Semiconductor Radiation Detectors

Technology and Applications

Devices, Circuits, and Systems

Series Editor
Krzysztof Iniewski
Emerging Technologies CMOS Inc.
Vancouver, British Columbia, Canada

PUBLISHED TITLES:

Advances in Imaging and Sensing
Shuo Tang and Daryoosh Saeedkia

Analog Electronics for Radiation Detection
Renato Turchetta

Atomic Nanoscale Technology in the Nuclear Industry
Taeho Woo

Biological and Medical Sensor Technologies
Krzysztof Iniewski

Building Sensor Networks: From Design to Applications
Ioanis Nikolaidis and Krzysztof Iniewski

Cell and Material Interface: Advances in Tissue Engineering, Biosensor, Implant, and Imaging Technologies
Nihal Engin Vrana

Circuits and Systems for Security and Privacy
Farhana Sheikh and Leonel Sousa

Circuits at the Nanoscale: Communications, Imaging, and Sensing
Krzysztof Iniewski

CMOS: Front-End Electronics for Radiation Sensors
Angelo Rivetti

CMOS Time-Mode Circuits and Systems: Fundamentals and Applications
Fei Yuan

Design of 3D Integrated Circuits and Systems
Rohit Sharma

Diagnostic Devices with Microfluidics
Francesco Piraino and Šeila Selimović

Electrical Solitons: Theory, Design, and Applications
David Ricketts and Donhee Ham

Electronics for Radiation Detection
Krzysztof Iniewski

PUBLISHED TITLES:

Electrostatic Discharge Protection: Advances and Applications
Juin J. Liou

Embedded and Networking Systems: Design, Software, and Implementation
Gul N. Khan and Krzysztof Iniewski

Energy Harvesting with Functional Materials and Microsystems
Madhu Bhaskaran, Sharath Sriram, and Krzysztof Iniewski

Gallium Nitride (GaN): Physics, Devices, and Technology
Farid Medjdoub

Graphene, Carbon Nanotubes, and Nanostuctures: Techniques and Applications
James E. Morris and Krzysztof Iniewski

High-Speed Devices and Circuits with THz Applications
Jung Han Choi

High-Speed Photonics Interconnects
Lukas Chrostowski and Krzysztof Iniewski

High Frequency Communication and Sensing: Traveling-Wave Techniques
Ahmet Tekin and Ahmed Emira

High Performance CMOS Range Imaging: Device Technology and Systems Considerations
Andreas Süss

Integrated Microsystems: Electronics, Photonics, and Biotechnology
Krzysztof Iniewski

Integrated Power Devices and TCAD Simulation
Yue Fu, Zhanming Li, Wai Tung Ng, and Johnny K.O. Sin

Internet Networks: Wired, Wireless, and Optical Technologies
Krzysztof Iniewski

Introduction to Smart eHealth and eCare Technologies
Sari Merilampi, Krzysztof Iniewski, and Andrew Sirkka

Ionizing Radiation Effects in Electronics: From Memories to Imagers
Marta Bagatin and Simone Gerardin

Labs on Chip: Principles, Design, and Technology
Eugenio Iannone

Laser-Based Optical Detection of Explosives
Paul M. Pellegrino, Ellen L. Holthoff, and Mikella E. Farrell

Low Power Emerging Wireless Technologies
Reza Mahmoudi and Krzysztof Iniewski

PUBLISHED TITLES:

Magnetic Sensors and Devices: Technologies and Applications
Laurent A. Francis, Kirill Poletkin, and Krzysztof Iniewski

Medical Imaging: Technology and Applications
Troy Farncombe and Krzysztof Iniewski

Metallic Spintronic Devices
Xiaobin Wang

MEMS: Fundamental Technology and Applications
Vikas Choudhary and Krzysztof Iniewski

Micro- and Nanoelectronics: Emerging Device Challenges and Solutions
Tomasz Brozek

Microfluidics and Nanotechnology: Biosensing to the Single Molecule Limit
Eric Lagally

MIMO Power Line Communications: Narrow and Broadband Standards, EMC, and Advanced Processing
Lars Torsten Berger, Andreas Schwager, Pascal Pagani, and Daniel Schneider

Mixed-Signal Circuits
Thomas Noulis

Mobile Point-of-Care Monitors and Diagnostic Device Design
Walter Karlen

Multisensor Attitude Estimation: Fundamental Concepts and Applications
Hassen Fourati and Djamel Eddine Chouaib Belkhiat

Multisensor Data Fusion: From Algorithm and Architecture Design to Applications
Hassen Fourati

MRI: Physics, Image Reconstruction, and Analysis
Angshul Majumdar and Rabab Ward

Nano-Semiconductors: Devices and Technology
Krzysztof Iniewski

Nanoelectronic Device Applications Handbook
James E. Morris and Krzysztof Iniewski

Nanomaterials: A Guide to Fabrication and Applications
Sivashankar Krishnamoorthy

Nanopatterning and Nanoscale Devices for Biological Applications
Šeila Selimović

Nanoplasmonics: Advanced Device Applications
James W. M. Chon and Krzysztof Iniewski

Nanoscale Semiconductor Memories: Technology and Applications
Santosh K. Kurinec and Krzysztof Iniewski

PUBLISHED TITLES:

Novel Advances in Microsystems Technologies and Their Applications
Laurent A. Francis and Krzysztof Iniewski

Optical, Acoustic, Magnetic, and Mechanical Sensor Technologies
Krzysztof Iniewski

Optical Fiber Sensors: Advanced Techniques and Applications
Ginu Rajan

Optical Imaging Devices: New Technologies and Applications
Ajit Khosla and Dongsoo Kim

Organic Solar Cells: Materials, Devices, Interfaces, and Modeling
Qiquan Qiao

Physical Design for 3D Integrated Circuits
Aida Todri-Sanial and Chuan Seng Tan

Power Management Integrated Circuits and Technologies
Mona M. Hella and Patrick Mercier

Radiation Detectors for Medical Imaging
Jan S. Iwanczyk

Radiation Effects in Semiconductors
Krzysztof Iniewski

Reconfigurable Logic: Architecture, Tools, and Applications
Pierre-Emmanuel Gaillardon

Semiconductor Devices in Harsh Conditions
Kirsten Weide-Zaage and Malgorzata Chrzanowska-Jeske

Semiconductor Radiation Detection Systems
Krzysztof Iniewski

Semiconductor Radiation Detectors, Technology, and Applications
Salim Reza and Krzysztof Iniewski

Semiconductors: Integrated Circuit Design for Manufacturability
Artur Balasinski

Smart Grids: Clouds, Communications, Open Source, and Automation
David Bakken

Smart Sensors for Industrial Applications
Krzysztof Iniewski

Soft Errors: From Particles to Circuits
Jean-Luc Autran and Daniela Munteanu

Solid-State Radiation Detectors: Technology and Applications
Salah Awadalla

Structural Health Monitoring of Composite Structures Using Fiber Optic Methods
Ginu Rajan and Gangadhara Prusty

PUBLISHED TITLES:

Technologies for Smart Sensors and Sensor Fusion
Kevin Yallup and Krzysztof Iniewski

Telecommunication Networks
Eugenio Iannone

Testing for Small-Delay Defects in Nanoscale CMOS Integrated Circuits
Sandeep K. Goel and Krishnendu Chakrabarty

Tunable RF Components and Circuits: Applications in Mobile Handsets
Jeffrey L. Hilbert

VLSI: Circuits for Emerging Applications
Tomasz Wojcicki

Wireless Medical Systems and Algorithms: Design and Applications
Pietro Salvo and Miguel Hernandez-Silveira

Wireless Technologies: Circuits, Systems, and Devices
Krzysztof Iniewski

Wireless Transceiver Circuits: System Perspectives and Design Aspects
Woogeun Rhee

FORTHCOMING TITLES:

3D Integration in VLSI Circuits: Design, Architecture, and Implementation Technologies
Katsuyuki Sakuma

Energy Efficient Computing: Devices, Circuits, and Systems
Santosh K. Kurinec and Krzysztof Iniewski

Nanoelectronics: Devices, Circuits, and Systems
Nikos Konofaos

Noise Coupling in System-on-Chip
Thomas Noulis

Radio Frequency Integrated Circuit Design
Sebastian Magierowski

X-Ray Diffraction Imaging: Technology and Applications
Joel Greenberg and Krzysztof Iniewski

Semiconductor Radiation Detectors

Technology and Applications

Edited by
Salim Reza

Managing Editor
Krzysztof Iniewski

CRC Press is an imprint of the
Taylor & Francis Group, an **informa** business

CRC Press
Taylor & Francis Group
6000 Broken Sound Parkway NW, Suite 300
Boca Raton, FL 33487-2742

First issued in paperback 2022

ISBN-13: 978-1-138-71034-4 (hbk)
ISBN-13: 978-1-03-233941-2 (pbk)
DOI: 10.1201/9781315200729

Publisher's Note

The publisher has gone to great lengths to ensure the quality of this reprint but points out that some imperfections in the original copies may be apparent.

Visit the Taylor & Francis Web site at
http://www.taylorandfrancis.com

and the CRC Press Web site at
http://www.crcpress.com

Contents

Preface

The foundation of the study of ionizing radiation was pioneered by three brilliant scientists: Wilhelm. C. Röntgen, Henri Becquerel, and Marie Curie. In November 1895, Professor Röntgen discovered a new kind of ray, later named as the x-ray, while studying electrical discharge through gases in a Crookes tube at the University of Würzburg. He noticed that the mysterious new ray has the ability to penetrate through opaque objects. A few months later, Henri Becquerel observed the emission of similar rays, naturally, from uranium salts. His observation eventually led to the discovery of the phenomenon called radioactivity. Then Marie Curie, a doctoral student of Becquerel, conducted extensive research on radioactive materials together with her husband Pierre. She is credited with the discovery of radium and polonium and is the only scientist who has won the Nobel Prizes in both physics and chemistry, to date.

Shortly after the discoveries of x-ray and radioactivity, scientists around the world became really curious and started to investigate the physics behind and the potential applications with great enthusiasm. Resources and efforts were put together in order to develop systems that can detect radiations. The earliest radiation detectors were the gas chambers. The incident ionizing radiations ionize the gas inside the chamber and create ion pairs. Then the charges generated by those ion pairs are measured under the influence of an externally applied electric field.

Substantially better energy resolution was achieved when the scientists started to use semiconductor detectors. The solid semiconductor materials have the ability to shorten the distance traveled by particles within them. So, it was possible to build durable detectors with smaller size and eventually better portability. During the 1960s, germanium had mostly been used as the detector material. Germanium detectors require the measurements to be conducted at low temperatures in order to reduce the thermally generated leakage current. This limitation in usability was a source of motivation to focus on building silicon-based detectors. Silicon has significantly larger bandgap when compared to germanium and thus can be used as sensitive material in radiation detectors to be operated at room temperature. Radiations can also be detected indirectly using a scintillation layer. The scintillation layer converts the incoming high-energy photons into visible light, which then can be detected by photodetectors.

The chapters in this book cover the physics and technologies behind modern semiconductor detectors for mainly x-ray radiation together with their applications. The contents of this book are carefully selected and arranged in such a way that it maintains a profound information flow.

Chapter 1: This chapter briefly shows the differences between the direct and the indirect conversion x-ray detectors and discusses the key properties of the photoconductive materials in direct conversion detectors in details. Two direct conversion detectors based on amorphous selenium (a-Se) and cadmium zinc telluride (CZT) for medical imaging applications are presented.

Chapter 2: Chromium compensated gallium arsenide (GaAs:Cr) is a relatively new detector material. It is possible to fabricate a thick, large area detector with high resistivity and with uniform and stable electric field throughout the sensitive volume using GaAs:Cr. In this chapter, the fabrication process, the material properties, and the performance of a GaAs:Cr based Medipix3RX detector are demonstrated.

Chapter 3: LAMBDA is a Medipix-based state-of-the-art x-ray imager, mainly developed for the experiments at synchrotrons. The authors of this chapter have presented the core technology and the applications of LAMBDA, together with the ongoing research on improving the system using edgeless technology and through-silicon via (TSV).

Chapter 4: In this chapter, the response function of a detector is defined and its importance is highlighted. The response functions of single-probe and multipixel CdTe detectors are estimated.

Chapter 5: Monte Carlo (MC) algorithm is widely used in order to study charge generation and charge transport inside semiconductor materials. MC algorithm-based modeling approaches for direct conversion x-ray detectors for medical imaging are discussed in this section of the book.

Chapter 6: The most important part in a detector system after the sensitive volume is the readout electronics, also known as the front-end electronics. It collects the charge generated inside the sensitive layer and processes it before sending it to the ADC, which then sends the digital information to a display or a storage unit. The authors of this chapter propose a novel inverter-based readout circuitry for radiation detectors.

Chapter 7: One of the most important applications of x-ray detectors is its usage in security scanning. This chapter continues the discussions about the readout electronics for x-ray detectors focused on the baggage-scanning applications. It also summarizes some of the popular imaging techniques and available detectors that are suitable for scanning applications.

Chapter 8: This chapter is devoted to the applications of photon-counting detectors (PCD) in high-resolution x-ray imaging, such as tomography, in a lab-scale environment. Other applications of PCDs in different fields of science are also listed.

Chapter 9: The principles of the cone beam computed tomography are discussed in this chapter along with its applications in medical imaging. The authors have reviewed the advantages and the challenges of the technology.

Chapter 10: The measurements of the polarization angle and the level of linear polarization using scattering polarimetry can play an important role in high-energy astrophysics. This chapter deals with the development and the performance of CdTe/CZT spectroscopic imagers for scattering polarimetry.

This book is written by internationally recognized experts in their respective fields from both academia and industry. The intended audiences are scientists and practicing engineers with some physics and electronics background. This book can also be used as a recommended reading and supplementary material in graduate-level courses. We wish all the readers an interesting journey through the existing and emerging technologies and the applications of semiconductor-based x-ray detectors.

Dr. Salim Reza
Dr. Krzysztof Iniewski

Series Editor

Krzysztof (Kris) Iniewski is managing R&D at Redlen Technologies, Inc., a startup company in Vancouver, British Columbia, Canada. Redlen's revolutionary production process for advanced semiconductor materials enables a new generation of more accurate, all-digital, radiation-based imaging solutions. Kris is also a founder of Emerging Technologies CMOS Inc. (http://www.etcmos.com), an organization of high-tech events covering communications, microsystems, optoelectronics, and sensors. In his career, Dr. Iniewski held numerous faculty and management positions at the University of Toronto, University of Alberta, SFU, and PMC-Sierra Inc. He has published over 100 research papers in international journals and conferences. He holds 18 international patents granted in the United States, Canada, France, Germany, and Japan. He is a frequent invited speaker and has consulted for multiple organizations internationally. He has written and edited several books for CRC Press, Cambridge University Press, IEEE Press, Wiley, McGraw-Hill, Artech House, and Springer. His personal goal is to contribute to healthy living and sustainability through innovative engineering solutions. In his leisure time, Kris can be found hiking, sailing, skiing, or biking in beautiful British Columbia. He can be reached at kris.iniewski@gmail.com.

Editor

Salim Reza is a research scientist at Mid Sweden University, Sweden. He has a licentiate and a PhD degree in radiation detectors and measurements. He has earned his PhD degree from Mid Sweden University and Deutsches Elektronen-Synchrotron (DESY), Hamburg, Germany. He is an expert in the phase-contrast x-ray imaging technique, Monte Carlo and finite element method-based simulation of radiation sensors, and developing radiation detection systems. He has worked in the development of the PERCIVAL detector system, intended to be used in the experiments at the European XFEL. He is also involved in the development and the characterization of the Medipix detector, developed by an international collaboration of scientists hosted by CERN. Improving work conditions and environment for academics is one of his major concerns. He has been elected as the vice-chairman in the national committee for The Swedish Association of University Teachers and Researchers, PhD section.

Contributors

Shiva Abbaszadeh
Department of Nuclear, Plasma, and Radiological Engineering
University of Illinois at Urbana-Champaign
Urbana, Illinois

Aldo Badano
Division of Imaging, Diagnostics and Software Reliability
Office of Science and Engineering Laboratories
Center for Devices and Radiological Health
Food and Drug Administration (FDA)
Silver Spring, Maryland

Julian Becker
Laboratory of Atomic and Solid State Physics
Cornell University,
and
Cornell High Energy Synchrotron Source (CHESS)
Cornell University,
Ithaca, New York

Ezio Caroli
Istituto Nazionale di Astrofisica (INAF)/Istituto di Astrofisica Spaziale (IASF)-Bologna
Bologna, Italy

Rui Miguel Curado da Silva
Laboratório de Instrumentação e Física Experimental de Partículas (LIP)
Departamento de Física
Universidade de Coimbra
Coimbra, Portugal

Stefano del Sordo
Istituto Nazionale di Astrofisica (INAF)/Istituto di Astrofisica Spaziale (IASF)-Palermo
Palermo, Italy

Jan Dudák
Institute of Experimental and Applied Physics
Czech Technical University
Prague, Czech Republic

Yuan Fang
Diagnostic X-ray Systems Branch
Division of Radiological Health
Office of In Vitro Diagnostics and Radiological Health
Center for Devices and Radiological Health
Food and Drug Administration (FDA)
Silver Spring, Maryland

Elias Hamann
Institute of Photon Science and Synchrotron Radiation
Karlsruhe Institute of Technology
Karlsruhe, Germany

Hiroaki Hayashi
Graduate School of Biomedical Sciences
Tokushima University
Tokushima, Japan

Kris Iniewski
Redlen Technologies
Vancouver, British Columbia, Canada

Yu Kuang
Department of Medical Physics
University of Nevada
Las Vegas, Nevada

Craig S. Levin
Department of Radiology
and
Department of Physics
and
Department of Electrical Engineering
and
Department of Bioengineering
Stanford University
Stanford, California

Jorge M. Maia
Departamento de Física
Universidade da Beira Interior
Covilhã, Portugal

Lampros Mountrichas
Electronics Laboratory of Physics Department
Aristotle University of Thessaloniki
Aristotle University Campus
Thessaloniki, Greece

Tianye Niu
Sir Run Run Shaw Hospital
Zhejiang University School of Medicine
and
Institute of Translational Medicine
Zhejiang University
Hangzhou, Zhejiang, China

Thomas Noulis
Electronics Laboratory of Physics Department
Aristotle University of Thessaloniki
Aristotle University Campus
Thessaloniki, Greece

David Pennicard
Deutsches Elektronen-Synchrotron (DESY)
Center for Free-Electron Laser Science (CFEL)
Hamburg, Germany

Milija Sarajlić
Deutsches Elektronen-Synchrotron (DESY)
Center for Free-Electron Laser Science (CFEL)
Hamburg, Germany

Stylianos Siskos
Electronics Laboratory of Physics Department
Aristotle University of Thessaloniki
Aristotle University Campus
Thessaloniki, Greece

Chris Siu
Department of Electrical and Computer Engineering Technology
British Columbia Institute of Technology (BCIT)
Burnaby, British Columbia, Canada

1

Direct Conversion Semiconductor Detectors for Radiation Imaging

Shiva Abbaszadeh and Craig S. Levin

CONTENTS

The detection and conversion of high-energy photons (such as x-ray, gamma-ray, and 511 keV annihilation photons) to electric charges can be divided into two classes: direct conversion and indirect conversion (Figure 1.1). Direct conversion detectors contain a photoconductive material, which converts high-energy photons directly into electrical charges. In contrast, indirect conversion detectors have a scintillation layer that converts the high-energy photons into optical photons. These optical photons are then converted into electrical charges using photodetectors, such as silicon photomultipliers (SiPMs). The indirect conversion detection is the most dominant technology for high-energy photon imaging due to the maturity of scintillation material and optical detection technology. Direct conversion semiconductors, especially compound semiconductors, have experienced slower development due to challenges in growth of high-purity, stable, and uniform crystals. However, there has been an ongoing interest in direct conversion for higher spatial resolution applications [1,2]. In this chapter, we discuss the key properties of photoconductive materials for high-performance direct conversion detectors for medical imaging applications. Two commercially available direct conversion photoconductors, amorphous selenium (a-Se) for large area x-ray imaging and cadmium zinc telluride (CZT) for high-resolution positron emission tomography (PET), are presented.

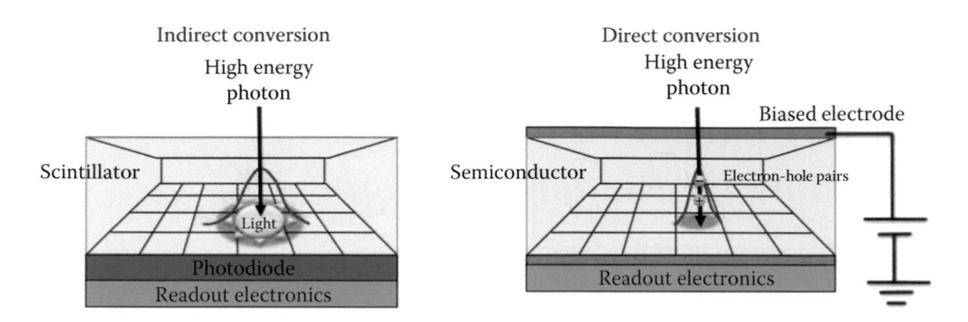

FIGURE 1.1
Schematic of cross-sectional view of (left) indirect conversion detector and (right) direct conversion detector. For indirect conversion, high-energy photons interact with a scintillator and generate visible photons, which are then absorbed by a photodetector (e.g., photodiode) and converted to charge, and then processed with readout electronics. Note that the light is emitted isotropically and spreads to neighboring pixels. For direct conversion, high-energy photons are converted directly into a collectable charge. Voltage is applied across the semiconductor, and the electric field within the semiconductor guides the generated charge to be collected by the pixel in the same planar spatial location where it is generated. This leads to a better spatial resolution since the electrode pattern pitch can be very fine. However, the energy of the photon is important, since if the size of the electron–hole cloud is bigger than the electrode pitch, the signal will be shared by neighboring pixels, reducing the spatial resolution.

1.1 Direct Conversion Photoconductor

A simple direct conversion detector (Figure 1.1) contains three main parts:

1. A photoconductive material, which converts high-energy photons directly into electrical charges.
2. Biasing electrodes, which are used to apply the bias and create the electric field within the photoconductive material. The electrical signal generated from the drift of photogenerated carriers is induced on the electrodes. They also define the position of charge collection generated by the incident photons. The electrodes have typically square pixelated or cross-strip configurations.
3. A readout circuit that reads the electrical signal from each pixel element and converts the signal chain to the digital domain to be transferred to a computer.

The photoconductive material can be complex in structure. It can be sandwiched between blocking layers (Figure 1.2) to prevent charge injection from the biasing electrodes and limit the dark current to the bulk properties of the photoconductive material. A hole-blocking layer is usually an n-type material that prevents the injection of holes from positively biased electrodes and allows conduction of the electrons being collected. An electron-blocking

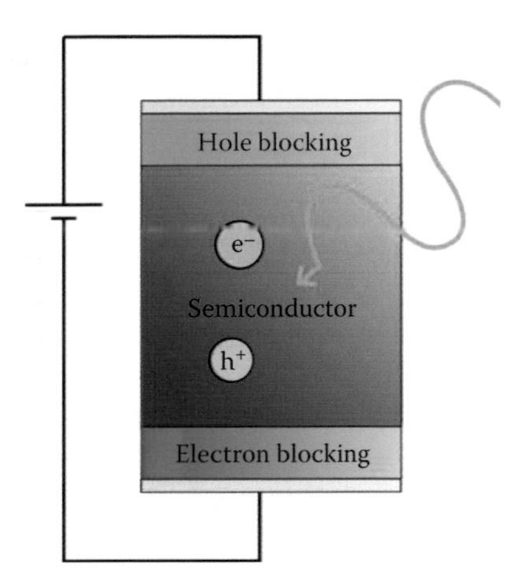

FIGURE 1.2
Schematic of a semiconductor detector sandwiched between electron and hole-blocking layers. A hole-blocking layer prevents injection of holes from positively biased electrodes and allows the photogenerated electrons to be collected. On the other hand, an electron-blocking layer prevents the injection of electrons from negatively biased electrodes and allows the photogenerated holes to pass through it. The blocking layers are used if the most dominant part of the dark current of the detector is due to injection of charge from biasing electrodes.

layer is usually a p-type material that prevents the injection of electrons from negatively biased electrodes and allows the conduction of holes [3]. A high-energy photon in medical imaging applications has an energy-depositing interaction within the photoconductive material either by photoelectric absorption or Compton scatter. The primary electrons produced from these interactions will ionize the material as they lose energy and generate more electron–hole pairs. A voltage is applied across the detector to drift the generated electron–hole pairs. The drift of the electrons and holes will induce a current in the detector as described by the Shockley–Ramo theory [4]. The charges generated at the electrodes due to the drift of electrons and holes are proportional to the number of electron–hole pairs generated and can be calculated by Hecht's equation [5]:

$$Q = Q_0 \left\{ \frac{\lambda_e}{L} \left[1 - \exp\left(-\frac{(L-x)}{\lambda_e} \right) \right] + \frac{\lambda_h}{L} \left[1 - \exp\left(-\frac{x}{\lambda_h} \right) \right] \right\}$$

where Q_0 is the initial ionization charge, L is the detector thickness, x is the distance between the point of interaction and cathode, and λ_e and λ_h are the carrier *Schubweg* for electrons and holes, respectively. The Schubweg is the average distance a charge carrier drifts under the applied electric field before it is permanently captured by traps or recombination centers [6]. It is determined by the product $\mu \cdot \tau \cdot E$, where μ is the mobility, τ is carrier lifetime, and E is the applied electric field. Due to the creation of a uniform electric field across the photoconductor, all of the generated holes or electrons (based on the polarity of the pixel electrode) will drift to the pixel underneath the position of generation. The lack of spatial dispersion effects (e.g., light scattering for scintillators) in the direct conversion process and the fine electrode pitch allow this method to have an increased spatial resolution compared to indirect conversion. However, the requirement of a high-voltage supply and the manufacturing of a uniform thick layer of photoconductor over a large area are drawbacks of direct conversion semiconductor detectors.

A high atomic number (Z) is desirable for the photoconductive material since it leads to a greater probability to absorb the photons incident on the detector. The photoelectric interaction probability increases proportional to Z^4/E^3, and the Compton scattering probability increases proportional to Z/E, where E is the energy of the incident photon [7]. Photoconductors with small atomic numbers need to be thicker in order to absorb the same amount of photons as materials with higher atomic numbers.

The amount of energy required to form a detectable electron–hole pair in a semiconductor is given by the electron–hole pair creation energy and is typically two to three times that of the semiconductor's bandgap energy [8]. As the amount of energy required to create an electron–hole pair decreases, the number of detectable pairs for a given photon energy increases, leading to a larger detected signal. Other important properties of a photoconductive material include low dark current and the possibility for large area fabrication.

The most commonly used semiconductor detector is silicon. However, due to its low Z and low density, silicon is not efficient in direct detection of high-energy photons such as x-rays, gamma-rays, and 511 keV annihilation photons. Therefore, materials with a higher atomic numbers have been under development as alternative semiconductor direct conversion detectors. It should be noted, however, that silicon has found use as a detector in an edge-on configuration for mammography, which is at the lower end of energies used for medical imaging [9,10]. Table 1.1 lists properties of numerous

TABLE 1.1

Selected Properties of Several Photoconductor Candidates for Direct Conversion Detectors

Material	Si	Ge	a-Se	CdZnTe	CdTe	HgI_2	PbI_2
Atomic number (Z)	14	32	34	48, 30, 52	48,52	80, 53	82, 53
Energy bandgap (eV)	1.12	0.67	2.2	1.5–1.7	1.44	2.1	2.3
Charge pair energy formation (W) (eV)	3.62	2.96	50	4.6	4.43	5	5.5
Mobility lifetime product (cm^2/V)	(h) >1 (e) ~1	(h) >1 (e) >1	(h) 10^{-6} (e) 10^{-8}	(h) 10^{-5} (e) 10^{-3}	(h) 10^{-4} (e) 10^{-3}	(h) 10^{-5} (e) 10^{-4}	(h) 10^{-6} (e) 10^{-7}

Source: Adapted from W. Zhao and J. A. Rowlands, *Proc. SPIE: Medical Imaging*, vol. 1651, pp. 134–143, 1992.

photoconductors. From Table 1.1, a-Se is a very mature technology that can be uniformly deposited up to a thickness of 1–2 mm [11]. A thickness of 200 μm a-Se is sufficient for absorption of the majority of photons having x-ray energy typical for mammography, and 1 mm thick a-Se is enough for chest x-ray imaging. For imaging photons with energy higher than 100 keV, the growth technology of CZT and cadmium telluride (CdTe) has progressed to meet this need, and crystals as thick as several centimeters are commercially available [12]. The crystal growth and purification of mercuric iodide (HgI_2) and lead iodide (PbI_2), which contain the high atomic number elements mercury (Z = 80) and lead (Z = 82), respectively, are still under research and development [13,14].

1.2 Direct Conversion Semiconductor Detectors in X-Ray Imaging

X-ray imaging is among the most important medical imaging procedures used today. The high energy of the radiation allows it to penetrate through the body and provide an image of the interior of the body, which is otherwise not visible by the human eye. A depiction of image acquisition using x-ray imaging is shown in Figure 1.3. The object, or patient, to be imaged is placed between the radiation source and the detector that acquires the image. The variation of x-ray attenuation within the body, due to varying tissue composition, causes the variation, or contrast, in the acquired image. An exposure

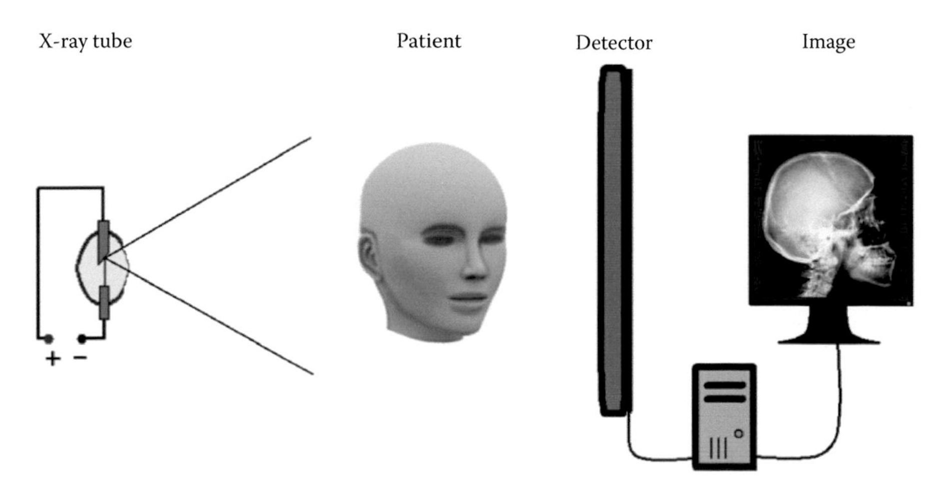

FIGURE 1.3
Radiographic image acquisition.

TABLE 1.2

Digital X-Ray Imaging System Properties

Clinical Task	Chest Radiography	Mammography	Fluoroscopy
Detector size	35 cm × 43 cm	18 cm × 24 cm	25 cm × 25 cm
Pixel size	200 μm × 200 μm	50 μm × 50 μm	250 μm × 250 μm
Number of pixels	1750 × 2150	3600 × 4800	1000 × 1000
Readout time	<5 s	<5 s	1/30 s
X-ray spectrum	120 kVp	30 kVp	70 kVp
Mean exposure	300 μR	12 mR	1 μR
Noise level	6 μR	60 μR	0.1 μR

Source: D. L. Y. Lee, L. K. Cheung and L. S. Jeromin, "New digital detector for projection radiography," in *Proc. SPIE: Medical Imaging*, pp. 237–249, 1995.

of uniform intensity exposes the body, and the intensity is modulated by the differential attenuation within the body. The modulated intensity that exits the body contains the internal structure information and is sensed by the detector to form the image. Therefore, adjacent regions having a greater difference in x-ray attenuation will have greater contrast.

There are several modalities used to meet the various needs in the field of medical imaging. X-ray imaging modalities include chest radiography, mammography, fluoroscopy, and computed tomography (CT). Each of these modalities serves a different purpose. Properties for the first three of these modalities are summarized in Table 1.2 [15]. Note that the x-ray spectrum is quoted in terms of kVp, which is the x-ray tube voltage, and the exposure is expressed in roentgens (R).

1.2.1 X-Ray System Performance Characteristics

In x-ray imaging, there is a need for high spatial resolution detectors in applications with small feature sizes, for example, microangiography, where we are looking at small blood vessels, and mammography, where microcalcifications are important since small microcalcifications are a hallmark of breast malignancy [16,17]. The spatial resolution in an x-ray system is characterized by measuring the modulation transfer function (MTF) of the system. The MTF is a measure of the contrast as a function of spatial frequency. As illustrated in Figure 1.4, if we have an input signal with components at all spatial frequencies, with the same intensity, and they are imaged with the imaging system in question, then if we take the ratio of the amplitude at the output compared to the amplitude at the input, we obtain the attenuation relationship as a function of frequency, where the high-frequency components are generally attenuated. This attenuation comes from the different sources of blurring in the system such as K-fluorescence reabsorption, Compton scattering, and primary photoelectron range, which is related to the electron–hole pair cloud diameter. One way to measure the MTF is using the so-called

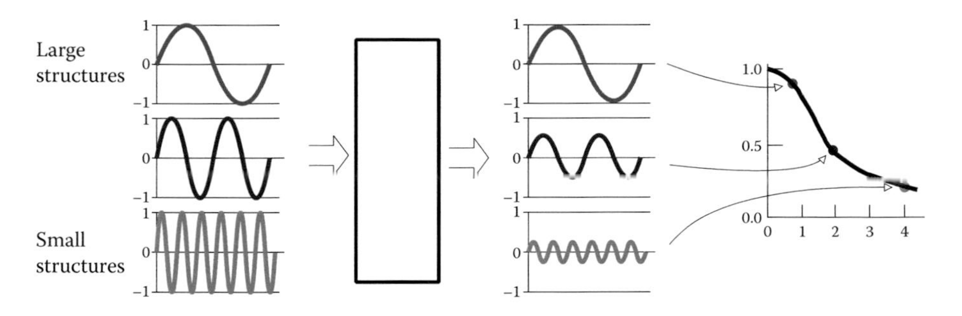

FIGURE 1.4
Conceptual depiction of the MTF. Large structures have low spatial frequency components, while small structures have high spatial frequency components. The imaging system may exhibit a different response to the various frequency components of the object being imaged. The relative signal of the output of the imaging system to its input as a function of frequency yields the MTF. Typically, higher frequency signals are more difficult to preserve due to various forms of blurring in the imaging system.

"edge-method" [18,19]. As was discussed earlier in the chapter, the direct conversion detector drifts the charge toward the pixel electrodes, and there is no blurring like there is in scintillators due to isotropic spreading of light. As a consequence, direct conversion semiconductor detectors provide a better MTF compared to their indirect conversion counterpart. We will compare the MTF of direct conversion imaging detectors with indirect conversion imaging detectors in the next section.

Another important characteristic of an x-ray detector is how well it preserves the signal-to-noise ratio (SNR) of the incident signal. Similar to the MTF, a metric for quantifying degradation in the signal quality as a function of spatial frequency can be computed by the ratio of the SNR at the output to that at the input of the imaging system, and taking the square of the result. This metric is known as the detective quantum efficiency (DQE) [20,21]. A DQE of unity perfectly preserves the SNR at the input of the imaging system; however, this is generally not achievable, even at zero spatial frequency, due to incomplete absorption of all incident photons (i.e., a probability less than unity for detecting incident photons). Noise sources in the imaging system will degrade the DQE, and thus the DQE is dependent on the incident exposure level. Typically the DQE will improve with increasing exposure as the noise becomes less dominated by the electronics (i.e., becomes quantum-noise limited).

1.2.2 a-Se Direct Conversion Detector

The a-Se photoconductor is the most highly developed photoconductor for large area x-ray imaging and is still the only commercially available material for flat-panel x-ray detectors [22–25]. a-Se has an acceptable x-ray absorption coefficient for low x-ray energy, good charge transport properties, and low dark current. Typically, a-Se detectors are operated at an electric field strength of ~10 V/μm to get an acceptable x-ray sensitivity and acceptable levels of lag and ghosting [26,27]. Figure 1.5 compares the MTF of an a-Se detector with another detector technology, demonstrating that a-Se is capable of providing a very high spatial resolution. In addition, in order to make a-Se reach its highest charge collection efficiency, higher electric fields are required [28–30]. A higher electric field within the a-Se layer improves the detector performance by increasing carrier motilities and photogeneration efficiency that leads to better dose efficiency [31,32]. However, maintaining low dark current (≤10 pA/mm^2) at high electric fields is challenging [33]. A high dark current reduces the dynamic range of the device. It also increases noise levels, thus degrading SNR and can lead to crystallization of the detector due to joule heating [34]. Previous studies of dark current on a simple metal/a-Se/metal structure have shown that dark current is dependent on time and voltage and changes by the nature of the positively biased metal contact employed [35].

Juska et al. first observed avalanche multiplication in a-Se in 1980 while they were studying the photogeneration efficiency and electron and hole mobility of a-Se at high electric fields [28]. They measured transient photocurrent in a-Se (20–200 μm) sandwiched between two insulating layers using the time-of-flight technique. It was observed that at fields higher than 80 V/μm, the photogeneration efficiency increased sharply. Photogeneration efficiency larger than unity is due to photocurrent multiplication phenomena. Photocurrent multiplication in Juska's structure has been attributed to the avalanche phenomena, which entails the generation of secondary

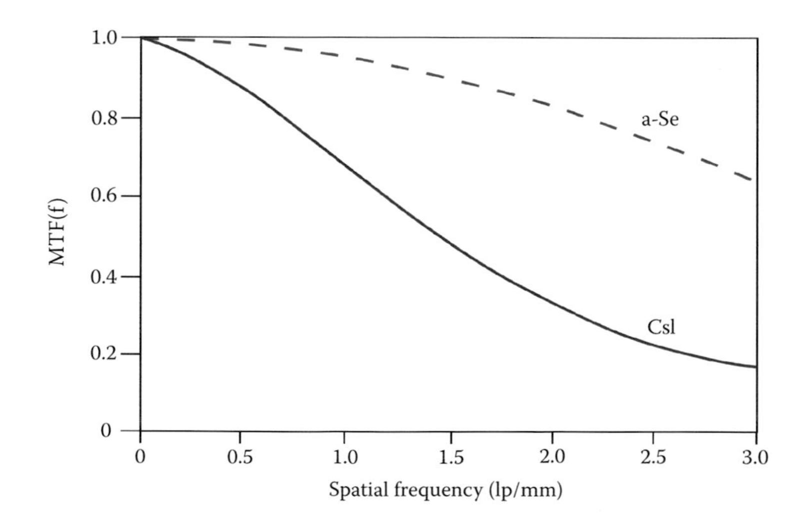

FIGURE 1.5
MTF of a-Se and CsI, a common indirect conversion material, x-ray detector technologies. (Adapted from M. Spahn, *Eur. Radiol.*, vol. 15, pp. 1934–1947, 2005.)

electron–hole pairs by impact ionization at high electric fields. Soon after that, a-Se was used in commercial ultrasensitive high-gain avalanche rushing photoconductor (HARP) TV camera tubes [36]. The basic structure of the HARP camera is shown in Figure 1.6.

In order to apply a high electric field across the a-Se, a multilayer vertical a-Se structure with proper electron and hole-blocking material has been used. In the HARP structure, a-Se, which can vary in thickness, is sandwiched between cerium oxide CeO_2 (several tens of nanometers) and the porous di-antimony trisulfide Sb_2S_3 (about 100 nm) that act as blocking layers for holes and electrons, respectively. Cerium oxide is an n-type wide bandgap material that prevents injection of holes from the anode by forming a large potential barrier to holes. The di-antimony trisulfide layer has a large number of electron traps that become filled and form a negative space-charge barrier that stops injection of electrons from cathode [37]. The selenium layer used is doped with arsenic and tellurium to suppress crystallization and to increase the sensitivity for red light, respectively [37]. A thin region of the selenium layer next to the signal electrode was doped with a small amount of lithium fluoride to decrease the white blemishes [38]. A transparent indium tin oxide (ITO) or tin dioxide (SnO_2) electrode, deposited on a glass substrate, provides a contact for a positive electrostatic bias to be applied. However, there is no physical contact or electrode at the other side of the structure. This side is kept at the cathode potential of the electron gun by a scanning electron beam. This free surface is capable of supporting a latent charge image. Research has been ongoing to improve the readout component of the HARP camera to establish a more practical method, and recently there is research to adapt the HARP technology for x-ray imaging [39–44] due to the potential benefits of avalanche multiplication.

In addition to the HARP structure, different hole-blocking layers and their behavior as a function of the electric field were investigated [45,46] to

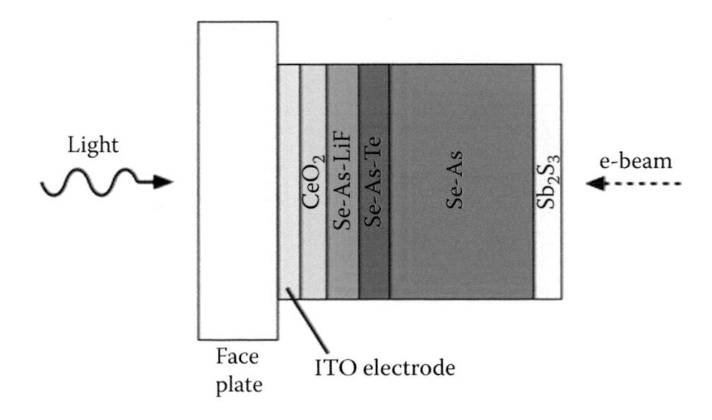

FIGURE 1.6
Schematic of HARP camera structure. (Adapted from K. Tanioka, *Nucl. Instrum. Methods A*, vol. 608, pp. S15–S17, 9/1, 2009.)

develop an a-Se detector capable of working at electric fields above the typical 10 V/μm. The role of the hole-blocking layer is essential to maintain low dark current at high electric fields since the mobility of holes in a-Se is 20–30 times larger than the electron mobility. It was found that a thin layer (<1 μm) of polyimide (PI) permits operation at high electric fields greater than 80 V/μm while maintaining a dark current density below 200 pA/mm^2. PI is commonly used as an insulator and when used as a blocking layer should create a significant potential barrier for holes. To avoid electric field reduction in the a-Se layer due to charge accumulation, the electrons should flow through the PI/a-Se interface and within the PI layer. Detectors using a PI layer utilize a simple fabrication process based on widely available semiconductor materials that can be easily integrated into current large area digital imager manufacturing processes, and the process is compatible with both thin-film transistor (TFT) and complementary metal–oxide–semiconductor (CMOS) technologies. It was initially suspected that charge (specifically electron) accumulation would occur at the PI/a-Se interface and would degrade the device performance over time by reducing the internal electric field of the device. However, it was demonstrated that photocurrent is quite stable over time using the time-of-flight experiment [47] and pulsed and long light exposure [46], suggesting that charge accumulation does not significantly impact the internal field within the a-Se layer. Figure 1.7 shows an x-ray image of a 25–50 μm wire diameter aortic stent that was acquired using a CMOS-PI-selenium prototype [48].

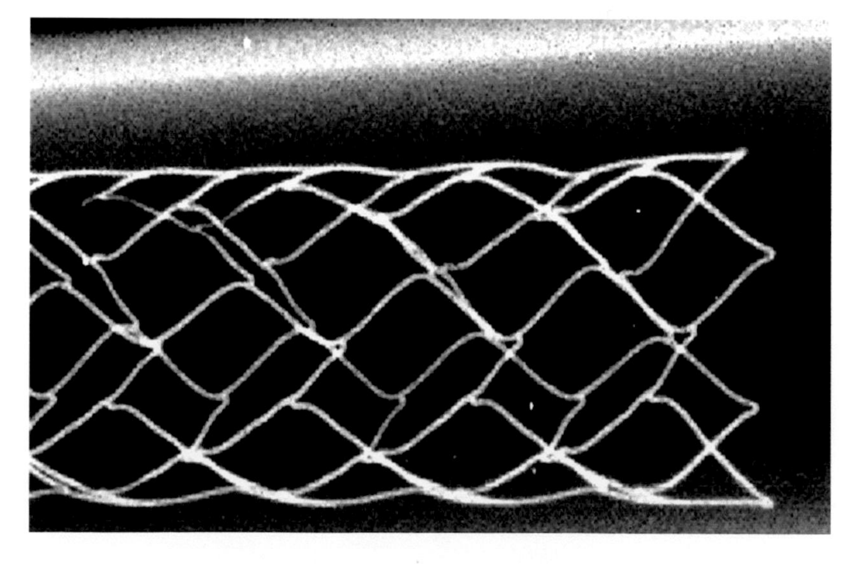

FIGURE 1.7
25–50 μm wire diameter aortic stent in a glass vial imaged by a prototype detector using PI/a-Se on CMOS with 25 μm pixel size. The x-ray source was operated at 30 kVp. (From C. C. Scott et al., "Amorphous selenium direct detection CMOS digital x-ray imager with 25 micron pixel pitch," in *SPIE Medical Imaging*, p. 90331G, 2014. With permission.)

1.3 Direct Conversion Semiconductor Detectors in PET

PET is a common imaging modality in nuclear medicine. Nuclear medicine uses radioactive tracers labeled to a molecule that is specifically engineered to assess specific biological processes within the body. As a consequence, it is capable of imaging the functions within the body as opposed to imaging the anatomy as in conventional x-ray imaging. The type of tracer used in PET is a positron-emitting radiotracer, which undergoes annihilation with an electron in the tissue and produces two photons with energy of 511 keV emitted in opposite directions, as shown in Figure 1.8. If these 511 keV photons are detected by two position sensitive detectors, the location of the annihilation positron is somewhere along the line of response (LOR) between these detectors. Through iterative image reconstruction algorithms that involve repeated forward- and back-projecting the data through the system LORs, the biodistribution of the tracer can be imaged in three dimensions.

1.3.1 PET System Performance

The sensitivity of a PET system to be able to detect small concentrations of radiotracer is determined by the PET system geometry, characteristics of the radiation detectors, and system data acquisition. Important performance parameters of a PET system are

1. *Energy resolution*: The energy resolution of a detector system is defined by its ability to and, it depends on the detector and data acquisition system. It is quantified by the width of the photopeak of the energy spectrum of the incident photons. Different components of noise such as quantum noise of the detector, signal propagation within the

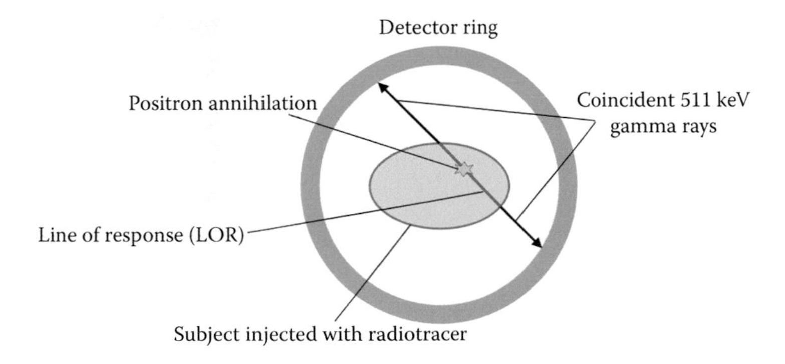

FIGURE 1.8
Schematic of a ring PET scanner where radiation detectors surround the subject. A LOR is recorded along the line that connected the detection elements where the two 511 keV annihilation photons are detected.

detector, input capacitance of the readout amplifier, thermal noise of the resistors, dark current of the detector, etc., worsen the energy resolution of the detector. The energy resolution in percentage is defined as the photopeak full width at half maximum (FWHM) divided by the peak energy of the photopeak $\left(\frac{FWHM_{keV}}{E_{keV}} \times 100\%\right)$.

2. *Spatial resolution*: The spatial resolution of the PET system is determined by the physical characteristics of the detectors and the basic physics of positron decay. The width of the pixel elements in the detector and capability of the detector in detecting the depth of interaction (DOI) are very important in defining the spatial resolution of the PET system. If we assume we were able to reduce the pixel size to zero, the range of positron in tissue and the acollinearity of annihilation photons fundamentally limit the spatial resolution of the PET system [49,50].
3. *Time resolution*: The time resolution of the PET system is affected by how fast the detector elements respond and the properties of the readout electronics. Better timing resolution will lead to less ambiguity for identifying coincident annihilation photons from background processes.

The majority of preclinical and clinical PET systems make use of indirect conversion of 511 keV photons and use a scintillator attached to a photomultiplier tube, avalanche photodiode, or SiPM for detection [51,52]. Imaging low concentration and small structures is a driving force in both academia and industry to develop PET systems with ultrahigh spatial resolution (≤ 1 mm^3) and high 511 keV photon detection efficiency. To reach such a resolution, different scintillator segmented detectors with crystals having dimensions as small as 0.8 and 0.5 mm have been investigated [53,54]. In addition, high-density semiconductor materials such as CZT have gained particular interest due to the following reasons:

1. In a semiconductor detector, the submillimeter pixel element is achieved with a metal deposition process on the semiconductor material, which eliminates the need for saw cutting of scintillator crystals into tiny elements, which would then require assembly into arrays and a labor-intensive process. Due to direct detection of charge carriers across a uniform electric field within the semiconductor detector, there is little or no spatial blurring.
2. DOI information for 3D positioning of photon interaction can be easily extracted with semiconductor detectors due to charge trapping within the semiconductor material. The ratio of charge induced at the anode and cathode represents the depth of photon interaction in the material [55,56]. This property eliminates the

complex scintillation detector designs that have been proposed such as double-sided detectors or different scintillator materials used to encode the DOI [57].

3. The electron–hole pair creation energy in the semiconductor is approximately two or three times that of the bandgap. With relatively low electron–hole pair creation energy (and therefore a lot of electron–hole pairs created) and direct detection of charge carriers, the energy resolution of semiconductor detectors is much better than that of scintillator-based detectors.
4. The high-energy resolution and accurate 3D position sensing of semiconductor detectors make the system capable of Compton kinematics that allow estimating the incoming incident angle of the photons that undergo Compton scatter [58]. This information can be exploited for a number of purposes, including geometric rejection of random and scatter coincidences, or acceptance and positioning of normally rejected events, and increasing system spatial resolution by more accurately estimating the first point of interaction in the detector in multi-interaction photon events [59].

1.3.2 CZT Direct Conversion Detector

Cadmium telluride and CZT are the most highly developed photoconductors for gamma ray and 511 keV annihilation photon detection for imaging. The manufacturing of CZT has progressed over the past 30 years and has become quite reliable. Different manufacturers, such as Redlen Technology (Victoria, Canada) and Imarad Imaging System Ltd. (Rehovot, Israel), have been manufacturing CZT since the 1990s. Several laboratories have investigated the spatial, energy, and timing resolution of individual CZT crystals (with different dimensions and electrode design) as a detector technology for PET system [60–63]. The electrode design of these detectors is typically either square pixelated anodes with a planar cathode or cross-strip electrodes (orthogonal anode and cathode strips). The cross-strip electrodes are attractive since they significantly reduce the number of readout channels at the system level.

Advances in CZT crystal growth and manufacturing led to a substantial improvement in mobility-lifetime product and achievable energy resolution. In order to preserve the intrinsic high-energy resolution of CZT PET detectors as they are scaled up to a system, the design of the detector's electrodes, interconnections, and its readout electronics are very important. He et al. have been developing CZT detectors for more than a decade for various imaging applications in the field of nuclear security. In one study, they have evaluated two $1.5 \times 1.5 \times 1\ cm^3$ pixelated CZT detectors and a data acquisition for a PET system [64]. The detector energy resolution was 1% FWHM at 662 keV. They also demonstrated that timing resolution of 10 ns FWHM is achievable on a 1 cm thick crystal using a cathode waveform digitizer. The time of occurrence of the event in the CZT detector was determined by the rising edge of the cathode signal using digital waveform sampling. Vaska et al. developed pixelated CZT detector modules with different thickness and reported 10 ns FWHM time resolution for a 1.4 mm thickness biased at 100 V in coincidence with BaF_2 (indirect conversion detector) and 21 ns FWHM time resolution for a 7.5 mm thickness biased at 1000 V [65]. It was confirmed by both groups that waveform analysis improves the time measurement over the leading edge discrimination of a shaped cathode or anode signal. The time-amplitude walk caused by leading-edge triggering limits the time resolution. Special electronics and proper calibration are very important for optimizing the time resolution to make it limited by CZT intrinsic properties and remove the effect of the electronics.

Cross-strip CZT detectors have also been developed and investigated by several groups [56,60,66]. Drezet et al. developed 16 × 20 × 0.9 mm^3 CZT detectors with 0.9 mm wide anode strips on a pitch of 1 mm with 3.9 mm cathode strips on a 4 mm pitch orthogonal to anode strips [60]. Timing resolution between two CZT detectors with an applied bias of 500 V was reported to be 2.6 ns. Matteson et al. [67] and Gu et al. [55] developed larger volume CZT detector modules. This design utilizes two 40 × 40 × 5 mm^3 CZT crystals stacked in an anode–cathode–cathode–anode configuration to form a single 40 × 40 × 10 mm^3 CZT detector module. Figure 1.9 shows the detector module and the cross section of the CZT crystal used in that design. Anode strips are 0.1 mm

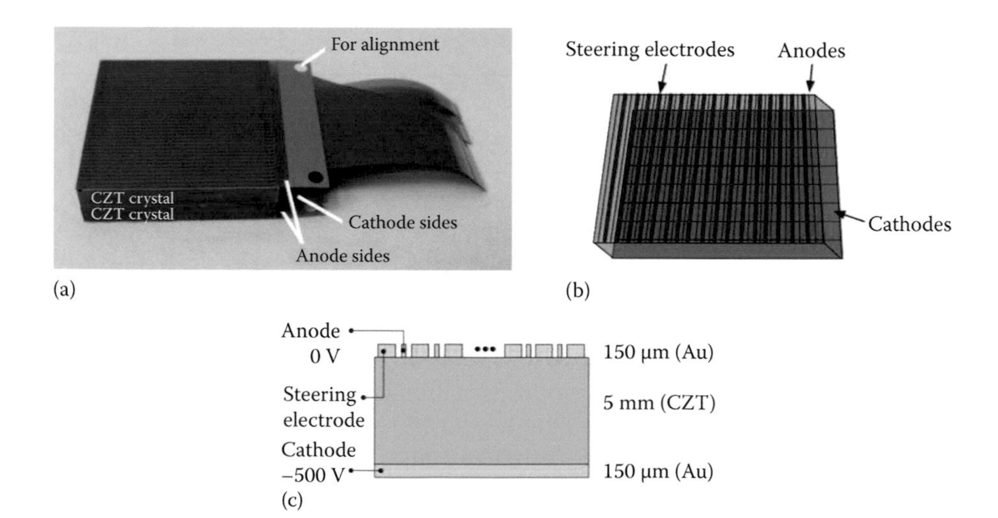

FIGURE 1.9
Two CZT crystals are assembled to flexible circuit and stacked based on anode–cathode–cathode–anode (ACCA) configuration to form a CZT module (4 cm × 4 cm × 1 cm). (b) Schematic of a CZT crystal with cross-strip pattern showing anodes, cathodes, and steering electrodes. (c) Drawing of cross section of the CZT crystal.

wide with 1 mm pitch, and cathode strips are 4.9 mm wide with 5 mm pitch. Gold (Au, 150 nm thick) was deposited on the polished surface of CZT crystals for both anode and cathode electrodes. Conductive silver epoxy was used as the bonding material to assemble the flexible circuit to the Au electrodes. Due to the small width of the anode compared to thickness of the crystal (5 mm), there is the so-called "small pixel effect," which means that the drifting electrons have to be in close proximity to the anodes in order to induce significant charge on those anodes. Based on the Shockley–Ramo theorem, the small pixel effect blinds the electrode to the charge trapping within the bulk of the crystal. The CZT detector design also used steering electrodes placed between anode strips to improve charge collection. It was shown that the steering electrode biased above 80 V with respect to the anode's bias was sufficient for complete charge collection [56,66]. The detector module showed an energy resolution of 3.9 ± 0.19% FWHM at 511 keV, and the point spread function (PSF) in the direction orthogonal to the anode strips was 0.78 ± 0.1 mm FWHM including the 250 μm diameter of the point source. A system using these detector modules is currently under construction [68]. The geometry of this system is arranged into a box as illustrated in Figure 1.10, and the detectors are oriented in an "edge-on" configuration such that the annihilation photons encounter 4 cm of CZT along the direction perpendicular to the axis. The first full-system data acquisition characterization of this system supporting readout of 94 CZT detector modules was tested with 12 detector modules with different crystal quality. Results showed the subsystem-wide energy resolution as being 7.43 ± 1.02% FWHM over 468 channels. The global time resolution of the system based on six CZT crystals in coincidence with six other CZT crystals was 37 ± 5 ns FWHM.

From the discussion and examples presented in this section, it can be concluded that the spatial and energy resolution of CZT is superior to those of

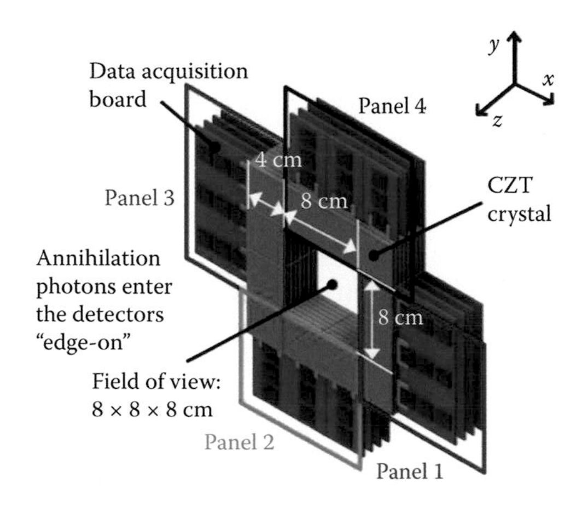

FIGURE 1.10
Schematic of a CZT-based PET detector system for small animal imaging.

current state-of-the-art scintillator-based detectors. The best timing resolution reported for 1 cm thick CZT crystal was 10 ns. Although 10 ns timing resolution is not acceptable for most PET applications, it is appropriate for small animal PET imaging since the random fraction for small animal imaging is relatively low. Since CZT is capable of Compton collimation, random coincidences may also be rejected by examining their incidence angle into the detectors. In addition, novel methods such as refractive index modulation by probing the Pockels effect in CdTe or CZT crystals are under investigation to dramatically improve PET coincidence time resolution [69]. If such an innovative method is achieved, there exists the possibility to combine the technique with conventional CZT detectors to simultaneously achieve superior spatial, energy, and timing resolution.

1.4 Summary

In summary, this chapter discussed direct conversion detector technology. This technology differs from that of indirect conversion by the conversion of incident radiation to charge (as opposed to light), which is collected with the aid of an applied bias that drifts the generated charge toward the electrodes. From a high level, the detector consists of a conversion material, biasing electrodes, and readout circuitry. Direct conversion detectors can achieve an increased spatial resolution compared to indirect conversion detectors due to the fine electrode pitch and lack of dispersion effects. On the other hand, drawbacks include the need of a high-voltage supply and manufacturing a uniform thick layer, potentially over a large area. The ultimate choice of conversion material depends on the application, since different radiation energies are used for different imaging modalities, and a material's probability to absorb photons is dependent on factors such as atomic number.

For x-ray imaging, the MTF can be used for characterizing the spatial resolution of the system and the DQE can be used to quantify the signal quality in terms of SNR. The numerous sources that affect the spatial resolution include K-fluorescence reabsorption, Compton scattering, and primary photoelectron range, while noise sources such as electronic noise affect the DQE. a-Se is a common direct conversion detector material for mammography and low x-ray energy applications. Benefits of a-Se include good charge transport properties, low dark current, and high spatial resolution. In addition, avalanche multiplication has been observed in a-Se, which could potentially be leveraged for x-ray imaging. The blocking layer also plays an important role, as this layer is essential to maintain low dark current and prevent charge injection at high electric fields.

For PET imaging, three important parameters of the system performance are energy resolution, spatial resolution, and time resolution. Although typically indirect conversion detectors are used for PET imaging, direct conversion detectors gained interest due to several factors including a pixel element defined by metal deposition (no need for cutting crystals), DOI that enables 3D positioning, high-energy resolution, and capability for Compton kinematics. CZT was highlighted as a highly developed photoconductor for high-energy imaging applications. Several groups have investigated CZT for PET imaging and have shown high energy and spatial resolution; although the time resolution observed was limited, it is still appropriate for small animal PET imaging. The detector electrodes, interconnections, and readout electronics also play large roles in a full system designed using CZT.

References

1. M. Overdick et al., "Status of direct conversion detectors for medical imaging with X-rays," *IEEE Trans. Nucl. Sci.*, vol. 56, pp. 1800–1809, 2009.
2. J. D. Kuttig et al., "Comparative investigation of the detective quantum efficiency of direct and indirect conversion detector technologies in dedicated breast CT," *Phys. Med.*, vol. 31, pp. 406–413, 2015.
3. H. Khlyap, *From Semiclassical Semiconductors to Novel Spintronic Device.* Bentham Science Publishers, Sharjah, 2013.
4. S. Ramo, "Currents induced by electron motion," *Proc. Ire*, vol. 27, pp. 584–585, 1939.
5. K. Hecht, "For the mechanism of the photoelectric primary current in insulating crystals," *Zeits. Phys.*, vol. 77, pp. 235–245, 1932.
6. S. Kasap, H. Ruda and Y. Boucher, *Cambridge Illustrated Handbook of Optoelectronics and Photonics.* Cambridge University Press, Cambridge, 2009.
7. G. F. Knoll, *Radiation Detection and Measurement.* John Wiley & Sons, New Jersey, 2010.
8. W. Que and J. A. Rowlands, "X-ray photogeneration in amorphous selenium: Geminate versus columnar recombination," *Phys. Rev. B*, vol. 51, pp. 10500–10507, 1995.
9. M. Lundqvist et al., "Measurements on a full-field digital mammography system with a photon counting crystalline silicon detector," *Med. Imaging*, pp. 547–552, 2003.
10. E. Fredenberg et al., "Energy resolution of a photon-counting silicon strip detector," *Nucl. Instrum. Methods Phys. Res. A*, vol. 613, pp. 156–162, 2010.
11. S. Kasap et al., "Amorphous selenium and its alloys from early xeroradiography to high resolution X-ray image detectors and ultrasensitive imaging tubes," *Phys. Stat. Sol. (B)*, vol. 246, pp. 1794–1805, 2009.
12. Z. He et al., "Position sensitive single carrier CdZnTe detectors," in *Nuclear Science Symposium, 1996. Conference Record, 1996 IEEE*, pp. 331–335, 1996.
13. R. Street et al., "Comparison of PbI2 and HgI2 for direct detection active matrix x-ray image sensors," *J. Appl. Phys.*, vol. 91, pp. 3345–3355, 2002.
14. W. Zhao and J. A. Rowlands, "Large-area solid state detector for radiology using amorphous selenium," *Proc. SPIE: Med. Imaging*, vol. 1651, pp. 134–143, 1992.

15. D. L. Y. Lee, L. K. Cheung and L. S. Jeromin, "New digital detector for projection radiography," *Proc. SPIE: Med. Imaging*, pp. 237–249, 1995.
16. U. Bick and F. Diekmann, *Digital Mammography*. Springer Science & Business Media, New York, 2010.
17. S. Rudin et al., "New light-amplifier-based detector designs for high spatial resolution and high sensitivity CBCT mammography and fluoroscopy," *Med. Imaging*, pp. 61421R, 2006.
18. I. Cunningham and B. Reid, "Signal and noise in modulation transfer function determinations using the slit, wire, and edge techniques," *Med. Phys.*, vol. 19, pp. 1037–1044, 1992.
19. E. Samei, M. J. Flynn and D. A. Reimann, "A method for measuring the presampled MTF of digital radiographic systems using an edge test device," *Med. Phys.*, vol. 25, pp. 102–113, 1998.
20. I. A. Cunningham, "Handbook of medical imaging," Anonymous, pp. 79–159, 2000.
21. M. Z. Kabir, M. W. Rahman and W. Shen, "Modelling of detective quantum efficiency of direct conversion x-ray imaging detectors incorporating charge carrier trapping and K-fluorescence," *IET Circuits, Devices Syst.*, vol. 5, pp. 222–231, 2011.
22. G. Belev and S. O. Kasap, "Amorphous selenium as an X-ray photoconductor," *J. Non-Cryst. Solids*, vol. 345, pp. 484–488, 2004.
23. S. A. Mahmood et al., "Dark current in multilayer amorphous selenium x-ray imaging detectors," *Appl. Phys. Lett.*, vol. 92, pp. 223506-1–223506-3, 2008.
24. R. Keshavarzi et al., "Performance of a prototype 32x32 pixel indirect x-ray imager based on a lateral selenium passive pixel sensor," *Proc. SPIE: Med. Imaging*, vol. 8313, p. 83135O, 2012.
25. K. S. Karim and S. Abbaszadeh, "Radiation detector system and method of manufacture," US9269838, Granted Feb. 2016.
26. W. Zhao and J. Rowlands, "X-ray imaging using amorphous selenium: Feasibility of a flat panel self-scanned detector for digital radiology," *Med. Phys.*, vol. 22, pp. 1595–1604, 1995.
27. O. Tousignant et al., "Spatial and temporal image characteristics of a real-time large area a-se x-ray detector," *Med. Imaging*, pp. 207–215, 2005.
28. G. Juška and K. Arlauskas, "Impact ionization and mobilities of charge carriers at high electric fields in amorphous selenium," *Phys. Stat. Sol. (A)*, vol. 59, pp. 389–393, 1980.
29. S. Abbaszadeh, N. Allec and S. K. Karim, "Characterization of low dark-current lateral amorphous-selenium metal–semiconductor–metal photodetectors," *IEEE Sensors*, vol. 13, pp. 1452–1458, 2013.
30. M. Yunus, "Monte Carlo Modeling of the Sensitivity of X-Ray Photoconductor," University of Saskatchewan, April 2005.
31. G. Juška and K. Arlauskas, "Impact ionization and mobilities of charge carriers at high electric fields in amorphous selenium," *Phys. Stat. Sol. (A)*, vol. 59, pp. 389–393, 1980.
32. K. Tsuji et al., "Impact ionization process in amorphous selenium," *J. Non-Cryst. Solids*, vol. 114, pp. 94–96, 1989.
33. S. O. Kasap and G. Belev, "Progress in the science and technology of direct conversion X-ray image detectors: The development of a double layer a-Se based detector," *J. Optoelectron. Adv. Mat.*, vol. 9, pp. 1–10, 2007.

34. O. Bubon et al., "Electroded avalanche amorphous selenium (a-Se) photosensor," *Curr. Appl. Phys.*, vol. 12, pp. 983–988, 2012.
35. R. E. Johanson et al., "Metallic electrical contacts to stabilized amorphous selenium for use in X-ray image detectors," *J. Non-Cryst. Solids*, vol. 227–230, pp. 1359–1362, 1998.
36. K. Tanioka et al., "An avalanche-mode amorphous Selenium photoconductive layer for use as a camera tube target," *IEEE Electron Device Lett.*, vol. 8, pp. 392–394, 1987.
37. E. Maruyama, "Amorphous built-in-field effect photoreceptors," *Jpn. J. Appl. Phys.*, vol. 21, pp. 213–223, 1982.
38. K. Tanioka, "High-gain avalanche rushing amorphous photoconductor (HARP) detector," *Nucl. Instrum. Methods A*, vol. 608, pp. S15–S17, 9/1, 2009.
39. M. M. Wronski, "Development of a Flat Panel Detector with Avalanche Gain for Interventional Radiology." Thesis, 2009.
40. M. M. Wronski et al., "A solid-state amorphous selenium avalanche technology for low photon flux imaging applications," *Med. Phys.*, vol. 37, pp. 4982–4985, 2010.
41. A. Reznik et al., "Applications of avalanche multiplication in amorphous selenium to flat panel detectors for medical applications," *J. Mater. Sci.-Mater. El.*, vol. 20, pp. 63–67, 2009.
42. A. Reznik et al., "Avalanche multiplication in amorphous selenium and its utilization in imaging," *J. Non-Cryst. Sol.*, vol. 354, pp. 2691–2696, 2008.
43. T. Watabe et al., "New signal readout method for ultrahigh-sensitivity CMOS image sensor," *IEEE Trans. Electron Devices*, vol. 50, pp. 63–69, 2003.
44. J. R. Scheuermann et al., "Charge transport model in solid-state avalanche amorphous selenium and defect suppression design," *J. Appl. Phys.*, vol. 119, p. 024508, 2016.
45. F. Nariyuki et al., "New development of large-area direct conversion detector for digital radiography using amorphous selenium with a C60-doped polymer layer," *Proc. SPIE: Med. Imaging*, vol. 7622, p. 762240, 2010.
46. S. Abbaszadeh et al., "Investigation of hole-blocking contacts for high-conversion-gain amorphous selenium detectors for x-ray imaging," *IEEE Trans. Electron Devices*, vol. 59, pp. 2403–2409, 2012.
47. S. Abbaszadeh et al., "Enhanced detection efficiency of direct conversion X-ray detector using polyimide as hole-blocking layer," *Scientif. Rep.*, vol. 3, pp. 3360-1–7, 2013.
48. C. C. Scott et al., "Amorphous selenium direct detection CMOS digital x-ray imager with 25 micron pixel pitch," *SPIE Med. Imaging*, p. 90331G, 2014.
49. C. S. Levin and E. J. Hoffman, "Calculation of positron range and its effect on the fundamental limit of positron emission tomography system spatial resolution," *Phys. Med. Biol.*, vol. 44, p. 781, 1999.
50. S. R. Cherry, "The 2006 Henry N. Wagner Lecture: Of mice and men (and positrons)—Advances in PET imaging technology," *J. Nucl. Med.*, vol. 47, pp. 1735–1745, 2006.
51. V. C. Spanoudaki and C. S. Levin, "Photo-detectors for time of flight positron emission tomography (ToF-PET)," *Sensors*, vol. 10, pp. 10484–10505, 2010.
52. E. Roncali and S. R. Cherry, "Application of silicon photomultipliers to positron emission tomography," *Ann. Biomed. Eng.*, vol. 39, pp. 1358–1377, 2011.
53. J. R. Stickel, J. Qi and S. R. Cherry, "Fabrication and characterization of a 0.5-mm lutetium oxyorthosilicate detector array for high-resolution PET applications," *J. Nucl. Med.*, vol. 48, pp. 115–121, 2007.

54. M. Nitta et al., "The X'tal cube PET detector of isotropic (0.8 mm) 3 crystal segments," in *2014 IEEE Nuclear Science Symposium and Medical Imaging Conference (NSS/MIC)*, pp. 1–3, 2014.
55. Y. Gu et al., "Study of a high resolution, 3-D positioning cross-strip cadmium zinc telluride detector for PET," in *Nuclear Science Symposium Conference Record, 2008. NSS'08. IEEE*, pp. 3596–3603, 2008.
56. Y. Gu et al., "Study of a high-resolution, 3D positioning cadmium zinc telluride detector for PET," *Phys. Med. Biol.*, vol. 56, p. 1563, 2011.
57. M. Ito, S. J. Hong and J. S. Lee, "Positron emission tomography (PET) detectors with depth-of-interaction (DOI) capability," *Biomed. Eng. Lett.*, vol. 1, pp. 70–81, 2011.
58. G. Chinn, A. M. Foudray and C. S. Levin, "A method to include single photon events in image reconstruction for a 1 mm resolution PET system built with advanced 3-D positioning detectors," in *Nuclear Science Symposium Conference Record, 2006. IEEE*, pp. 1740–1745, 2006.
59. C. S. Levin, "Promising new photon detection concepts for high-resolution clinical and preclinical PET," *J. Nucl. Med.*, vol. 53, pp. 167–170, 2012.
60. A. Drezet et al., "CdZnTe detectors for the positron emission tomographic imaging of small animals," in *Nuclear Science Symposium Conference Record, 2004 IEEE*, pp. 4564–4568, 2004.
61. L. J. Meng and Z. He, "Exploring the limiting timing resolution for large volume CZT detectors with waveform analysis," *Nucl. Instrum. Methods Phys. Res. A*, vol. 550, pp. 435–445, 2005.
62. P. Vaska et al., "A prototype CZT-based PET scanner for high resolution mouse brain imaging," in *Nuclear Science Symposium Conference Record, 2007. NSS'07. IEEE*, pp. 3816–3819, 2007.
63. S. Komarov et al., "Investigation of the limitations of the highly pixilated CdZnTe detector for PET applications," *Phys. Med. Biol.*, vol. 57, p. 7355, 2012.
64. F. Zhang et al., "Feasibility study of using two 3-D position sensitive CZT detectors for small animal PET," in *Nuclear Science Symposium Conference Record, 2005 IEEE*, p. 4, 2005.
65. P. Vaska et al., "Initial performance of the RatCAP, a PET camera for conscious rat brain imaging," in *Nuclear Science Symposium Conference Record, 2005 IEEE*, pp. 3040–3044, 2005.
66. Y. Gu and C. Levin, "Study of electrode pattern design for a CZT-based PET detector," *Phys. Med. Biol.*, vol. 59, p. 2599, 2014.
67. J. L. Matteson et al., "Charge collection studies of a high resolution CZT-based detector for PET," in *Nuclear Science Symposium Conference Record, 2008 IEEE*, pp. 503–510, 2008.
68. S. Abbaszadeh et al., "Characterization of a sub-assembly of 3D position sensitive cadmium zinc telluride detectors and electronics from a sub-millimeter resolution PET system," *Phys. Med. Biol.*, vol. 61, p. 6733, 2016.
69. L. Tao, H. M. Daghighian and C. S. Levin, "A promising new mechanism of ionizing radiation detection for positron emission tomography: Modulation of optical properties," *Phys. Med. Biol.*, vol. 61, p. 7600, 2016.
70. M. Spahn, "Flat detectors and their clinical applications," *Eur. Radiol.*, vol. 15, pp. 1934–1947, 2005.

2

GaAs:Cr as Sensor Material for Photon Counting Pixel Detectors

Elias Hamann

CONTENTS

2.1 Introduction

In recent decades, photon counting x-ray pixel detectors have found their way into many scientific fields like particle physics, scattering and diffraction, (pre-) clinical research, materials science, at synchrotrons, and also increasingly at laboratory x-ray setups. Here, they more and more replace or work concurrently with classical detection media like films or indirect converting, scintillator-based detector systems. The conversion of x-rays to charge carriers and thus electrical signals directly in the (semiconductor) sensors, as well as several technological developments, led to features like high-dynamic ranges, high signal-to-noise ratios (and thus dose efficiency), and frame rates up to several kilohertz. Pixel sizes in the range of several tens of microns ensure a spatial resolution sufficiently high for many applications like (pre-)clinical studies and nondestructive testing. Several chips, for instance the Medipix3RX [1], additionally offer the possibility of setting more than one energy threshold, which allows making use of the spectral composition of the radiation. Further, two of the biggest problems in small pixel detectors, namely charge sharing and the influence of characteristic x-rays from the sensor material, were recently tackled by advanced hit allocation and charge summing functionalities [2,3]. This has opened up new ways in rather novel techniques like spectral imaging for material decomposition (e.g., separating contrast agents) or allows the filtering of higher harmonics at synchrotron beamlines. A comprehensive review about the current status, the challenges, as well as potentials of photon-counting detectors can be found in Ref. [4], and, exemplarily for the LAMBDA detector system, in Chapter 3.

Most photon-counting pixel detectors are so-called *hybrid* detectors, where a suitable semiconductor sensor is attached to dedicated readout electronics by flip-chip techniques like bump bonding. In this way, the sensor material can be chosen to best fit the requirements imposed by the application. The most widely used semiconductor sensor material is silicon (Si) since it is available in large wafer sizes at relatively low cost and is nearly defect-free. However, due to the relative low absorption and the increasing influence of the (unwanted) Compton effect for energies above 30 keV, the (dose) efficient use of Si is limited. Since many applications like medical CT and material testing require higher x-ray energies (100 keV and more) in order to penetrate large and/or heavy samples, the search for suitable semiconductor sensor materials thus has become an important task besides the constant development and improvement of the readout electronics itself.

This chapter presents the characterization and application of high-resistivity, chromium compensated gallium arsenide (GaAs:Cr), a semiconductor sensor material that recently has (re-)gained increasing attention. As will be shown below, this material shows detector-grade properties and thus is well suitable to be used for x-ray detectors. The chapter is divided in three parts: First, a detailed characterization of the material properties of GaAs:Cr with respect to its use as a sensor material for photon-counting x-ray pixel detectors (homogeneity, resistivity, charge carrier properties) is performed. Then, the fabrication steps of GaAs:Cr Medipix detector assemblies as well as the functionality of the Medipix3RX photon-counting readout electronics are described. Finally, the performance of these detectors with respect to (spectral) x-ray imaging is investigated. Interested readers are kindly advised to read the PhD thesis [5] written by the author for more detailed information.

2.2 High Resistivity, Chromium Compensated Gallium Arsenide

Since the absorption of a material strongly depends on its atomic number Z, so-called *high-Z* materials like germanium, gallium arsenide, or cadmium (zinc) telluride have been thoroughly investigated in the last decades in order to find a material suitable for high-energy x-ray applications [6]. Figure 2.1 shows a comparison of the absorption properties of silicon, gallium arsenide (similar to germanium), and cadmium telluride in several available thicknesses. Cadmium (zinc) telluride (CdTe, CZT) exhibits high-absorption capabilities for photon energies up to 150 keV and thus has gained particular interest, e.g., for medical applications. However, detectors based on thick sensors may suffer from charge trapping and detrimental effects like charge sharing and long-ranged sensor fluorescence radiation, the last two leading to crosstalk between neighboring pixels. Further, CdTe and CZT can show temporal instabilities, especially under high fluxes and when blocking Schottky contacts are employed [7–10], in the worst case leading to (temporary) device failure. Germanium (Ge), like Si, can be produced in large, almost defect-free volumes, but has to be cooled due to the small bandgap (0.67 eV, compared to 1.12 eV in Si, 1.43 eV in GaAs, and 1.5 eV in CdTe), which results in higher leakage currents at room temperature.

Gallium arsenide (GaAs) has been considered promising for radiation detection since more than half a decade [11], but did not have sustaining success. Its atomic numbers 31 and 33, embracing germanium with $Z = 32$, lead to an efficient absorption of x-ray energies up to ~60 keV (see Figure 2.1), which makes it a perfect intermediate sensor material between Si and CdTe/CZT and thus suitable for applications like nondestructive testing, mammography, or dental

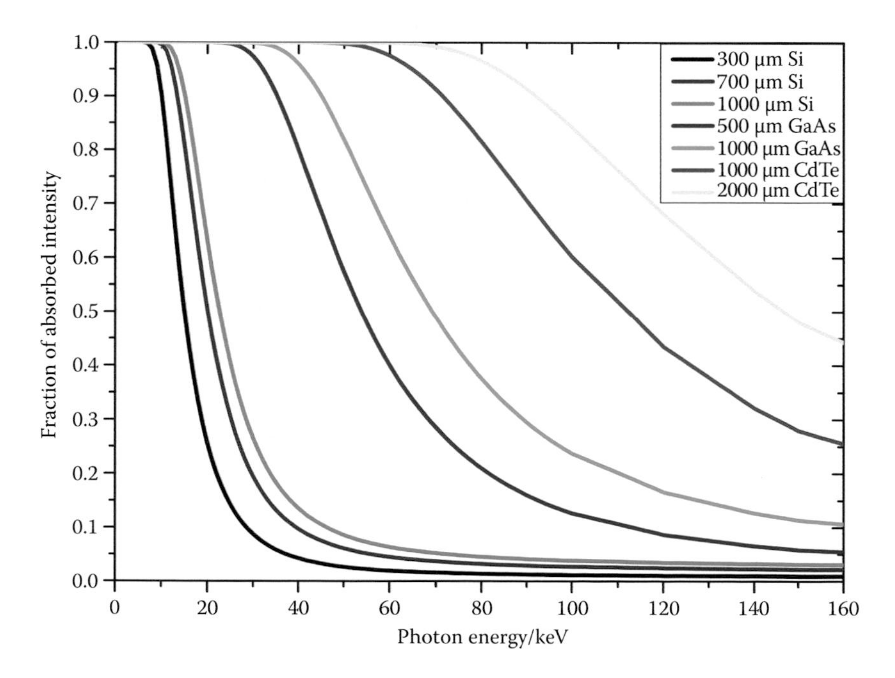

FIGURE 2.1
Absorption properties versus photon energy for Si, GaAs, and CdTe in several available thicknesses.

imaging. For these energies, the contribution of inelastic (Compton) scattering is also very small, which means that almost all photon interaction takes place via the desired photo effect. Further, characteristic x-rays produced in the sensor are rather short-ranged as compared to CdTe (~40 μm for Ga fluorescences as compared to ~120 μm for characteristic Cd radiation, both denoting 1/e absorption lengths excluding coherent scattering), thus facilitating spectroscopic imaging.

Most of the proposed and investigated detector technologies have been based on either epitaxial or melt-grown bulk material and a considerable amount of publications were written, especially around the 1990s and early 2000s, which arguably can be considered as the most productive era for GaAs-based radiation detectors (see, e.g., Refs. [6,12–15] and many more).

However, the thickness of the active layers was mainly limited either due to the growth technique (epitaxy) or the need of creating depletion zones (e.g., by diode structures or by applying Schottky barriers) to increase resistivity, or a combination of both. This often led to large parts of the sensor being insensitive to radiation, thus preventing efficient use of the sensor volume for the absorption of radiation. Furthermore, persistent technological problems with this material have impeded the routine application of GaAs for x-ray imaging detectors, mainly caused by short ranges and temporal instabilities of the electrical field in the sensor. Hence, only very few successful studies of imaging detectors based on epitaxial or semi-insulating GaAs are reported (e.g., in Refs. [16,17]), but eventually most of these devices did not go beyond prototype status. As a result, the number of publications abated somewhat during the first 10 years after the millennium change.

However, it was also in the early 2000s that a research group from Tomsk, Russia, introduced a novel technique to fabricate detector-grade GaAs sensors. This method is based on high-temperature chromium (Cr) diffusion into n-type GaAs wafers grown by the liquid encapsulated Czochralski (LEC) method, which allows the production of relatively thick and large (up to 1 mm thickness and 4 in. diameter) GaAs:Cr wafers with high resistivity and fully active volumes, i.e., without the need for blocking contacts leading to confined depletion regions [18,19]. Further, it was found that the electric field in this material is uniform and stable throughout the whole wafer volume and that electronic properties were sufficiently high to guarantee good charge collection efficiency (CCE). However, despite these very promising properties, GaAs:Cr remained somewhat unnoticed and thus unstudied with respect to x-ray imaging detectors for almost a decade and has regained increasing attention only within the last few years [20–24].

In the next subchapters, a detailed material characterization of GaAs:Cr with respect to homogeneity, resistivity, and charge carrier properties is given.

2.2.1 Homogeneity

For x-ray imaging applications, the uniformity of the detector material is an important point. Especially for computed tomography (CT), inhomogeneities in the sensor material can result in image artifacts and information loss on potentially important sample features. If these inhomogeneities remain constant in time and do not distort the image quality too much, they can be removed from the final image by standard techniques like *flatfield correction*. In contrast to that, time-dependent variations in x-ray sensitivity or leakage currents of the sensor are much more difficult to correct for (if possible at all) and result in degraded image quality [25]. If additionally spectroscopic information is desired, e.g., for spectral CT, crystal defects like impurities, dislocations, or grain and twin boundaries are reported to have a negative effect on the energy resolution since they lead to local variations of material properties like resistivity and charge carrier mobility [26,27]. Thus, a good sensor optimally should consist of a monolithic piece of defect-free semiconductor material. Whereas this is realizable for elemental semiconductors like Si or Ge, it is very difficult to obtain for almost all compound semiconductors grown from the melt (like GaAs or CdTe), mainly due to limitations and problems related to the crystal growth. Sensors made from compound semiconductors thus often contain crystal defects like dislocations, inclusions, or grain boundaries, which are also subjects of intensive analysis, e.g., in order to correlate them to features later visible in flood illumination images of the detectors made from them [9,28–29].

One convenient method to investigate the crystalline quality of semiconductor wafers or, respectively, the presence of defects in the material is white beam synchrotron x-ray topography (WBXT) [30–32]. Here, a polychromatic synchrotron beam is impinging on the sample, and certain Bragg reflections of choice can be recorded with photographic films or digital synchrotron cameras for further analysis. Crystal defects then are visible as contrast variations on the topograms due to local changes in the reflectivity (i.e., distortions of the crystal lattice) around the defects. In order to study the crystal quality of GaAs:Cr, WBXT measurements of a 500 μm thick, 40 mm diameter wafer were conducted at the TopoTomo beamline of the ANKA synchrotron [33] at the Karlsruhe Institute of Technology (KIT). The diffracted beam was recorded with a standard synchrotron camera system consisting of a scintillating screen, visible light optics, and a CCD. Wafer mapping was realized by laterally shifting the sample through the 5×5 mm^2 beam and stitching the single images in a postprocessing step. A 15×15 mm^2 region of interest (ROI) from the central part of the final topogram of the wafer is shown in Figure 2.2. Besides the prominent horizontal stripes, which stem from a non-uniform illumination of the scintillating screen, several topological features are visible, namely a relatively dense cellular network of dislocations (typical of LEC grown GaAs [34,35]) as well as a few elongated dislocation agglomerations called *lineages*. As expected, these features in the sensor material give rise to variations in x-ray sensitivity, resulting in count rate variations in x-ray images (see Figure 2.5a), which have to be corrected for in order to restore image homogeneity. As shown further below, this is possible with the help of a standard flatfield correction (cf. Figure 2.5b), which significantly

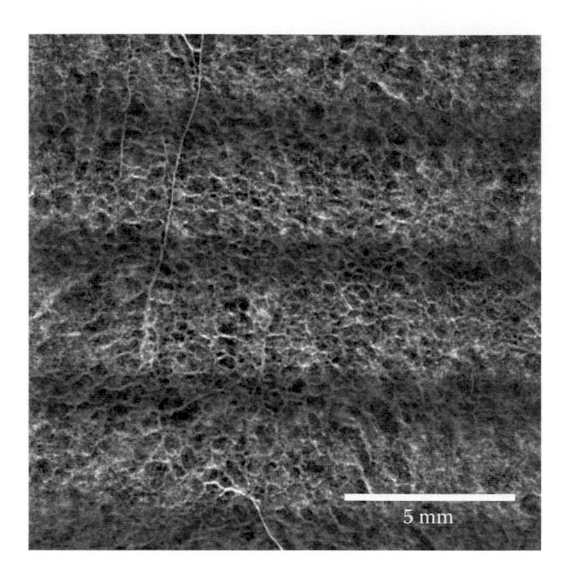

FIGURE 2.2
White Beam synchrotron x-ray Topography (15 × 15 mm^2 ROI) of a GaAs:Cr Wafer.

improves the image quality. However, these structures seem to be rather stable in time under x-ray illumination even at higher fluxes*—in contrast to dislocations and grain boundaries in CdTe, which are found to change in size and shape with time [9]—and thus even longer measurements like CT scans are possible without frequent interruptions for image correction, e.g., periodic bias voltage reset, as often applied to CdTe sensors [36,37].

2.2.2 Resistivity

The resistivity of a sensor material is important since it determines the maximum applicable bias voltage at an acceptable level of leakage current, which acts as one source of electronic noise. Furthermore, the charge carrier velocity and thus the drift time of the charges to the collecting electrodes directly depend on the electric field in the material, which, as a consequence, usually should be as high and uniform as possible.† Hence, a high resistivity is required for detector-grade sensor materials to guarantee a high CCE since shorter drift times lead to reduced charge trapping probability, a reduction of charge sharing, and thus to better spectroscopic performance.

The resistivity and linearity of the current–voltage characteristics of a 1 mm thick GaAs:Cr sensor already bump-bonded to a Medipix3RX readout chip

* This is at least true when using photon counting detectors with leakage current compensation; only very recently first observations of temporal effects like signal "afterglow" were reported using charge integrating readout electronics.

† Note that there is a peculiar nonlinear behavior of GaAs, which results in a maximum charge carrier velocity at a certain electric field, above which the velocity is decreasing again and eventually converges to a constant value.

(see also Section 2.3) were measured by applying negative bias voltages to the backside contact and measuring the current flowing through the sensor. Guard rings surrounding the active area of the sensor ensured that only currents flowing through the sensor volume were measured. The current–voltage (*I*–*V*) characteristics of the sensor as well as the calculated resistivity at each voltage are shown in Figure 2.3a and b, respectively. In agreement

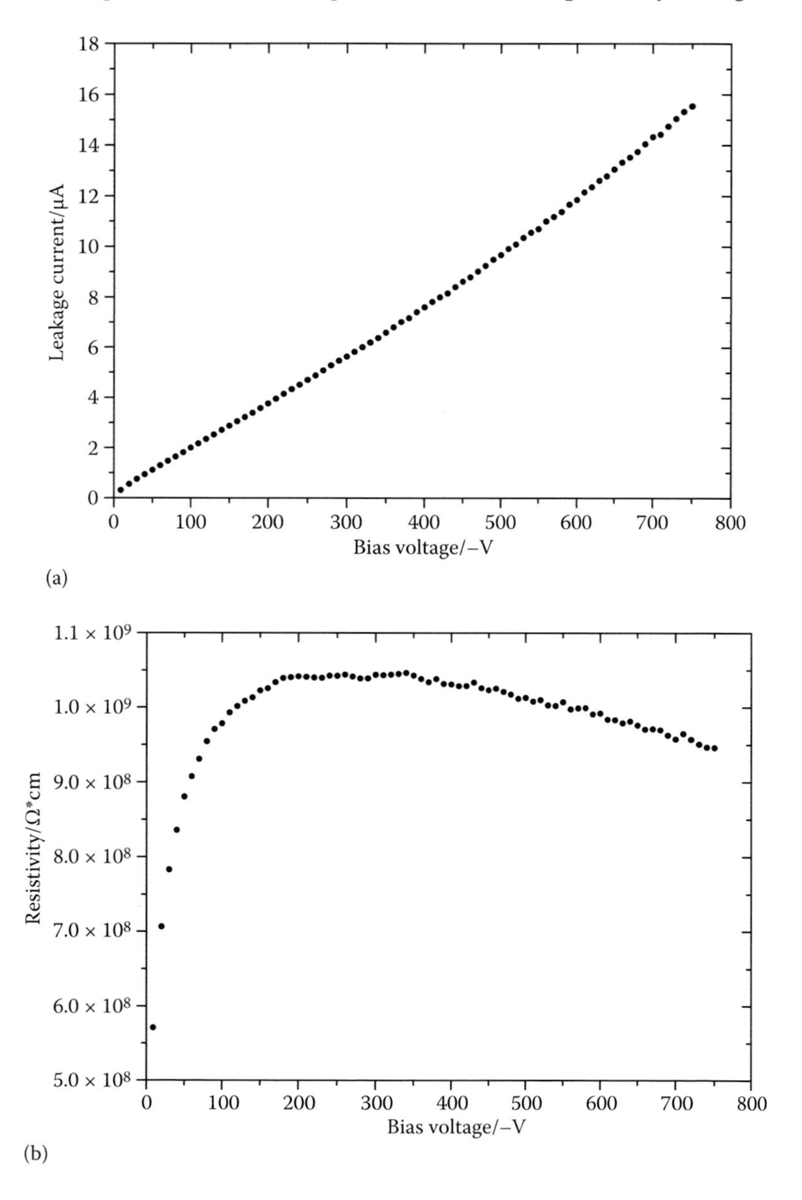

FIGURE 2.3
(a) Current–voltage characteristics and (b) calculated resistivity for a 1 mm GaAs:Cr sensor.

with Refs. [5,19], the leakage currents are rising linearly up to a certain bias voltage, confirming an ohmic behavior of the material in this voltage range (here up to approximately –500 V). Above this value, the dependence becomes superlinear, probably due to self-amplification effects from charge carrier scattering, the nonlinear voltage dependence of the electron velocity, or charge carrier injection from the contacts. Further, due to the relatively small influence from the metal contacts (and their effect on the band structure close to the metal–semiconductor interface), it can be assumed that the whole bias voltage drops across the sensor only and that the overall resistivity is thus determined by the GaAs:Cr bulk properties only [38–40]. The resistivity was found to be in the range of $1 \times 10^9\ \Omega$ cm, which is in good agreement with previously found values [19,41], and is even almost as high as the values reported for CdTe. This allows applying bias voltages of several hundred volts to maximize the CCE at sufficiently low leakage currents. Further, as shown in Ref. [42], the resistivity and electric field in GaAs:Cr are constant throughout the whole wafer thickness, resulting in fully active sensor volumes and ensuring a good absorption for higher x-ray energies.

2.2.3 Charge Carrier Properties

Both mobility and lifetime of charge carriers, and especially their product (often called $\mu\tau$-product), are further important parameters in semiconductors to be used as x-ray sensors. Multiplied by the electric field strength E, $\mu\tau{*}E$ gives the mean free ($1/e$) path length λ of the charge carriers in the material before trapping or recombination takes place. Thus, the $\mu\tau$-product determines the maximum thickness a sensor should have for a given or desired CCE. Usually, in semiconductors, the $\mu\tau$-product of one charge carrier type is much larger than of the other. However, since pixel detectors are so-called *single-polarity* devices due to the *small pixel effect*, which strongly suppresses the influence of one charge carrier type on the signal generation [43–45], it is sufficient that the $\mu\tau$-product of either electrons or holes is high. In GaAs:Cr, like in CdTe, the $\mu\tau$-product of the electrons is much higher than the one of holes.

One method used to measure the $\mu\tau$-product of a given charge carrier type consists of irradiating an electrode of a detector with monoenergetic alpha particles (e.g., as emitted by radionuclides) and measuring the corresponding pulse height in dependence of the applied bias voltage. Alpha particles are usually used since, on the one hand, a large number of charge carriers are generated, of which a sufficiently high number should arrive at the opposite electrode even under low-bias voltages, and on the other hand, it ensures that the interaction takes place directly underneath the contact, allowing only one charge carrier type to drift through the whole sensor. The resulting measurement curve then can be fitted with the so-called *Hecht equation* [46], which directly yields the $\mu\tau$-product as one of the fit parameters.* In

* Note that neither mobility nor lifetime can be determined separately with this method alone.

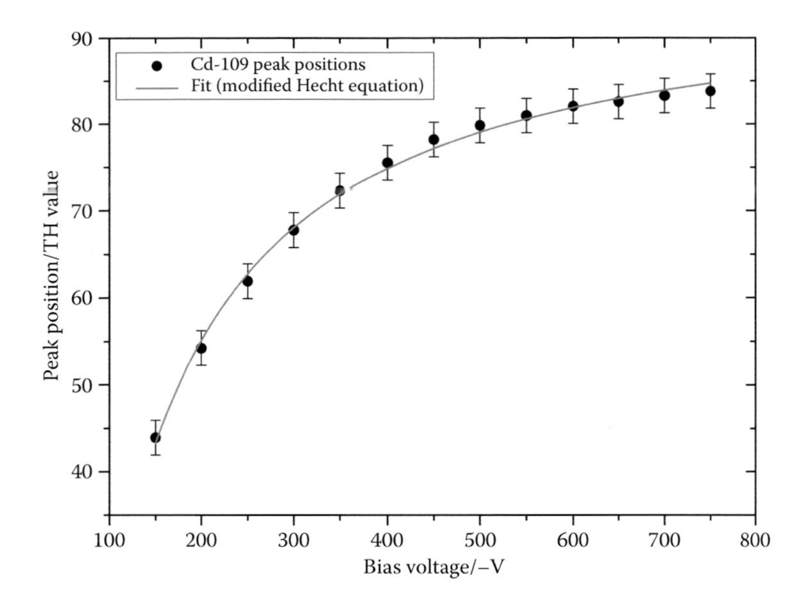

FIGURE 2.4
Cd-109 peak positions (in units of threshold steps) versus bias voltage measured with a 1 mm GaAs:Cr Medipix3RX detector. By fitting the data using a modified Hecht equation, the electron $\mu\tau$-product can be extracted.

this work, however, the electron $\mu\tau$-product of a 1 mm thick GaAs:Cr sensor, bump-bonded to a Medipix3RX chip, was measured using a monoenergetic gamma source (Cd-109, emitting at ~22 keV). This low-energy gamma source was chosen since, as reported in Refs. [47,48], alpha particles cannot be used to reliably determine the electron $\mu\tau$-product of GaAs:Cr due to the relatively long electron-hole plasma dissolving time and the resulting underestimation of the results. The peak positions of the gamma spectra for several bias voltages were obtained by threshold scans,* and the resulting plot is shown in Figure 2.4. A fit using a modified Hecht equation [5] resulted in a value of $\mu\tau_e \approx 1 \times 10^{-4}$ cm²/V, which is slightly larger, but close to the values presented in Ref. [41]. Whereas for 500 µm thick sensors this results in a high CCE† of 92%, for x-rays interacting directly at the backside electrode (i.e., the planar electrode opposite to the pixelated sensor surface), this value already drops to ~85% for a 1 mm thick sensor, possibly leading to low-energy tailing and a reduced energy resolution when recording gamma ray spectra due to partial charge trapping and/or recombination. This means that, an application-specific compromise has to be found between good spectroscopic performance (thinner sensor) and high absorption efficiency (thicker sensor).

* Unlike all other measurements presented in this work, the threshold scans were performed in high gain mode in order to improve the dynamic range for the $\mu\tau$-measurements.

† CCE values are calculated assuming an electrical field of 6 kV/cm and an electron mobility of 4000 cm²/Vs [41].

In conclusion, both resistivity and electron $\mu\tau$-product of the investigated GaAs:Cr wafers/sensors are found to be relatively high. Together with the reported stability [22] and uniformity of the electric field [19], this shows an excellent, detector-grade material quality and proves the suitability to be used as sensor material for x-ray pixel detectors.

This being said, the next section will focus on the description of the Medipix3RX detector assemblies with GaAs:Cr sensors, before in the final section of the chapter the performance of the assemblies with respect to (spectral) x-ray imaging applications is investigated.

2.3 GaAs:Cr Medipix3RX Detectors: Fabrication and Functionality

2.3.1 Device Fabrication

After the high-temperature chromium diffusion into the low-resistivity, n-type LEC wafers, further processing steps are necessary for the production of a fully assembled detector device. First, the wafers are polished in order to remove parts with higher Cr concentration close to the wafer surfaces, leading to a more homogeneous distribution of the Cr atoms in the wafer. In a following step, intentionally ohmic metal contacts are deposited on both sides of the wafer, with a planar contact structure on the (cathode) backside and a pixelated layout on the electron-collecting anode side, which later is connected to the readout chip. The materials used for the contacts are thin layers of vanadium and nickel for the anode and cathode, respectively. These materials generate only relatively small potential barriers (~0.8 V [48,49]) at the metal–semiconductor transition, which means that the overall resistivity can be assumed to be mainly determined by the bulk resistivity of the GaAs:Cr wafer. Besides the metal contacts, a surrounding metal guard ring structure can be deposited on the pixelated side in order to remove parasitic leakage currents and inhomogeneities in the electric field coming from the sensor edges.

After the processing of the wafer, it is diced in order to obtain sensors of the desired size (slightly larger than 14×14 mm^2 in the case of Medipix detectors) and eventually attached to the readout chip via flip-chip processes (in our case, bump bonding), which connect the segmented anode contacts with the pixel passivation openings of the readout chip. This is one of the most crucial steps in the detector processing chain, since the flip-chip yield is directly related to the number of functional pixels of the detector assembly. In the case of GaAs pixel detectors, this has often been an issue and considerable amounts of malfunctioning pixels are often found in x-ray images presented in earlier literature. Further, defects in the flip-chip connections (like pores, cracks, or voids) can also potentially influence the charge transport from the sensor to the readout chip and thus the signal generation in the pixel electronics.

Another important point during flip-chipping is the temperature under which this process takes place. Normally, relatively high temperatures are needed to melt solder bumps in order to stick to the readout chip pixels as well as to the sensor metal contacts. However, since silicon (the material the readout chip is mainly made of) and gallium arsenide exhibit different thermal expansion coefficients, this may lead to stresses in the sensor when the assembly is cooled down after the flip-chip process and thus poses an increased risk of instabilities of the solder connections.

In order to reduce thermal stresses during the flip-chip process, a low-temperature bump-bonding process optimized for high-Z sensor materials has been developed at the Freiburg Materials Research Center (FMF, University of Freiburg, Germany) [50]. This technique has earlier proven to produce a high pixel yield for CdTe Medipix detectors, and consequently has also been applied to connect the diced GaAs:Cr sensors to the Medipix3RX readout chips used in this work. After the flip-chip process, the bump bonds are no longer available for visible inspection, and one has to rely on other methods for investigating the bonding quality. Synchrotron Radiation Computed Laminography (SRCL, [51,52]), which is a modification of CT and is particularly suitable for laterally extended, flat samples, has shown to be a useful tool to nondestructively investigate the flip-chip connections of high-Z hybrid pixel detectors. Such measurements would go beyond the scope of this work, but as described in Ref. [53], the quality of the low-temperature flip-chip process used to connect GaAs:Cr sensors to Medipix chips is very good. By means of SRCL, sensor and readout chips were found to be perfectly aligned and no structural damages of the solder bumps (cracks, pores, etc.) were visible. This results in a very high-pixel yield (>99.99%) and only very few nonconnected pixels visible in flood images, such as shown further below in Figure 2.5b. As compared to earlier publications, this also shows a significant improvement in the flip-chip technology for GaAs detectors and a step toward routinely producing high-quality detector devices.

As a last step in the assembly processing chain, the chip/sensor unit is attached to a suitable carrier board, which connects the detector assembly to the readout electronics, and the data and control signal lines between the carrier board and the chip are established with wire-bond connections.

2.3.2 Medipix3RX Readout Chips

After having described the fabrication process of the GaAs:Cr Medipix3RX detector assemblies, this section shall shortly review the functionality of this particular readout chip. A more detailed description of the chip can be found in Ref. [1].

The Medipix3RX is a single photon-counting readout chip fabricated in 0.13 μm CMOS technology. It has 256 × 256 square pixels with a pitch of

55 μm. As all readout chips from the Medipix family, it features the so-called Krummenacher preamplifier layout [54], which can handle both negative and positive input charge polarities. This offers the possibility to use different sensor materials, depending on the desired application and which polarity offers the better charge carrier properties. This preamplifier layout also features an active leakage current compensation controlled by the so-called *Ikrum* DAC, which at the same time is responsible for the analog pulse shaping time [55].

The photon-counting principle simply means that, as opposed to energy integration, a counter in a pixel is increased by one, if the electrical signal coming from an interaction in the sensor exceeds a certain energy threshold globally set for all pixels of the detector. In this way, the events are weighted all equally and not according to their energy. Further, the possibility of setting an energy threshold also allows cutting off the electronic noise completely, resulting in single photon sensitivity, large dynamic ranges, and increased signal-to-noise ratios optimally only limited by Poisson statistics.

The Medipix3RX was specifically designed to overcome one of the most limiting factors of small pixel detectors, namely charge sharing, in which the charge cloud is spread (at least partially) over several pixels upon arrival at the readout electronics. The resulting signal is thus "shared" between these pixels and crosstalk between them occurs. Whereas this effect can be helpful for particle detection (e.g., alpha particles, heavy ions, or in high-energy physics), it is detrimental for (spectral) x-ray imaging since it worsens spatial and especially energy resolution of the detector. Hence, besides the "normal" operation mode, in which all pixels act individually (the so-called *single pixel mode* [SPM]), the Medipix3RX chip features an event-by-event based interpixel communication functionality (the so-called *charge summing mode* [CSM]), which recovers the full charge signal from all participating pixels of each event and assigns it to the pixel in which the largest fraction of the charge has been collected. If the overall charge signal then exceeds the threshold, the counter in this pixel (and this pixel only) is increased. Despite an increase in noise due to the analog summing of the charge signals, the CSM functionality has shown to significantly increase energy resolution and to reduce the influence of the energy threshold on the effective spatial resolution (see also Sections 2.4.3 and 2.4.4 and Refs. [3,56]).

By design, two energy thresholds are available per pixel of the Medipix3RX chip, each connected to a 12-bit counter. In SPM, both thresholds can be set individually, whereas in CSM, one threshold is needed for the hit allocation circuitry. In both modes, however, the two counters can be combined to one 24-bit counter to increase the dynamic range (in SPM, this is only possible when omitting one of the two thresholds). If only one in four pixels (in 2 × 2 submatrices) of the chip is bump-bonded to the sensor, the so-called *color mode* can be employed. In this mode, which then exhibits 110 μm pixel pitch, all available thresholds of the four physical pixels can be combined and used within the new "superpixel" (i.e., eight thresholds are then available in SPM and four in CSM).

In the frame of this work, two GaAs:Cr Medipix3RX detector assemblies were used: one with a 500 μm thick sensor and one with a 1 mm thick sensor. Since the 500 μm thick sensor has already been characterized thoroughly in our previous work [3], the main focus here lies on the investigation of the 1 mm thick sensor. All measurements were performed at room temperature using the USB2 *FitPix* interface [57] and the *Pixelman* software package [58], both being developed by the Institute of Experimental and Applied Physics (IEAP) of the CTU Prague (Czech Republic). Prior to the measurements, a threshold adjustment procedure (so-called *equalization*) using the pixels' electronic noise level was undertaken in order to reduce the inherent pixel-to-pixel threshold mismatch. Unless noted otherwise, all measurements were done with the 1 mm GaAs:Cr Medipix3RX assembly in CSM, employing low-gain mode, low-noise settings (for more details, see Ref. [1]), and a bias voltage of −650 V applied to the planar backside contact, with the threshold set just above the electronic noise level.

2.4 GaAs:Cr Medipix3RX Detectors: Imaging Performance

2.4.1 Image Quality

Twenty-one flood images were taken using a microfocus x-ray tube with a tungsten target, operated at 80 kV acceleration voltage. One of these images, normalized with respect to the mean count value, is shown in Figure 2.5a, together with the pixel-count histogram (Figure 2.5c). Since no sample was in the beam, it is clear that the variations in x-ray sensitivity solely come from the features in the sensor material itself (mainly dislocation networks, as discussed in Section 2.2.1), disturbing the image homogeneity due to their local influence on the charge transport properties. The slight asymmetry toward higher counts stems from the higher counting sensor edges, caused by residual edge effects, which are not removed by the guard ring. The small darker spot at the central left part of the image corresponds to the silver conductive glue with which the high-voltage wire connection was attached to the backside contact.

In order to correct for these image inhomogeneities, the remaining 20 images were summed, normalized, and then used to flatfield correct the first image. The result is shown in Figure 2.5b, again together with the corresponding histogram (Figure 2.5d). Obviously, the homogeneity of the corrected image is largely increased, and the influences of the material inhomogeneities are successfully removed. Quantitatively, the standard deviation of the count variations is reduced by a factor of 10. Note that a minimum number of flatfield frames are needed to optimize the SNR of the corrected image. Indeed, as shown in Ref. [3], with a sufficiently large number of

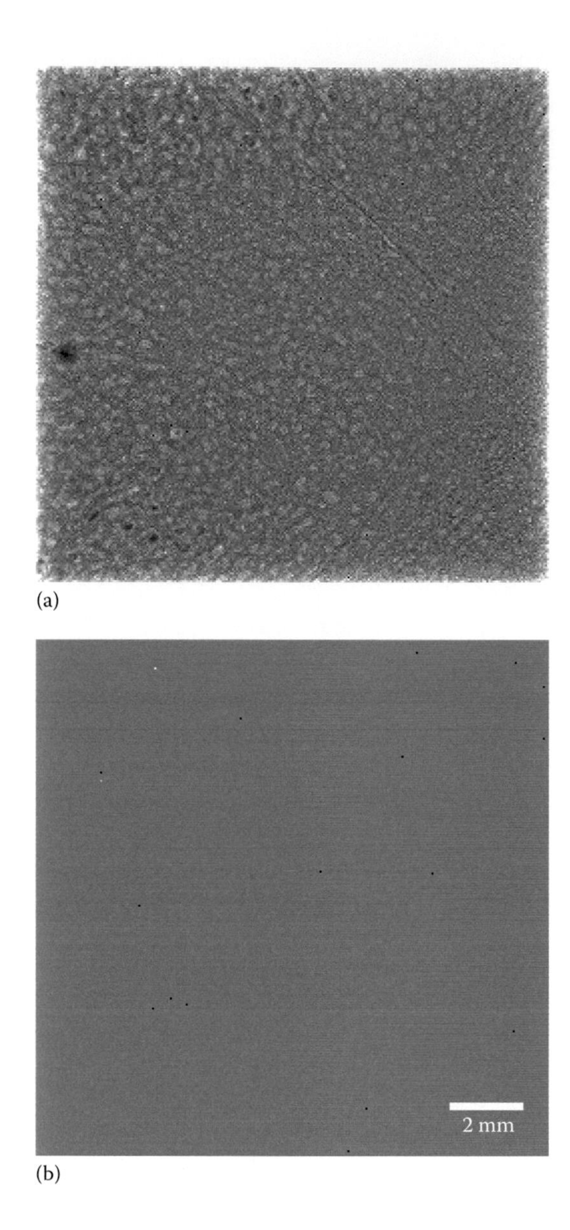

FIGURE 2.5
(a) Normalized flood image recorded with the 1 mm GaAs:Cr Medipix3RX assembly; (b) flat-field corrected image from (a). *(Continued)*

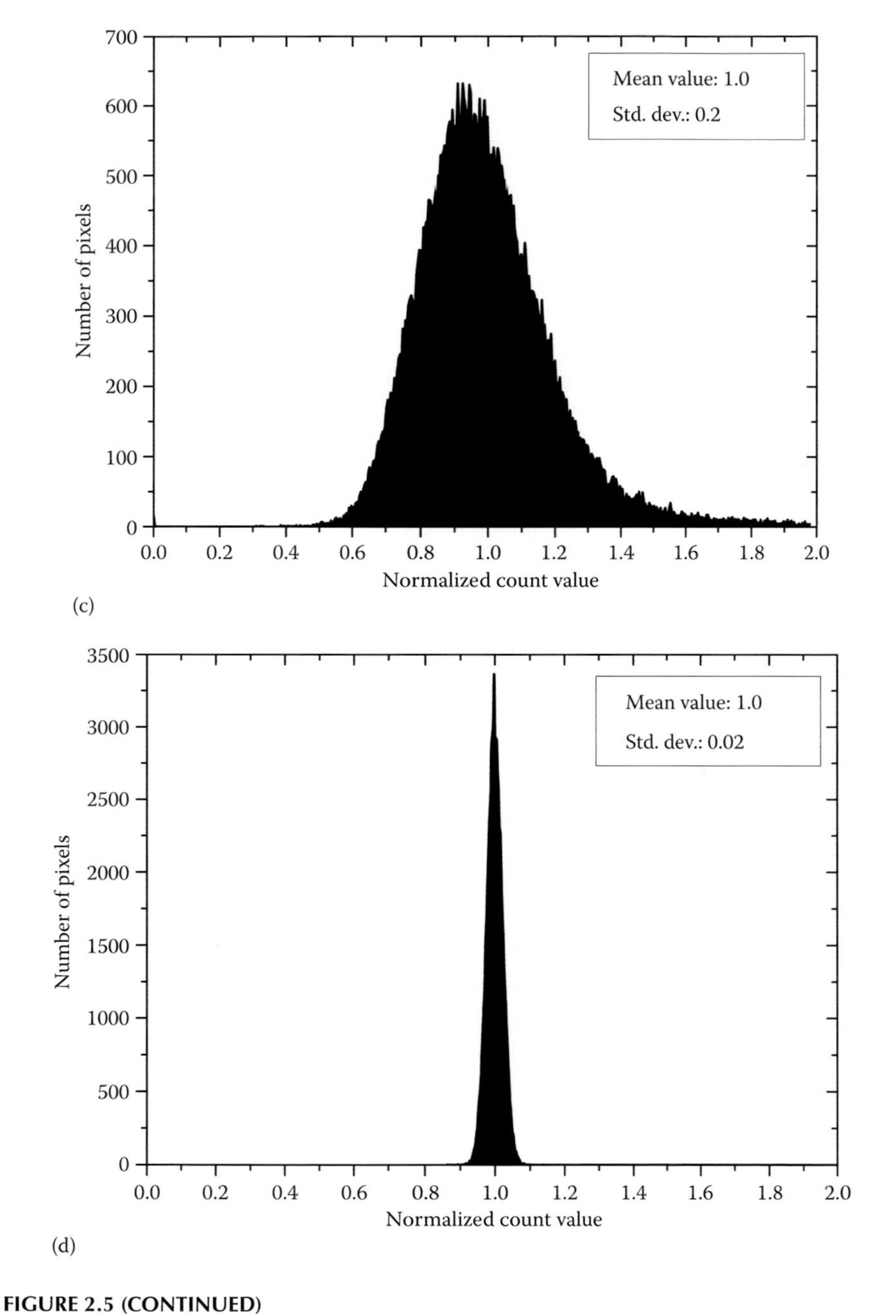

FIGURE 2.5 (CONTINUED)
(c and d) corresponding count histograms (normalized) of (a) and (b), respectively. (a) and (b) are shown in a linear gray scale from 0 to 2. *(Continued)*

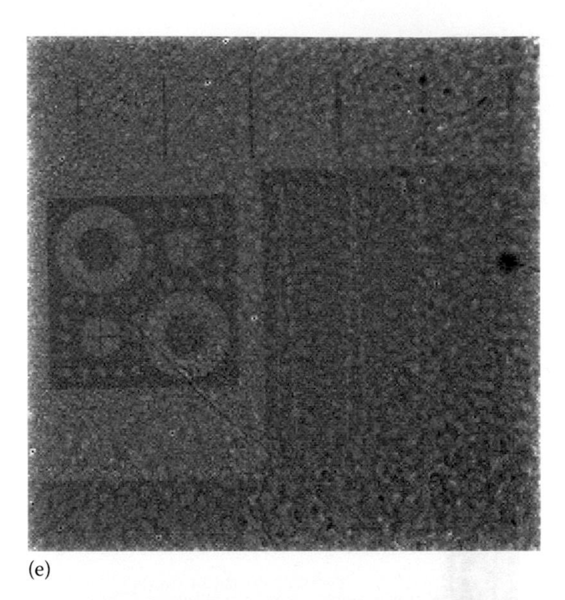

(e)

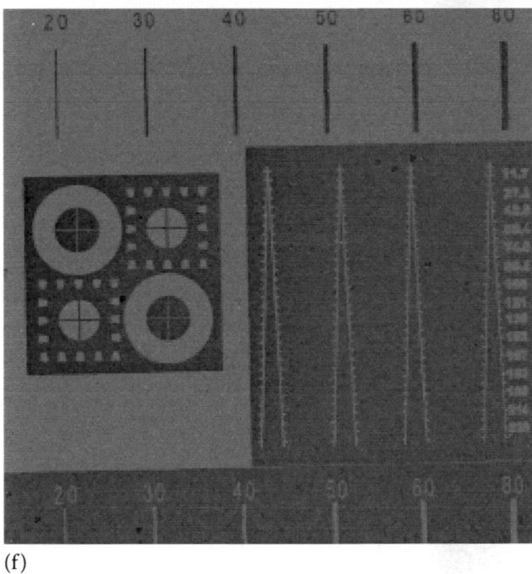

(f)

FIGURE 2.5 (CONTINUED)
(e) Raw image of a thin gold test pattern; (f) flatfield corrected image from (e) in linear gray scale from 0.5 to 1.5.

flatfield images, an SNR very close to the theoretical value as predicted by pure Poisson statistics is reachable with GaAs:Cr Medipix detectors.

The effect of the flatfield correction on x-ray images of a specimen (in this case, a thin gold test pattern deposited on a membrane) is shown in Figure 2.5e and f.

Whereas in the raw image the strong influence of the dislocation pattern is again visible, somewhat "hiding" many details of the specimen, the flatfield correction procedure successfully removes these distortions and reveals even fine details of the test pattern.

Figure 2.6a–d shows a collection of x-ray images (flatfield corrected and defective pixels removed; for details, see the figure caption), again demonstrating

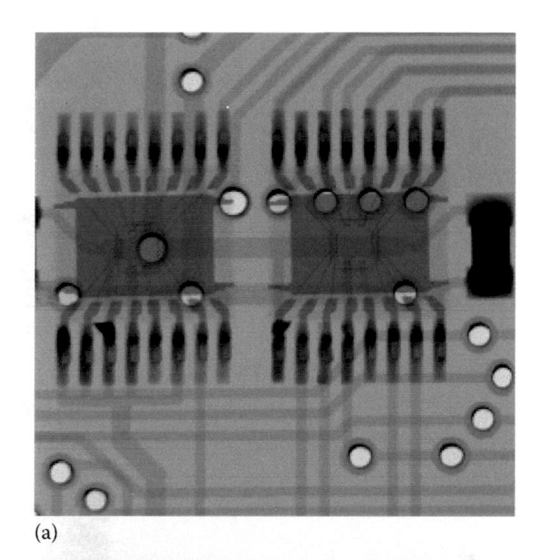

(a)

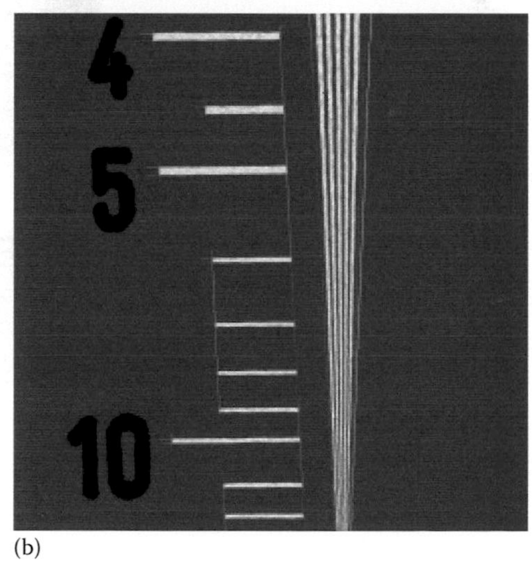

(b)

FIGURE 2.6
X-ray images of (a) two voltage converters on a PCB (Magnification M ≈ 2, linear gray scale from 0 to 1.2). (b) A lead resolution pattern (M ≈ 1, linear gray scale from 0.5 to 2.5, acquired with 0.05 mm Pb filter to reduce beam hardening). *(Continued)*

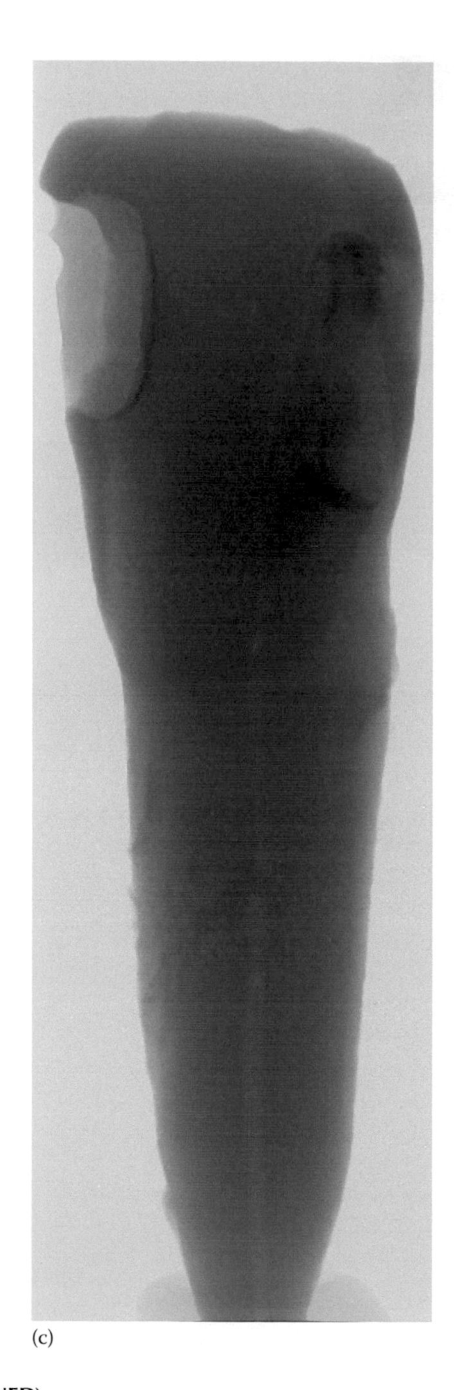

(c)

FIGURE 2.6 (CONTINUED)
(c) Human tooth with two fillings (sample provided by owner, 4 × 1 tiles, M ≈ 2, linear gray scale from 0 to 1.2).
(Continued)

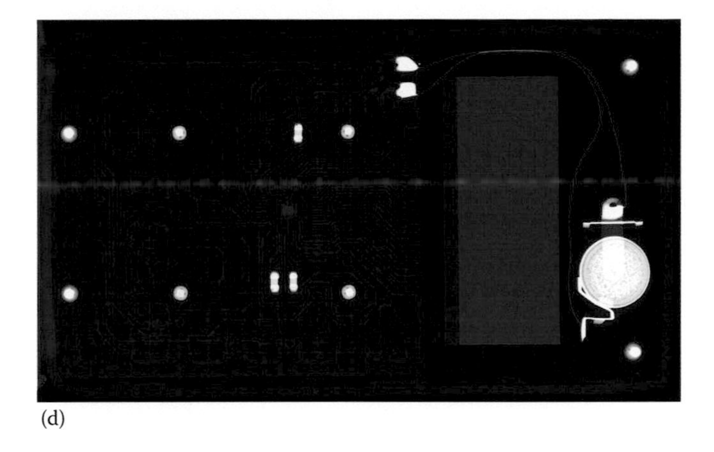

(d)

FIGURE 2.6 (CONTINUED)
(d) Pocket calculator (13 × 8 tiles, M ≈ 1.2, neg. log. false color scale [ImageJ 1.51 h *Fire* lookup table; http://imagej.nih.gov/ij] from 0 to 2.8). All acquired with an 80 kVp spectrum with 1 mm Al filter (except [b]).

the efficiency of the flatfield procedure and the good image quality that can be obtained with GaAs:Cr Medipix detectors. In Figure 2.6a, even small-sized features like the wire bonds of the IC are visible. This already indicates a high-spatial resolution of the detector, which will be investigated in more detail in Section 2.4.3. It has to be noted, however, that for denser and/or thicker samples, a pure flatfield correction might not suffice to obtain the desired image quality since the absorption properties of such samples might change the spectral composition of the transmitted x-ray beam (known as *beam hardening*). In this case, residual artifacts can arise in the flatfield corrected image due to the different interaction depths of different photon energies of the spectrum, mainly caused by the low-energy photons present in the flatfield images but not in the (highly absorbing) specimen image. Thus, adequate filtering of the beam in order to remove the lower-energy parts from the spectrum (as done in Figure 2.6b) and/or a beam hardening correction procedure [59] should be performed to maintain a high image quality.

2.4.2 Long-Term Stability and High-Flux Performance

For applications like (pre-)clinical CT, the deviation of the count rate linearity (i.e., the ratio of output to input count rate) should not be too pronounced at high photon fluxes in order to reliably reconstruct the real event rate observed in each pixel. The count rate behavior of Medipix detectors can be described with the so-called *paralyzable* model, which means that the pixel electronics and thus the counters are blocked during the processing of a single event. Further, this dead-time is extended if a second event occurs before the first event has been fully processed (so-called *pulse pileup*). It can thus be expected that the output count rate (OCR) increases slower than the input count rate (ICR) above a certain flux due to event losses, and eventually even decreases with further increasing count rate up to complete detector paralyzation (i.e., zero count rate). The count rate behavior of the GaAs:Cr detector was studied in a laboratory setup using a microfocus tungsten x-ray tube and an unfiltered 80 kVp spectrum. Photon fluxes were varied by different tube anode current settings. At each flux, 100 frames were recorded with 24-bit counter settings to avoid saturation. From these frames, the mean value and standard deviation of the median count rates* were then calculated and plotted versus the real incoming flux. The latter was calculated from the first seven data points, for which the count rate increase can be assumed to be linear (i.e., OCR = ICR). The results are shown in Figure 2.7a together with a fit using the paralyzable model (OCR = ICR*exp[-ICR*τ]), yielding a detector dead-time of $\tau = 4.9 \pm 0.2$ μs. The flux $\Phi_{0.9} = 8.3 \times 10^6$ 1/(s * mm^2), at which the measured OCR falls below 90% of the hypothetical linear count rate (light gray line in Figure 2.7a), is also indicated. These values are in the same order of what has previously been shown with Medipix3RX detectors using CdTe [56] and Si [60] sensors, confirming that the readout electronics and not the sensor material is the limiting factor. Despite flux rates of >10^8 1/(s * mm^2) used in medical CT applications are way beyond the detector's linear regime, it has to be pointed out that the observed dead-time as well as the value for $\Phi_{0.9}$ are specific for the used setup and chip settings and can be further optimized for high-flux applications (however, at cost of a loss in energy resolution), particularly by adjusting the *Ikrum* DAC [55,60].

Besides count rate linearity, a stable performance of the detector over time, optimally also under high-flux x-ray irradiation, is required. In order to investigate the long-term stability, the detector was illuminated with the x-ray tube (80 kVp) at three different fluxes/anode currents (100, 500, and 2000 μA). One frame each 5 s was consecutively taken for 1 h measurement time per current. The median count rate per frame versus time (i.e., frame number) is shown in Figure 2.7b, indicating an excellent long-term stability of the detector also at higher fluxes, where the OCR already significantly deviates from the ICR. A very slight decrease in the count rate during the 1 h measurements (visible, e.g., at the 2000 μA measurement) results from an effective increase in the threshold position due to a slight temperature increase in the x-ray cabinet during the measurements (as also observed with a Medipix3RX chip and a CdTe sensor in Ref. [25]), and thus can be attributed to effects from the readout chip rather than from the sensor material.

* Median counter values were used in order to remove the influence of defective (e.g., "hot") pixels on the measurements, possibly leading to biased results.

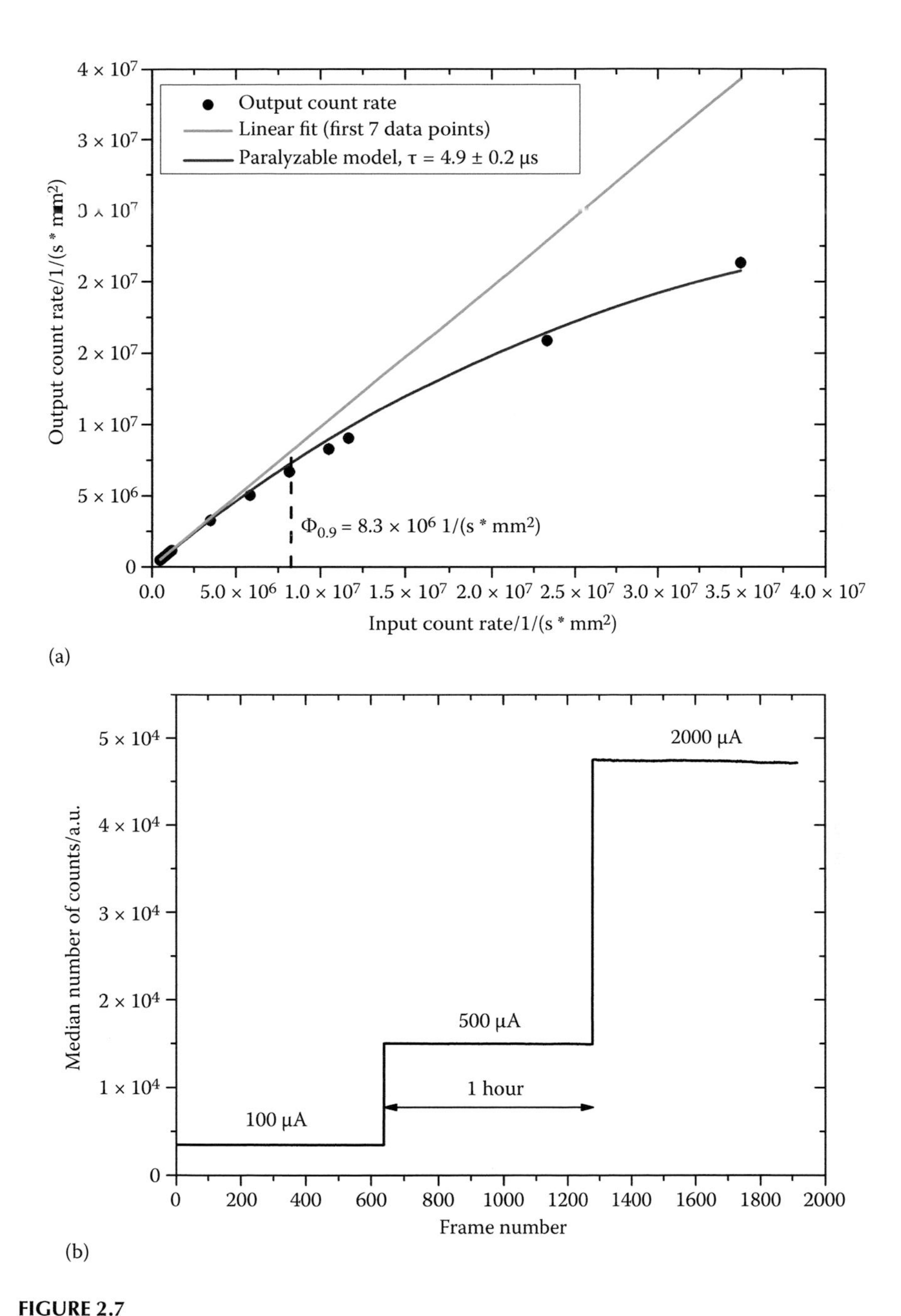

FIGURE 2.7
(a) Measured OCR versus real ICR with ideal linearized count rate (light gray line) and fit according to the paralyzable detector model (dark gray line). $\Phi_{0.9}$ is indicated with the black dashed line. (b) Count rate stability measurements at three different fluxes; 1 h measurement time per flux.

2.4.3 Spatial Resolution

The spatial resolution of a detector is certainly one of the most important properties since it determines the smallest specimen feature size that will be visible in the resulting image. Whereas for indirect converting detectors the so-called *point spread function* (PSF; the detector's response to a point source) often is rather Gaussian-shaped and smeared out over several pixels, direct converting photon-counting semiconductor pixel detectors are more likely to exhibit a rather box-shaped PSF [61,62], which is mainly determined by the physical pixel size. However, it was also shown that the level of the energy threshold can have an influence on the detector's *modulation transfer function* (MTF)* [63]. Charge sharing effects can lead to image blur (due to multiple counting of single events by neighboring pixels), and thus the threshold level relative to the photon energy decides if such events are present or omitted in the image. This effect is depicted in Figure 2.8. Here, the presampling MTF was measured for a 500 μm thick GaAs:Cr Medipix3RX detector in SPM and CSM using the *slanted-edge* method [3,62,64]. Images of a tilted tungsten edge were recorded at an x-ray energy of 40 keV (monochromatic synchrotron radiation) and for several threshold positions E_{th} (set below/above Ga and As fluorescence energies and below/above half the photon energy). The influence of the threshold position on the MTF and thus on the spatial resolution in the SPM measurements, corresponding to the "normal" behavior for many photon-counting detectors, is evident (Figure 2.8a). However, note that the effective increase in spatial resolution (or, in other words, reduction of the effective pixel size) only comes at the cost of a reduced detective quantum efficiency (DQE) due to an increasing number of events being omitted with higher and higher threshold levels.

In CSM, the situation is completely different (Figure 2.8b). Due to the interpixel communication and charge allocation algorithms in this mode, each event is attributed with its full energy information to one pixel at most. Thus, the MTF remains almost constant for all threshold levels, as long as they are reasonably far away from the photon energy itself. Further, despite the MTF values are not as high as in the SPM case, it was shown that a very good DQE is maintained in CSM even at higher threshold levels [2].

These results also confirm that the sensor material is not the limiting factor for the spatial resolution of GaAs:Cr Medipix3RX detectors, given the MTF values at the Nyquist frequency f_{Ny} determined by the native pixel size (55 μm, corresponding to f_{Ny} = 9.1 lp/mm). However, one could argue that the spatial resolution of thicker sensors might decrease due to a higher probability of larger charge cloud spread. Thus, qualitative measurements of a resolution pattern were performed with the 1 mm detector, shown in Figure 2.6b. The contrast at spatial frequencies close to the Nyquist frequency is still

* The MTF is the Fourier transform of the *line-spread-function* (LSF), which itself is the derivative of the easily measurable *edge spread function* (ESF).

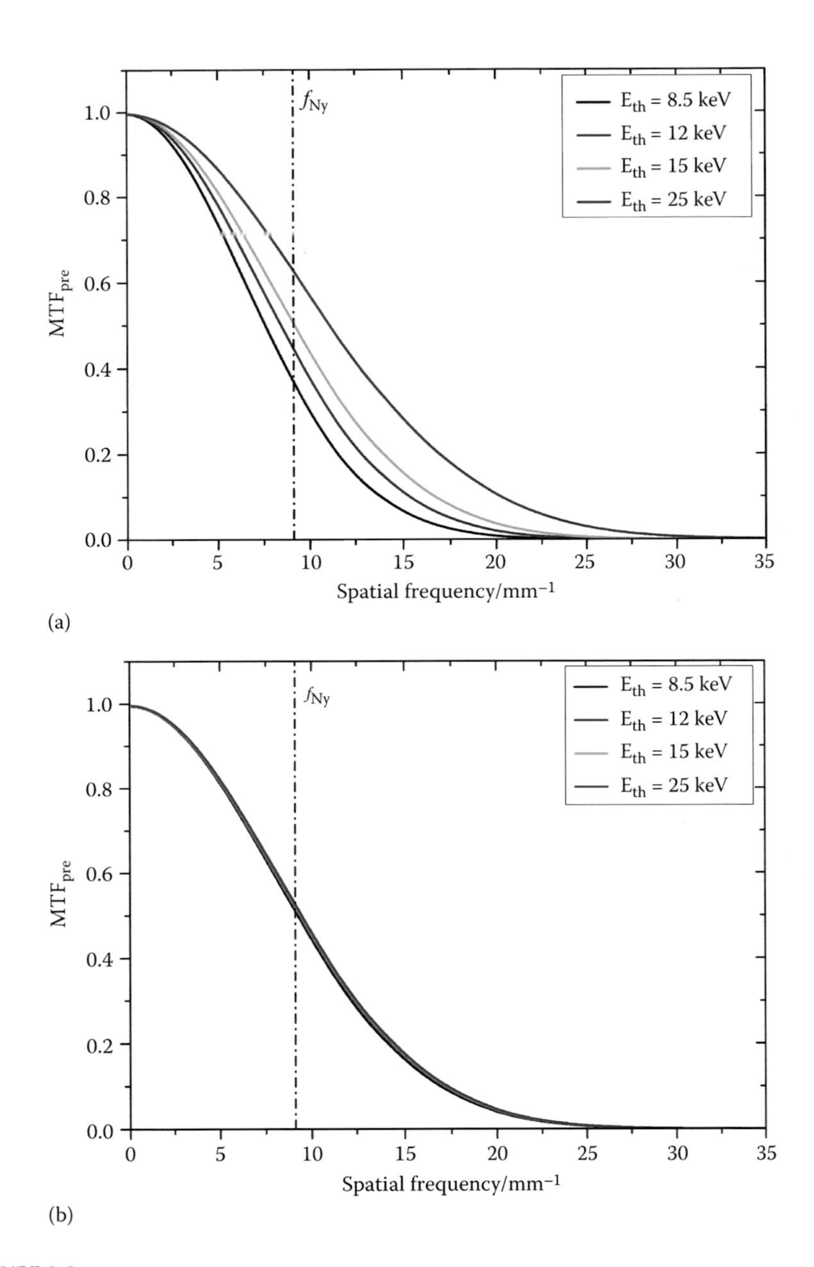

FIGURE 2.8
Presampling MTF measured with the 500 μm GaAs:Cr detector using the slanted edge method. (a) Measurements in SPM and (b) measurements in CSM. The Nyquist frequency is indicated by the dot-dashed black line.

considerably high, so the conclusions drawn from the 500 μm thick sensor should also remain valid for the 1 mm sensor.

However, note that a possibly disadvantageous feature that arises for detectors with box-shaped PSFs (and that is mostly absent for detectors with a blurred PSF) can be deduced from Figure 2.8 and is seen in Figure 2.6b, namely *aliasing*. This effect is characterized by nonzero contrast at certain sample frequencies higher than the Nyquist frequency, which, in principle, cannot be reliably measured by the detector anymore. Further, aliasing can depend on the sample orientation, which, in turn, can hamper diagnostic significance in medical imaging. Here, pixel binning or using a detector with possibly lower spatial resolution and/or more Gaussian-shaped PSF would be an alternative solution.

2.4.4 Spectroscopic Properties

One of the advantages of photon-counting semiconductor pixel detectors is that the possibility of setting one (or more) energy threshold(s) allows making use of the energy information contained in the recorded x-ray spectrum for the so-called *spectral x-ray imaging* or *spectral CT*. This feature can be used to differentiate certain materials that a specimen is composed of or to identify and separate tissues and contrast agents used in medical imaging [4,65–67]. For certain methods, even quantitative results (like concentrations of contrast agent solutions) are possible with dedicated algorithms [2,25].

For spectral imaging, a sufficiently good energy resolution of the detector is needed. The energy resolution is usually given in terms of the full width at half maximum (FWHM) of energy peaks measured using monochromatic radiation. Whereas some readout electronics like the Timepix chips [68] allow a direct measurement of the energy of an incoming photon (e.g., in terms of the time span the charge pulse is above the energy threshold), the measurement of energy spectra with photon-counting detectors is usually realized with threshold scans. Since the pixel electronics count all events whose energy is higher than the corresponding threshold level, the successive scanning of the threshold yields an integrated energy spectrum. The differentiation of such curves with respect to the threshold value then yields the desired energy spectrum. When several spectra are recorded at different photon energies (for example, by using different radionuclides or monochromatic synchrotron radiation), the detector, respectively, the threshold, can be calibrated. This helps to later position the energy threshold at a desired level, for example, closely below and above an absorption edge of a certain material one would like to detect (as used in K-edge subtraction imaging).

In order to calibrate the energy threshold and to investigate the energy resolution of the 1 mm GaAs:Cr Medipix3RX detector, threshold scans were performed using monochromatic synchrotron radiation (energy bandwidth $\Delta E/E \approx 10^{-2}$) and two radionuclides, namely Cd-109 (emitting at 22.1 and 88 keV) and Am-241 (59.5 keV), both with negligible bandwidth. Since the tremendous improvement in spectroscopic performance of Medipix3RX detectors in CSM as compared to SPM has already been shown impressively for CdTe sensors [56] and the 500 μm thick GaAs:Cr sensor [3], only measurements in CSM were conducted in the frame of this work. For data processing, defective pixels were first removed from the frames before the mean count value was calculated for each threshold position of the respective scan. The derivative of these plots was then smoothed and finally normalized to the peak height. The resulting energy spectra are shown in Figure 2.9a.

The synchrotron measurements were then used to calibrate the energy threshold by plotting the peak position versus the corresponding photon energy and applying a standard linear regression. The result is shown in Figure 2.9b together with the fit parameters, generally proving a good linearity. However, as also can be seen in Figure 2.9b, the energy calibration is more accurate for the lower energies, whereas it deviates from the linear behavior for energies above ~40 keV, shifting the peaks to slightly higher energy/threshold levels compared to the theoretical values calculated with the fit parameters. This behavior, which has already been observed with the 500 μm thick sensor [5], most probably comes from nonlinearities in the pixel preamplifiers at higher input charges, as well as from the CCE variations discussed further below.

The energy resolution of the detector in terms of measured FWHM values, which are listed in Table 2.1 for all measurements, varies between 3.4 and 8.6 keV. These are relatively high values (i.e., reduced energy resolution) compared to what is generally possible with semiconductor pixel detectors. On the other hand, these are good results given the very small pixel size and as compared to measurements without employing the CSM functionality. Mainly, the relatively large FWHM values come from the increased noise during the analog summation of charges from neighboring pixels in the CSM circuitry and the low energy tailing discussed below. Nevertheless, with the given energy resolutions and calibration results, it should still be possible to set the energy threshold to the desired values with sufficient accuracy also at higher x-ray energies, especially since the deviations from linearity are in the order of (or even less than) the energy resolution of the detector. This is a first and important step toward spectral x-ray imaging, and first examples will be shown further below in this section.

When looking at the spectra in Figure 2.9a, two more aspects are worth being noted. First, the fact that there is almost no photopeak visible at the energy of the Ga and As fluorescences (around 10 keV) again confirms the high effectiveness of the CSM in combination with GaAs:Cr sensors, given the mean free path of the fluorescence photons of $\lambda \leq 40$ μm in the sensor and the effective charge collection area (i.e., the *dynamical* pixel size*) of 3×3 pixels,

* Note that the difference between the *dynamically* increased charge collection area of the Medipix3RX chip and physically larger pixel sizes has already been investigated in Ref. [56] for CdTe sensors.

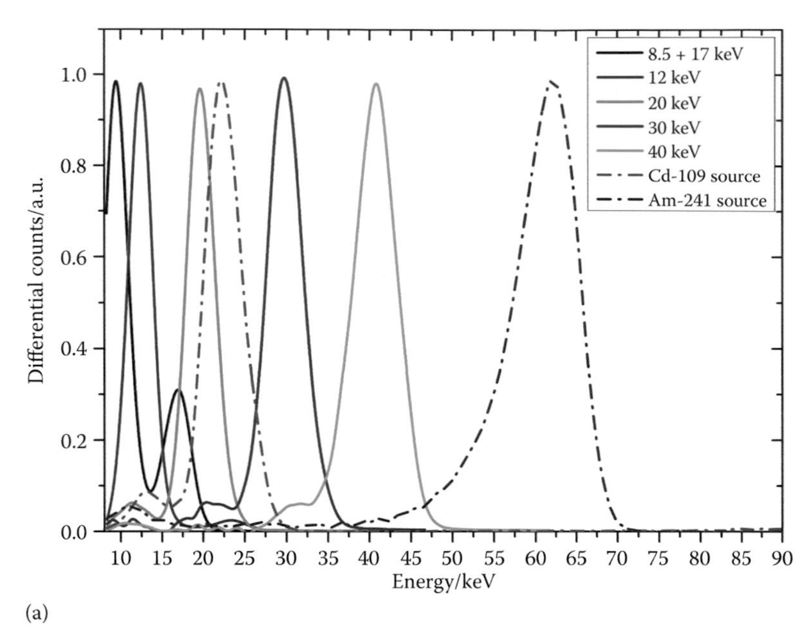

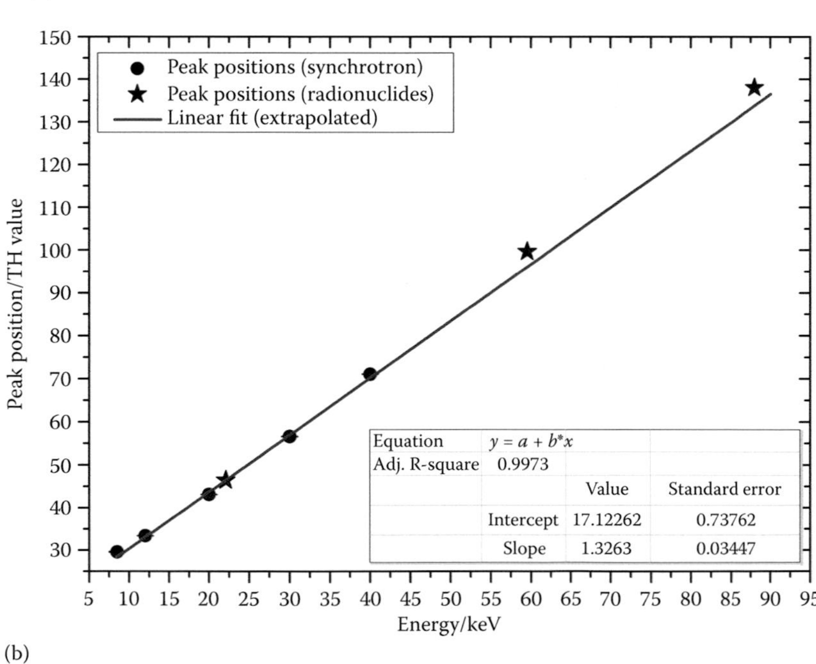

FIGURE 2.9
(a) Energy spectra recorded with the 1 mm GaAs:Cr detector (using a monochromatic synchrotron beam and radionuclides). (b) Calibration of the energy threshold by linear regression using the synchrotron measurements.

TABLE 2.1

Energy Resolution of the 1 mm Detector Assembly (in Terms of FWHM) at Several Photon Energies (Measured Using Monochromatic Synchrotron Beam and Radionuclides)

Photon Energy (keV)	Energy Resolution (FWHM; keV)
8.5	3.4
12	3.3
20	4.0
22.1[a]	5.1
30	5.1
40	6.3
59.5	8.6

[a] With influences from photons coming from the 25 keV k_β decay.

corresponding to 165 × 165 μm². Thus, only very low escape peaks, in which the fluorescence photon leaves the sensor and thus cannot be captured by the CSM, are visible at energies around 10 keV below the nominal photon energy.

Second, it can be seen that the peaks are rather Gaussian-shaped for photon energies below 30 keV, whereas for the higher energies, they are asymmetrically broadened toward lower energies. There are two reasons for this phenomenon: First, the increasing energy resolution at higher energies eventually leads to a merging of the main energy peak with the escape peak. Second, whereas the penetration depths for the lower-energy photons are rather limited, i.e., confined to the sensor volume underneath the planar backside contact (onto which the sensor was illuminated), the photon interaction probability is more homogeneously spread across the whole sensor depth for the higher photon energies. Given the finite electron mobility-lifetime product, this also means that the CCE is more widespread and shifted toward higher values as compared to the low-energy photons, which on average have to travel larger distances through the sensor before reaching the pixel electronics. This leads to the observed low-energy tailing and apparent shifting of the peaks toward higher than nominal energies as discussed above.

2.4.5 Exemplary Application: Spectral CT

As a first example of possible spectral imaging applications, CT measurements of a contrast agent phantom containing capillaries filled with aqueous gadolinium and iodine solutions (50 and 250 μmol/ml each) were performed. In analogy to the procedure described in Refs. [2] and [25], scans at four different threshold positions were acquired, namely at 28, 48, 68, and 88 keV, while employing a 120 kVp tube spectrum with 1 mm Al filtering. After the tomographic reconstruction (using the *Octopus* software package, Inside Matters NV, Gent, Belgium), two-thirds of the slices of each scan were used

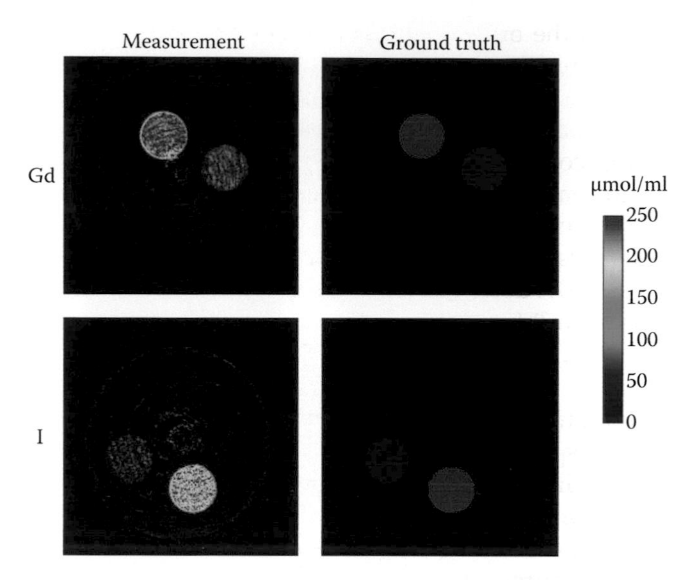

FIGURE 2.10
Concentrations of Gd and I contrast agents in µmol/ml. (Left) Results from spectral CT measurements after application of a trained algorithm. (Right) Ground truth. Color code corresponds to the ImageJ 1.51h *physics* lookup table.

as training data for an algorithm based on a regression model. The remaining slices were then processed with the trained algorithm to quantitatively recover the contrast agent concentration. The resulting images are shown in Figure 2.10 along with the ground truth. Despite the deviation of the measurements from the ground truth (mainly due to the relatively low absorption of the 1 mm GaAs:Cr sensor above 40 keV and the resulting higher noise level in the CT scans at the two highest thresholds), these results impressively demonstrate the possibilities of such algorithms in combination with photon-counting detectors for spectral x-ray imaging. Needless to say, there exist several other methods to perform (material resolved) spectral CT, but the results shown here could already be a first step toward the routine use of such detectors and algorithms for clinical applications [69,70].

2.5 Conclusions and Outlook

In this chapter, it was shown that chromium-compensated gallium arsenide (GaAs:Cr) can be considered as a promising sensor material for photon-counting pixel detectors for x-ray imaging applications with photon energies up to ~60 keV, which is a perfect intermediate regime between routinely used sensor materials like silicon and cadmium (zinc) telluride. GaAs:Cr shows a high resistivity in the order of 10^9 Ω cm and an electron mobility-lifetime product of >10^{-5} cm^2/V, both confirming detector-grade material quality. In combination with a Medipix3RX chip, it was found that the quality of x-ray images is very good, but adequate correction methods like flatfield and/or beam hardening correction or filtering of the x-ray spectrum are mandatory in order to remove sensitivity variations due to defects in the crystal structure (mainly dislocation networks). Count rate stability is excellent over time spans of hours even at higher fluxes (if adequate temperature stability is ensured), and a 10% deviation from count rate linearity was found at a flux of ~8×10^6 1/(s * mm^2) for the given experimental settings. The spatial resolution when using GaAs:Cr sensors was found to be limited by the physical pixel size of the readout chip, leading to good detail recognizability of feature sizes in the range of tens of micrometers (however, aliasing might be present for higher spatial frequencies). The spectroscopic performance of the detector assembly in the CSM is very good due to the effective elimination of charge sharing and sensor fluorescences, with energy resolutions lying in the range a few kiloelectron volts. As a first example, the detector was used for spectral CT measurements showing promising results.

Future steps could include a more in-depth characterization of the material properties, namely the dislocation networks and possible time dependencies due to charge trapping/detrapping as well as the production of larger detectors with cutting-edge readout chips (see, for instance, Chapter 3).

Acknowledgments

The author would like to thank A. Tyazhev from the Functional Electronics Laboratory, Tomsk State University for providing the GaAs:Cr wafers, A. Fauler and M. Fiederle (both FMF Freiburg) for the production of the detector assemblies, M. Zuber (KIT) for valuable help during the measurements and data analysis, and the KIT institutes IPS and IBPT for the provision of beamtime. This work was partially funded by the BMBF project GALAPAD2 under grant no. 05K16VK6.

References

1. R. Ballabriga, J. Alozy, G. Blaj, M. Campbell, M. Fiederle, E. Frojdh, E. Heijne, X. Llopart, M. Pichotka, S. Procz, L. Tlustos and W. Wong, "The Medipix3RX: A high resolution, zero dead-time pixel detector readout chip allowing spectroscopic imaging," *J Instrum*, Bd. 8, Nr. 02, p. C02016, 2013.

2. T. Koenig, M. Zuber, E. Hamann, A. Cecilia, R. Ballabriga, M. Campbell, M. Ruat, L. Tlustos, A. Fauler, M. Fiederle and T. Baumbach, "How spectroscopic x-ray imaging benefits from inter-pixel communication," *Phys Med Biol*, Bd. 59, Nr. 20, pp. 6195–6213, Sep. 2014.
3. E. Hamann, T. Koenig, M. Zuber, A. T. A. Cecilia, O. Tolbanov, S. Procz, A. Fauler, T. Baumbach and M. Fiederle, "Performance of a Medipix3RX Spectroscopic pixel detector with a high resistivity gallium arsenide sensor," *IEEE Trans Med Imag*, Bd. 34, Nr. 3, pp. 707–715, Mar. 2015.
4. K. Taguchi and J. S. Iwanczyk, "Vision 20/20: Single photon counting x-ray detectors in medical imaging," *Med Phys*, Bd. 40, Nr. 10, p. 100901, 2013.
5. E. Hamann, "Characterization of high resistivity GaAs as sensor material for photon counting semiconductor pixel detectors," PhD dissertation, University of Freiburg, 2013, URN: urn:nbn:de:bsz:25-opus-93550.
6. D. McGregor and H. Hermon, "Room-temperature compound semiconductor radiation detectors," *Nucl Instrum Meth A*, Bd. 395, Nr. 1, pp. 101–124, 1997.
7. H. Toyama, A. Higa, M. Yamazato, T. Maehama, R. Ohno and M. Toguchi, "Quantitative analysis of polarization phenomena in CdTe radiation detectors," *Jpn J Appl Phys*, Bd. 45, Nr. 11, pp. 8842–8847, Nov. 2006.
8. D. S. Bale and C. Szeles, "Nature of polarization in wide-bandgap semiconductor detectors under high-flux irradiation: Application to semi-insulating Cd(1-x)Zn(x)Te," *Phys Rev B*, Bd. 77, Nr. 3, 035205, Jan. 2008.
9. M. Ruat and C. Ponchut, "Defect signature, instabilities and polarization in CdTe X-ray sensors with quasi-ohmic contacts," *J Instrum*, Bd. 9, Nr. 04, pp. C04030–C04030, Apr. 2014.
10. A. Ruzin, "On polarization of compensated detectors," *IEEE Trans Nucl Sci*, Bd. 63, Nr. 2, pp. 1188–1193, Apr. 2016.
11. W. Harding, C. Hilsum, M. Moncaster, D. Northrop and O. Simpson, "Gallium arsenide for gamma-ray spectroscopy," *Nature*, Bd. 187, p. 405, 1960.
12. R. Irsigler, J. Andersson, J. Alverbro, J. Borglind, C. Fröjdh, P. Helander, S. Manolopoulos, V. O'shea and K. Smith, "Evaluation of 320 × 240 pixel LEC GaAs Schottky barrier x-ray imaging arrays, hybridized to CMOS readout circuit based on charge integration," *Nucl Instrum Meth A*, Bd. 434, Nr. 1, pp. 24–29, 1999.
13. R. Irsigler, J. Andersson, J. Alverbro, Z. Fakoor-Biniaz, C. Fröjdh, P. Helander, H. Martijn, D. Meikle, M. Östlund, V. O'Shea and K. Smith, "320 × 240 GaAs pixel detectors with improved x-ray imaging quality," *Nucl Instrum Meth A*, Bd. 460, Nr. 1, pp. 67–71, 2001.
14. S. Amendolia, A. Annovazzi, A. Bigongiari, M. Bisogni, F. Catarsi, F. Cesqui, A. Cetronio, M. Chianella, P. Delogu, M. Fantacci, D. Galimberti, D. Lanzieri, S. Lavanga, M. Novelli, G. Passuello, M. Pieracci, M. Quattrocchi, V. Rosso, A. Stefanini, A. Testa and L. Venturelli, "A prototype for a mammographic head and related developments," *Nucl Instrum Meth A*, Bd. 518, Nr. 1, pp. 382–385, 2004.
15. L. Tlustos, M. Campbell, C. Fröjdh, P. Kostamo and S. Nenonen, "Characterisation of an epitaxial GaAs Medipix2 detector using fluorescence photons," *Nucl Instrum Meth A*, Bd. 591, Nr. 1, pp. 42–45, 2008.
16. A. Annovazzi, S. Amendolia, A. Bigongiari, M. Bisogni, F. Catarsi, F. Cesqui, A. Cetronio, F. Colombo, P. Delogu, M. Fantacci, A. Gilberti, C. Lanzieri, S. Lavagna, M. Novelli, G. Passuello, G. Paternoster, M. Pieracci, M. Poletti, M. Quattrocchi,

V. S. A. Rosso, A. Testa and L. Venturelli, "A GaAs pixel detectors-based digital mammographic system: Performances and imaging tests results," *Nucl Instrum Meth A*, Bd. 576, Nr. 1, pp. 154–159, 2007.
17. A. Zwerger, A. Fauler, M. Fiederle and K. Jakobs, "Medipix2: Processing and measurements of GaAs pixel detectors," *Nucl Instrum Meth A*, Bd. 576, Nr. 1, pp. 23–26, 2007.
18. G. Ayzenshtat, N. Bakin, D. Budnitsky, E. Drugova, V. Germogenov, S. Khludkov, O. Koretskaya, L. Okaevich, L. Porokhovnichenko, A. Potapov, K. Smith, O. Tolbanov, A. V. M. Tyazhev and A. Vorobiev, "GaAs structures for x-ray imaging detectors," *Nucl Instrum Meth A*, Bd. 466, Nr. 1, pp. 25–32, 2001.
19. A. Tyazhev, D. Budnitsky, O. Koretskay, V. Novikov, L. Okaevich, A. Potapov, O. Tolbanov and A. Vorobiev, "GaAs radiation imaging detectors with an active layer thickness up to 1 mm," *Nucl Instrum Meth A*, Bd. 509, Nr. 1, pp. 34–39, 2003.
20. L. Tlustos, G. Shelkov and O. P. Tolbanov, "Characterisation of a GaAs(Cr) Medipix2 hybrid pixel detector," *Nucl Instrum Meth A*, Bd. 633, pp. S103–S107, 2011.
21. E. Hamann, A. Cecilia, A. Zwerger, A. Fauler, O. Tolbanov, A. Tyazhev, G. Shelkov, H. Graafsma, T. Baumbach and M. Fiederle, "Characterization of photon counting pixel detectors based on semi-insulating GaAs sensor material," *J Phys: Conf Ser*, Bd. 425, Nr. 6, p. 062015, 2013.
22. M. Veale, S. Bell, D. Duarte, M. French, M. Hart, A. Schneider, P. Seller, M. Wilson, V. Kachkanov, A. Lozinskaya, V. Novikov, O. Tolbanov, A. Tyazhev and A. Zarubin, "Investigating the suitability of GaAs:Cr material for high flux X-ray imaging," *J Instrum*, Bd. 9, Nr. 12, pp. C12047–C12047, Dec. 2014.
23. D. Budnitsky, A. Tyazhev, V. Novikov, A. Zarubin, O. Tolbanov, M. Skakunov, E. Hamann, A. Fauler, M. Fiederle, S. Procz, H. Graafsma and S. Ryabkov, "Chromium-compensated GaAs detector material and sensors," *J Instrum*, Bd. 9, Nr. 07, pp. C07011–C07011, Jul. 2014.
24. D. Pennicard, S. Smoljanin, B. Struth, H. Hirsemann, A. Fauler, M. Fiederle, O. Tolbanov, A. Zarubin, A. Tyazhev, G. Shelkov and H. Graafsma, "The LAMBDA photon-counting pixel detector and high-Z sensor development," *J Instrum*, Bd. 9, Nr. 12, pp. C12026–C12026, Dec. 2014.
25. M. Zuber, E. Hamann, R. Ballabriga, M. Campbell, M. Fiederle, T. Baumbach and T. Koenig, "An investigation into the temporal stability of CdTe-based photon counting detectors during spectral micro-CT acquisitions," *Biomed Phys Eng Express*, Bd. 1, Nr. 2, p. 025205, Aug. 2015.
26. M. Amman, P. Luke and J. Lee, "CdZnTe material uniformity and coplanar-grid gamma-ray detector performance," in *IEEE Nuclear Science Symposium Conference Record, 1999*, 1999.
27. M. Amman, J. S. Lee and P. N. Luke, "Electron trapping nonuniformity in high-pressure-Bridgman-grown CdZnTe," *J Appl Phys*, Bd. 92, Nr. 6, pp. 3198–3206, 2002.
28. E. Hamann, A. Cecilia, P. Vagovic, D. Hanschke, J. Butzer, D. Greiffenberg, A. Fauler, T. Baumbach and M. Fiederle, "Applications of Medipix2 single photon detectors at the ANKA synchrotron facility," *2010 IEEE Nuclear Science Symposium Conference Record*, Oct. 2010.

29. E. Frojdh, C. Frojdh, E. N. Gimenez, D. Krapohl, D. Maneuski, B. Norlin, V. O'Shea, H. Wilhelm, N. Tartoni, G. Thungstrom and R. Zain, "Probing defects in a small pixellated CdTe sensor using an inclined mono energetic x-ray micro beam," *IEEE Trans Nucl Sci*, Bd. 60, Nr. 4, pp. 2864–2869, Aug. 2013.
30. B. Raghothamachar, G. Dhanaraj, J. Bai and M. Dudley, "Defect analysis in crystals using x-ray topography," *Microsc Res Tech*, Bd. 69, Nr. 5, pp. 343–358, 2006.
31. A. E. Bolotnikov, S. Babalola, G. S. Camarda, Y. Cui, R. Gul, S. U. Egarievwe, P. M. Fochuk, M. Fuerstnau, J. Horace, A. Hossain, F. Jones, K. Kim, O. Kopach, B. McCall, L. Marchini, B. Raghothamachar, R. Taggart, G. Yang, L. Xu and R. B. James, "Correlations between crystal defects and performance of CdZnTe detectors," *IEEE Trans Nucl Sci*, Bd. 58, Nr. 4, pp. 1972–1980, Aug. 2011.
32. A. Cecilia, E. Hamann, C. Haas, D. Greiffenberg, A. Danilewsky, D. Haenscke, A. Fauler, A. Zwerger, G. Buth, P. Vagovic, T. Baumbach and M. Fiederle, "Investigation of crystallographic and detection properties of CdTe at the ANKA synchrotron light source," *J Instrum*, Bd. 6, Nr. 10, pp. P10016–P10016, Oct. 2011.
33. A. Rack, T. Weitkamp, S. Bauer Trabelsi, P. Modregger, A. Cecilia, T. dos Santos Rolo, T. Rack, D. Haas, R. Simon, R. Heldele, M. Schulz, B. Mayzel, A. Danilewsky, T. Waterstradt, W. Diete, H. Riesemeier, B. Müller and T. Baumbach, "The micro-imaging station of the TopoTomo beamline at the ANKA synchrotron light source," *Nucl Instrum Meth B*, Bd. 267, Nr. 11, pp. 1978–1988, 2009.
34. T. Kamejima, F. Shimura, Y. Matsumoto, H. Watanabe and J. Matsui, "Role of Dislocations in Semi-Insulation Mechanism in Undoped LEC GaAs Crystal," *Jpn J Appl Phys*, Bd. 21, Nr. Part 2, No. 11, pp. L721–L723, Nov. 1982.
35. P. Rudolph, "Dislocation cell structures in melt-grown semiconductor compound crystals," *Cryst Res Technol*, Bd. 40, Nr. 1-2, pp. 7–20, 2005.
36. S. Basolo, J. Berar, N. Boudet, P. Breugnon, B. Chantepie, J. Clemens, P. Delpierre, B. Dinkespiler, S. Hustache, K. Medjoubi, K. Menouni, C. Morel, P. Pangaud and E. Vigeolas, "A 20kpixels CdTe photon-counting imager using XPAD chip," *Nucl Instrum Meth A*, Bd. 589, Nr. 2, pp. 268–274, May 2008.
37 V. Astromskas, E. N. Gimenez, A. Lohstroh and N. Tartoni, "Evaluation of polarization effects of e-collection Schottky CdTe Medipix3RX hybrid pixel detector," *IEEE Trans Nucl Sci*, Bd. 63, Nr. 1, pp. 252–258, Feb. 2016.
38. A. Tyazhev, D. Budnitsky and O. Tolbanov, "Electric field distribution in chromium compensated GaAs," in *Semiconducting and Insulating Materials, 2002. SIMC-XII-2002. 12th International Conference on*, 2002.
39. D. Budnitsky, A. Lychagin, L. Okaevich and O. Tolbanov, "Study of particularities for metal contact formation to the semiconductor high-resistance GaAs: Cr," in *IEEE International Siberian Conference on Control and Communications (SIBCON)*, 2005.
40. G. Ayzenshtat, M. Lelekov, V. Novikov, L. Okaevich and O. Tolbanov, "Charge transport in detectors on the basis of gallium arsenide compensated with chromium," *Semiconductors*, Bd. 41, Nr. 5, pp. 612–615, 2007.
41. M. Veale, S. Bell, D. Duarte, M. French, A. Schneider, P. Seller, M. Wilson, A. Lozinskaya, V. Novikov, O. Tolbanov, A. Tyazhev and A. Zarubin, "Chromium compensated gallium arsenide detectors for x-ray and gamma-ray spectroscopic imaging," *Nucl Instrum Meth A*, Bd. 752, pp. 6–14, Jul. 2014.
42. A. Tyazhev, D. Budnitsky, D. Mokeev, V. Novikov, A. Zarubin, O. Tolbanov, G. Shelkov, E. Hamann, A. Fauler, M. Fiederle and S. Procz, "GaAs pixel detectors," *MRS Proceedings*, Bd. 1576, 001, 2013.

43. P. Sellin, "Modelling of the small pixel effect in gallium arsenide x-ray imaging detectors," *Nucl Instrum Meth A*, Bd. 434, Nr. 1, pp. 75–81, 1999.
44. Z. He, "Review of the Shockley-Ramo theorem and its application in semiconductor gamma-ray detectors," *Nucl Instrum Meth A*, Bd. 463, Nr. 1–2, pp. 250–267, May 2001.
45. M. D. Wilson, P. Seller, M. C. Veale and P. J. Sellin, "Investigation of the small pixel effect in CdZnTe detectors," *2007 IEEE Nuclear Science Symposium Conference Record*, 2007.
46. K. Hecht, "Zum Mechanismus des lichtelektrischen Primärstromes in isolierenden Kristallen," *Zeitschrift für Physik*, Bd. 77, Nr. 3–4, pp. 235–245, 1932.
47. G. Ayzenshtat, D. Budnitsky, O. Koretskaya, V. Novikov, L. Okaevich, A. Potapov, O. Tolbanov, A. Tyazhev and A. Vorobiev, "GaAs resistor structures for X-ray imaging detectors," *Nucl Instrum Meth A*, Bd. 487, Nr. 1, pp. 96–101, 2002.
48. A. Tyazhev, V. Novikov, O. Tolbanov, A. Zarubin, M. Fiederle and E. Hamann, "Investigation of the current-voltage characteristics, the electric field distribution and the charge collection efficiency in x-ray sensors based on chromium compensated gallium arsenide," *Hard X-Ray, Gamma-Ray, and Neutron Detector Physics XVI*, Sep 2014.
49. G. I. Ayzenshtat, M. A. Lelekov and O. P. Tolbanov, "Barrier-height measurement for a gallium arsenide metal-semi-insulator interface," *Semiconductors*, Bd. 41, Nr. 11, pp. 1310–1311, Nov. 2007.
50. M. Fiederle, H. Braml, A. Fauler, J. Giersch, J. Ludwig, G. Anton and K. Jakobs, "Development of flip-chip bonding technology for (Cd, Zn) Te," in *IEEE Nuclear Science Symposium Conference Record, 2003*, 2003.
51. L. Helfen, T. Baumbach, P. Mikulik, D. Kiel, P. Pernot, P. Cloetens and J. Baruchel, "High-resolution three-dimensional imaging of flat objects by synchrotron-radiation computed laminography," *Appl Phys Lett*, Bd. 86, Nr. 7, pp. 71915–71915, 2005.
52. F. Xu, L. Helfen, T. Baumbach and H. Suhonen, "Comparison of image quality in computed laminography and tomography," *Opt Express*, Bd. 20, Nr. 2, pp. 794–806, 2012.
53. A. Cecilia, E. Hamann, T. Koenig, F. Xu, Y. Cheng, L. Helfen, M. Ruat, M. Scheel, M. Zuber, T. Baumbach, A. Fauler and M. Fiederle, "High resolution 3D imaging of bump-bonds by means of synchrotron radiation computed laminography," *J Instrum*, Bd. 8, Nr. 12, pp. C12029–C12029, Dec. 2013.
54. F. Krummenacher, "Pixel detectors with local intelligence: An IC designer point of view," *Nucl Instrum Meth A*, Bd. 305, Nr. 3, pp. 527–532, 1991.
55. E. Hamann, T. Koenig, M. Zuber, A. Cecilia, A. Tyazhev, O. Tolbanov, S. Procz, A. Fauler, M. Fiederle and T. Baumbach, "Investigation of GaAs:Cr Timepix assemblies under high flux irradiation," *J Instrum*, Bd. 10, Nr. 01, pp. C01047–C01047, Jan. 2015.
56. T. Koenig, E. Hamann, S. Procz, R. Ballabriga, A. Cecilia, M. Zuber, X. Llopart, M. Campbell, A. Fauler, T. Baumbach and M. Fiederle, "Charge summing in spectroscopic x-ray detectors with high-Z sensors," *IEEE Trans Nucl Sci*, Bd. 60, p. 4713, 2013.
57. V. Kraus, M. Holik, J. Jakubek, M. Kroupa, P. Soukup and Z. Vykydal, "FITPix-fast interface for Timepix pixel detectors," *J Instrum*, Bd. 6, Nr. 01, p. C01079, 2011.

58. D. Turecek, T. Holy, J. Jakubek, S. Pospisil and Z. Vykydal, "Pixelman: A multiplatform data acquisition and processing software package for Medipix2, Timepix and Medipix3 detectors," *J Instrum*, Bd. 6, Nr. 01, p. C01046, 2011.
59. D. Vavrik and J. Jakubek, "Radiogram enhancement and linearization using the beam hardening correction method," *Nucl Instrum Meth A*, Bd. 607, Nr. 1, pp. 212–214, 2009.
60. E. Frojdh, R. Ballabriga, M. Campbell, M. Fiederle, E. Hamann, T. Koenig, X. Llopart, D. d. P. Magalhaes and M. Zuber, "Count rate linearity and spectral response of the Medipix3RX chip coupled to a 300μm silicon sensor under high flux conditions," *J Instrum*, Bd. 9, Nr. 04, pp. C04028–C04028, Apr. 2014.
61. T. Donath, S. Brandstetter, L. Cibik, S. Commichau, P. Hofer, M. Krumrey, B. Lüthi, S. Marggraf, P. Müller, M. Schneebeli, C. Schulze-Briese and J. Wernecke, "Characterization of the PILATUS photon-counting pixel detector for X-ray energies from 1.75 keV to 60 keV," *J Phys: Conf Ser*, Bd. 425, Nr. 6, p. 062001, Mar. 2013.
62. S. Ehn, F. M. Epple, A. Fehringer, D. Pennicard, H. Graafsma, P. Noël and F. Pfeiffer, "X-ray deconvolution microscopy," *Biomed Opt Express*, Bd. 7, Nr. 4, p. 1227, Mar. 2016.
63. L. Tlustos, R. Ballabriga, M. Campbell, E. Heijne, K. Kincade, X. Llopart and P. Stejskal, "Imaging properties of the Medipix2 system exploiting single and dual energy thresholds," *IEEE Trans Nucl Sci*, Bd. 53, Nr. 1, pp. 367–372, Feb. 2006.
64. E. Samei, M. J. Flynn and D. A. Reimann, "A method for measuring the presampled MTF of digital radiographic systems using an edge test device," *Med Phys*, Bd. 25, p. 102, 1998.
65. A. de Vries, E. Roessl, E. Kneepkens, A. Thran, B. Brendel, G. Martens, R. Proska, K. Nicolay and H. Grüll, "Quantitative spectral k-edge imaging in preclinical photon-counting x-Ray computed tomography," *Invest Radiol*, Bd. 50, Nr. 4, pp. 297–304, Aug. 2015.
66. P. M. Shikhaliev, "Photon counting spectral CT: Improved material decomposition with K-edge-filtered x-rays," *Phys Med Biol*, Bd. 57, Nr. 6, pp. 1595–1615, Mar. 2012.
67. R. Gutjahr, C. Polster, S. Kappler, H. Pietsch, G. Jost, K. Hahn, F. Schöck, M. Sedlmair, T. Allmendinger, B. Schmidt, B. Krauss and T. Flohr, "Material decomposition and virtual non-contrast imaging in photon counting computed tomography: An animal study," *Medical Imaging 2016: Physics of Medical Imaging*, Mar. 2016.
68. X. Llopart, R. Ballabriga, M. Campbell, L. Tlustos and W. Wong, "Timepix, a 65k programmable pixel readout chip for arrival time, energy and/or photon counting measurements," *Nucl Instrum Meth A*, Bd. 581, Nr. 1, pp. 485–494, 2007.
69. R. Gutjahr, A. F. Halaweish, Z. Yu, S. Leng, L. Yu, Z. Li, S. M. Jorgensen, E. L. Ritman, S. Kappler and C. H. McCollough, "Human imaging with photon counting-based computed tomography at clinical dose levels," *Invest Radiol*, Bd. 51, Nr. 7, pp. 421–429, Jul. 2016.
70. Z. Yu, S. Leng, S. M. Jorgensen, Z. Li, R. Gutjahr, B. Chen, A. F. Halaweish, S. Kappler, L. Yu, E. L. Ritman and C. H. McCollough, "Evaluation of conventional imaging performance in a research whole-body CT system with a photon-counting detector array," *Phys Med Biol*, Bd. 61, Nr. 4, pp. 1572–1595, Feb. 2016.

3

LAMBDA Detector—An Example of a State-of-the-Art Photon Counting Imaging System

David Pennicard, Julian Becker, and Milija Sarajlić

CONTENTS

3.1 Introduction

Ever since their discovery by the German physicist Wilhelm Röntgen in 1895, x-rays have been one of the most useful tools for determining the structure of matter. The brilliance* of x-ray sources has increased exponentially since then, leading to the establishment of research centers built around a dedicated accelerator and storage ring complex, commonly referred to as synchrotron light sources.

As of early 2017, there are approximately 50 of these synchrotron facilities around the globe [1].

This increase in x-ray brilliance increases the possibilities available to experimenters, but at the same time puts high demands on every component involved in an experiment, especially the photon detector. While early experiments were commonly recorded on photographic plates or film, these media are insufficient for most modern experiments due to their inherent limitations in sensitivity, dynamic range, and temporal resolution.

X-ray detector technology has evolved, improving in every aspect since its beginnings; however, before the advent of dedicated x-ray detector developments in the late 1980s, detectors were either general purpose or optimized for tasks other than synchrotron science.

Today, there are many research groups around the world developing the next generation of x-ray detectors for synchrotron sciences, and dedicated x-ray detectors are commercially available from multiple vendors, the LAMBDA system being one of them.

3.1.1 State-of-the-Art Detectors at Storage Ring Sources

Current state-of-the-art x-ray detectors for scientific applications commonly use hybrid pixel technology. A hybrid pixel detector consists of two pixelated layers bonded together by an array of fine solder bumps. In this way, the sensor layer, which stops x-rays and converts them to an electrical signal, can be optimized independently from the application specific integrated circuit (ASIC) layer, which is responsible for signal processing.

Hybrid pixel detectors have been around for more than two decades [2] and have become the working horses of scientific x-ray detectors at synchrotron sources. This was facilitated by their inherently excellent performance

* Brilliance defines a useful quantity in x-ray sciences: it is given by the number of photons falling within a certain wavelength range (typically 0.1%) per time, per solid angle per area. Optics can alter a beam but cannot increase its brilliance, so brilliance determines (for example) the photon flux that can ultimately be focused onto a small sample.

and the ever-increasing availability of those systems due to ongoing commercialization efforts by different companies.

The hybrid pixel technology offers many inherent advantages over other detector systems, such as x-ray detection, it is possible to achieve a point spread function on the order of the pixel size, a high efficiency, and a high signal per hit. Combined with suitable signal processing in the ASIC, it is then possible to detect single x-ray photons. Additionally, since each pixel has its own circuitry, this design has the potential for high-speed operation.

Most hybrid pixel detectors for x-ray detection can be separated into one of two categories: integrating systems, which sum up the signal arriving from the sensor over a given exposure interval; and photon counting systems, which process the incoming signal on the fly. Both approaches have their merits and limitations. Photon counting detectors are very effective at rejecting the noise associated with leakage current, and can achieve single photon sensitivity even during very long acquisition times. A photon counting design can allow energy discrimination of photons (at the very least, a minimum detection threshold), which can help distinguish useful signal from the background—for example, excluding fluorescence photons at lower energies than the primary beam. The main drawback of photon counting is that this approach relies on being able to distinguish individual photons over time. If two photons hit a pixel in quick succession, they may pile up and could be misinterpreted as a single hit, thus limiting the maximum count rate in the detector. In particular, this approach is unsuitable for experiments at free-electron lasers, where large numbers of photons may hit the detector simultaneously. For these situations, an integrating detector designed for high-dynamic range is more appropriate.

The Large Area Medipix3 Based Detector Array (LAMBDA) camera is a photon counting hybrid pixel detector, based on the Medipix3 chip [3,4], and was designed for high-end x-ray experiments, particularly at synchrotron sources. It achieves an extremely high image quality by combining this photon counting capability with a small pixel size of 55 μm. LAMBDA is a modular system that is tileable and systems with sizes ranging from 14 × 14 mm^2 (a single Medipix3 chip) up to 85 × 85 mm^2 (more than 2 million pixels) have been built. For fast and time-resolved experiments, it can be read out at up to 2000 frames per second with no time gap between images, no matter the size of the detector.

LAMBDA pixel detectors are available with different sensor layers for different x-ray energy ranges. For hard x-ray detection, the GaAs and CdTe LAMBDA systems replace the standard silicon sensor layer with a "high-Z" (high-atomic number) sensor. This provides high quantum efficiency at high x-ray energies (75% at 40 keV for GaAs, and 75% at 80 keV for CdTe), while retaining single-photon counting performance and the high frame rate of up to 2 kHz.

3.2 Medipix3 Chip

At the heart of the LAMBDA detector is the Medipix3 chip [5]. This ASIC was developed by an international collaboration of more than 20 institutes based at the European Centre for Nuclear Research (CERN). Each chip has 256 × 256 pixels of 55 μm size and can work with signals of either polarity, allowing it to be used with the most common types of silicon and high-Z sensors. The Medipix3 ASIC works using the photon counting principle that is also found in other ASICs like the Pilatus or XPAD chips.

Each pixel of the Medipix3 chip has built-in signal processing electronics for performing photon counting. When a photon hits the sensor layer, a pulse of current flows into that pixel's electronics. A two-stage amplifier, consisting of charge integrating and shaping stages, produces a signal pulse whose amplitude is proportional to the total charge in the pulse (and hence the photon's energy). This pulse is then compared to one or more user-definable thresholds. For each hit above threshold, a counter is incremented.

The Medipix3 chip features additional resources to facilitate scientific experiments. One of these features is the inclusion of two 12-bit counters per pixel. These can either be configured as a single 24-bit counter for extended dynamic range or configured in a "continuous read–write" mode. In continuous read–write mode, one counter is being read out while the other one is used for counting, and so there is no dead time between images for frame rates up to 2 kHz.

It is also possible to configure the chip such that the resources of each 2 × 2 block of pixels are shared. This means that the chip can be bonded to a sensor with 110 μm size to produce a detector with eight thresholds and counters per pixel. In this "color imaging" mode, the incoming photon flux is decomposed into up to eight distinct energy bins, resulting in eight different images for each acquisition. This spectroscopic information can, for example, be used for elemental identification in the imaged object.

Finally, the Medipix3 chip employs a patented charge sharing compensation circuitry. Once activated, charge split events within each 2 × 2 cluster of pixels are detected, summed together, and counted in the pixel having the highest individual collection from the event. This feature allows better energy discrimination and ensures that no double counting of photons occurs. Simulations have shown that this feature is particularly beneficial when working with high-Z materials [6].

These features of the Medipix3 chip can be switched on and off independently. Typically, experiments at synchrotrons are done with monochromatic beam, and require high spatial resolution and high count rate capability. So, these experiments typically use 55 μm pixel size, operate with a single threshold in continuous read–write mode, and have charge summing switched off (because charge summing reduces the maximum count rate). In contrast, x-ray tube experiments can fully exploit the color imaging capability of the chip by using 110 μm pixel size, multiple energy bins, and charge summing.

The Medipix3 chip went through substantial improvements since the first version of the chip. The most significant revision to the chip was the second revision, and the resulting chip is normally referred to as Medipix3RX [5]. This new version included a change in the charge-summing algorithm to improve image uniformity, lower pixel-to-pixel and chip-to-chip variation in performance, and improved continuous read–write operation. Unless otherwise stated, all the work described in this chapter was done with Medipix3RX.

3.3 LAMBDA Camera

3.3.1 Hardware Design

As shown above, the capabilities provided by the Medipix3 chip make it an excellent ASIC for experiments at synchrotron sources. However, many of these experiments require sensitive areas that are much larger than the approximately 14×14 mm^2 that a single chip provides. Furthermore, achieving 2000 frames per second readout of a large area system requires a high-data bandwidth, corresponding to just under 1.6 gigabit/s per chip.

The LAMBDA camera was designed to address these issues [7]. The main electronic components are shown in Figure 3.1 [8]. The detector head consists of one or more sensors bonded to Medipix3 chips and mounted on a ceramic

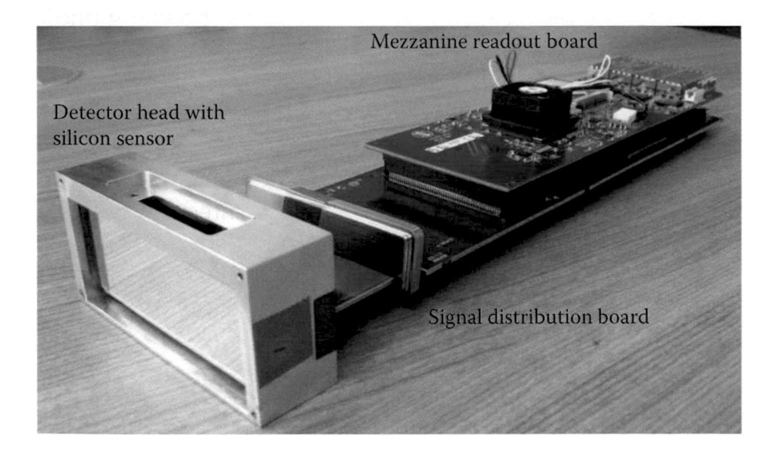

FIGURE 3.1
Single module LAMBDA system without housing. (Reproduced from Pennicard, D. et al. "LAMBDA—Large area Medipix3-based detector array." *Journal of Instrumentation* 7.11, C11009, 2012.)

circuit board. The ceramic board is designed to read out a 6 × 2 arrangement of Medipix3 chips, capitalizing on the fact that the Medipix3 chip is 3-side-buttable. When working with silicon, a single large sensor of 85 × 28 mm bonded to 12 Medipix3 chips is typically used. In the case of high-Z materials, two assemblies of 42 × 28 mm are typically used, because only smaller sensor wafers are available.

Wire bonds from the ceramic board to the Medipix3 chips allow input and output. The data output from each Medipix3 chip consists of eight differential signal pairs, with a data rate of 200 MHz for each pair. This means the full 12-chip detector head has 96 signal pairs, plus additional clock output signals to allow accurate sampling of the data. These I/O lines pass through the board to a 500-pin high-density connector on the back side. The ceramic board plugs into a "signal distribution" board, which provides power to the detector head and additional components needed to operate the detector. Mounted on the signal distribution board, there is a high-speed-readout mezzanine board, which has a Virtex-5 FPGA, RAM modules, two 10-Gigabit Ethernet optical links, and one 1-Gigabit Ethernet copper link. This mezzanine board receives commands from the control PC over the 1-Gigabit link. The FPGA controls the operation of the readout chips during imaging. It serializes the image data from the Medipix3 chips and sends the data back to the control PC at a high frame rate via the two 10-GbE links. Finally, the LAMBDA module can be mounted in a housing that provides cooling and shields the readout boards from x-rays during operation.

All the readout boards in the LAMBDA system are designed to fit behind the detector head, so multiple modules can be tiled to cover a larger area. To date, systems consisting of three modules have been constructed, giving a detection area of 85 × 85 mm (1536 × 1536 pixels). Single-module (750 kpixel) and three-module (2 Mpixel) LAMBDA systems are shown in Figure 3.2 [9].

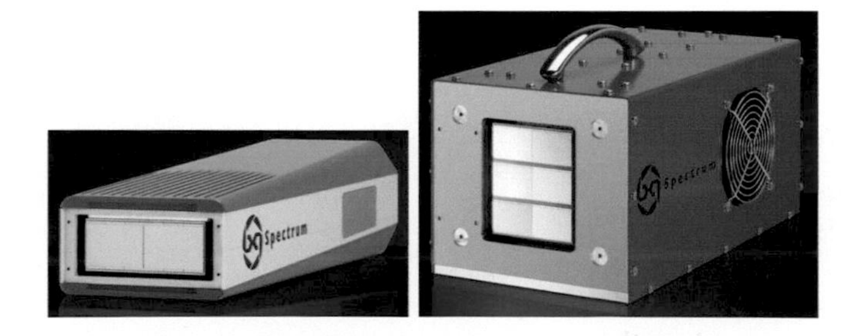

FIGURE 3.2
Comparison of different LAMBDA sizes; a single module of 1528 × 512 pixels and a three-module system of 1528 × 1536 pixels. (Reproduced from X-Spectrum, http://www.x-spectrum.de.)

3.3.2 High-Speed Readout

Acquiring images at 2000 frames per second corresponds to a data rate of 19.2 Gb/s for each 750K module. Achieving this data rate places special demands on the detector hardware, as described above, and also on the hardware and software of the control PC [10].

The data stream from the detector is sent across two 10-gigabit ethernet links. The links use the UDP protocol, which simply does one-way transmission with no resending of data. Using this simple protocol increases the maximum data rate and makes the FPGA firmware simpler to design, but this also places greater demands on the readout PC and software to ensure reliability. The readout PC receives the data using a 2 × 10 GbE network card. This PC is a multicore machine with a large amount of RAM, with four cores being dedicated to receiving the data from the links reliably and buffering the data in the RAM. The raw image data then need to be processed to produce a proper image. For example, the data output from each chip does not directly correspond to a series of 12-bit pixel counts; instead, data from different pixels are interleaved. (In the longer term, more of this processing could be done in the FPGA to reduce the load on the server.) After decoding, the images can be saved to disk.

The control software for the detector uses a library of LAMBDA control functions, which is then built into a Tango device server. Tango is a control system used at DESY and other synchrotron labs, which makes it possible to control and monitor the detector and a range of other systems at an x-ray beamline in a standardized and convenient way. The Tango server code is also responsible for writing output files using the HDF5 format. This is a format that allows a large series of images to be saved to a single file with extensive metadata. The format includes optional data compression, using an approach where metadata and images can be extracted from the file without needing to decompress the entire file. These features make it much more practical to perform high-speed experiments where very large numbers of images are required.

3.4 High-Z Materials for Hard X-Ray Detection

At photon energies of 20 keV and above, the stopping power of silicon becomes insufficient for many applications, as shown in Figure 3.3 [9]. For hard x-ray detection, LAMBDA can be equipped with different sensor materials. Since these alternate sensors are commonly made from elements with a high atomic number, they are referred to as "high-Z" sensors.

The Medipix3 chip is well-suited to work with sensor materials other than silicon [11]. It is important that the input polarity of the chip can be selected

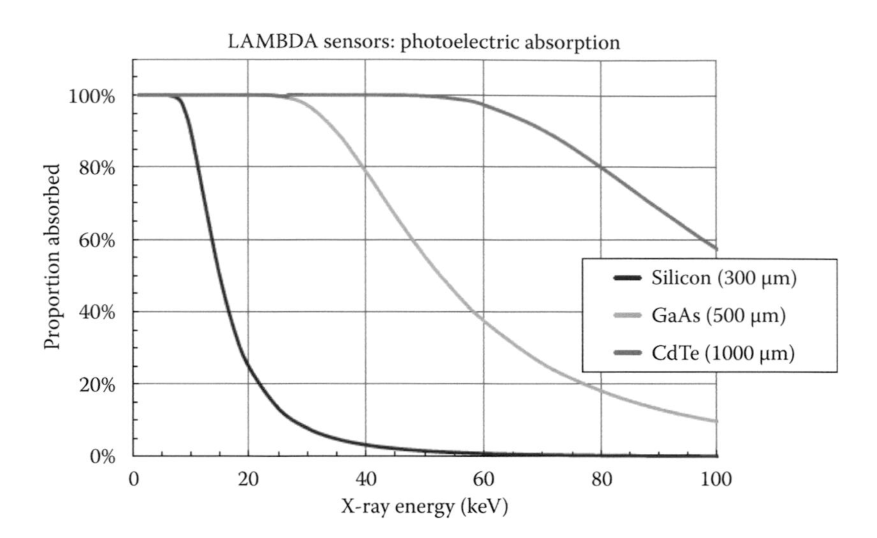

FIGURE 3.3
Proportion of x-rays absorbed as a function of photon energy by three commercially available sensor materials [12]. The absorption of germanium is almost identical to that of GaAs. (Reproduced from X-Spectrum, http://www.x-spectrum.de.)

since most high-Z sensors collect electrons, while most silicon sensors collect holes. It also has leakage current compensation capabilities, which allow the use of a wider range of materials, especially those with poor transport properties and/or large leakage currents. Lastly, the charge summing feature is particularly useful for high-Z sensors, because they typically have greater wafer thicknesses and hence experience more charge sharing.

While most high-Z materials are available in wafers, similar to silicon sensors, the size of these wafers is usually much smaller, commonly 3 or 4 in. (75 or 100 mm diameter). This results in a limitation of the maximum size of monolithic sensors, making them smaller than the largest monolithic silicon sensors. In the case of the LAMBDA system, this means that a sensor layout corresponding to 3 × 2 Medipix3 chips (42 × 28 mm or 1536 × 512 pixels) is the largest size that can be used. Larger modules like the LAMBDA 750K are tiled from two of these 3 × 2's next to each other on the same carrier board [13].

3.4.1 CdTe

Cadmium telluride is a well-established detector material, which is now commercially available in 3-in. wafers. These wafers provide both good resistivity and carrier lifetime, though they still have some non-uniformities. Since it is a more brittle material than GaAs and silicon, the risk of damaging the material during bump bonding is greater. The biggest benefit of CdTe is that it has substantially higher/larger quantum efficiency at higher energies. which makes it a natural choice for experiments at 80 keV and above. However, CdTe produces fluorescence photons around 25–30 keV, so if the incoming x-rays are in the range of about 25–50 keV, these fluorescence photons can both reduce the signal by escaping from the detector and blur the image by being reabsorbed in neighboring pixels. So, there are advantages to being able to choose different high-Z materials for different energy ranges. Additionally, CdTe can display changes in behavior with time and irradiation (polarization), which must be characterized, and where possible controlled, for example by choosing biasing conditions that minimize these effects [14].

3.4.1.1 X-Ray Response

Figure 3.4 shows x-ray test results from a CdTe LAMBDA module with a 3 × 2 chip layout as described above [13]. The sensor is 1000-μm-thick CdTe from the company Acrorad, which was then processed and bump bonded with a 55-μm pixel size by the University of Freiburg.

The flat-field response of the detector is shown in Figure 3.4a. This was measured using uniform illumination from a Mo target x-ray tube at 50 kV, a detector bias of −300 V, and a threshold setting of 10 keV. The flat-field image shows a pattern of lines superimposed over a more uniform response. These lines correspond to dislocations in the sensor material and show a high-counting region in the center of the line, and lower counts toward the edges. Additionally, in some places, there are "blobs" of insensitive pixels, typically surrounded by rings of pixels with increased counts. These clusters of insensitive pixels are mostly flooded with leakage current, rather than disconnected. Excluding dark and noisy pixels, the standard deviation of the pixel-to-pixel variation is 7.5%.

A USB stick was then imaged with the x-ray tube, and flat-field correction applied to the image. The result is shown in Figure 3.4b. Flat-field correction significantly improves the uniformity of the image, though the "blobs" of insensitive pixels cannot be corrected this way. There are 1116 dark pixels (0.28%) and 60 noisy pixels (0.015%), giving 99.70% functional pixels. This is a 0.06% lower yield than the GaAs module (see below), and since the insensitive pixels are more clustered, they are also more visible in the image. In the corrected images, the network of lines is faintly visible, indicating that the detector response has changed somewhat from the first to the second set of flat images. This change could be due to both the 10-min time gap between the measurement sets and also the irradiation of the sensor.

3.4.1.2 Stability over Time

One downside of CdTe is that its behavior can change over time due to polarization of the material. This CdTe module used ohmic pixel contacts, which typically suffer less from polarization than Schottky devices, but these effects can still occur over extended time periods or after irradiation [15].

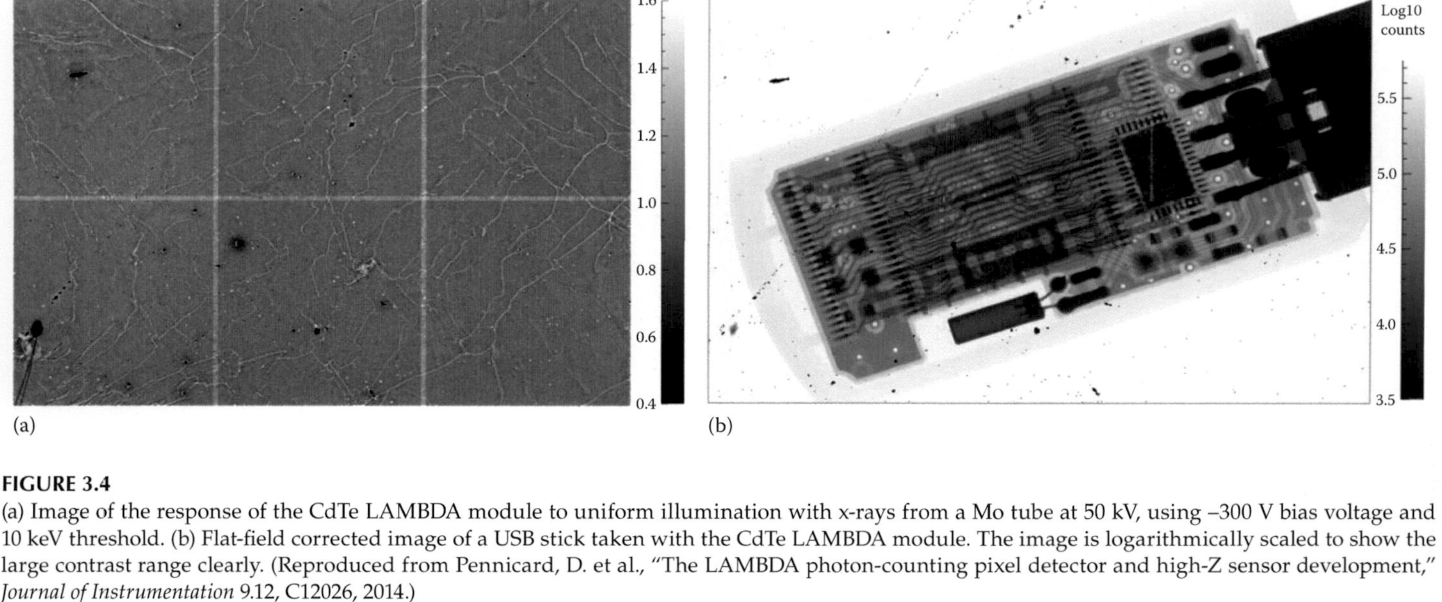

FIGURE 3.4
(a) Image of the response of the CdTe LAMBDA module to uniform illumination with x-rays from a Mo tube at 50 kV, using –300 V bias voltage and 10 keV threshold. (b) Flat-field corrected image of a USB stick taken with the CdTe LAMBDA module. The image is logarithmically scaled to show the large contrast range clearly. (Reproduced from Pennicard, D. et al., "The LAMBDA photon-counting pixel detector and high-Z sensor development," *Journal of Instrumentation* 9.12, C12026, 2014.)

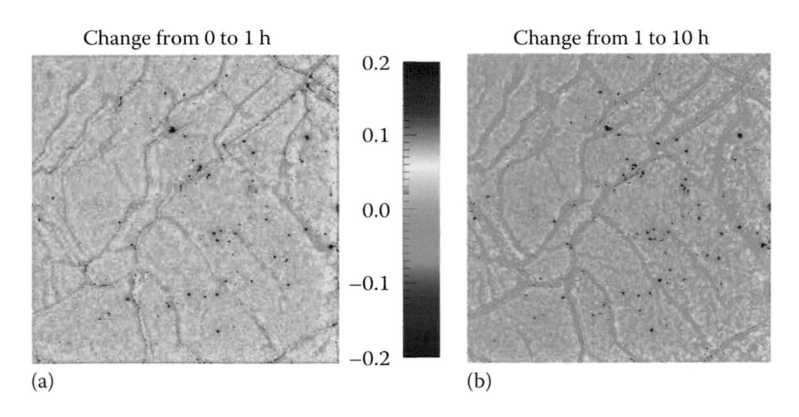

FIGURE 3.5
(a) Change in the flat-field response of the CdTe detector to 70-keV photons over the first hour of operation. The plot shows the flat-field response after 1 h minus the initial flat-field response, scaled so that 1 corresponds to the average response after 1 h. (b) Change in the flat-field response over the following 9 h, i.e., from 1 h into the measurement to 10 h. (Reproduced from Pennicard, D. et al., "The LAMBDA photon-counting pixel detector and high-Z sensor development," *Journal of Instrumentation* 9.12, C12026, 2014.)

To investigate this, during a beamtime at ESRF beamline ID15C, a long-term measurement was made with low, uniform illumination with 70-keV photons. (The measurement period was started immediately when the detector was powered on.)

Firstly, it was found that the overall count rate on the detector increased by 4% over the first hour of operation, before leveling out and becoming uniform over time. Additionally, changes in flat-field response over time were found. Figure 3.5 [13] shows the change in the flat-field image in a region of the sensor over the first hour of operation, then the following 9 h. Over the period from the start of the measurement to 1 h, there is not only an overall increase in counts but also an intensification of the dislocation lines, with the high-count region in the center increasing further and the low-count region at the edge decreasing. Then, over the period from 1 to 10 h, the overall count rate remains the same, but the intensity of the defect pattern is reduced somewhat. These results imply that a single flat-field measurement is not sufficient to fully correct for the non-uniformities in the detector over an extended period of time. The fact that the change is most rapid after initially switching on the high voltage also implies that rather than frequently resetting the bias, it is helpful to allow the detector to stabilize before making measurements, at least when working with ohmic contacts.

3.4.2 GaAs

Gallium arsenide is a widely used semiconductor material in applications such as optoelectronics. The high atomic number of this material makes it appealing for hard x-ray detection, and its high electron mobility potentially makes it a faster material than silicon. While it has lower x-ray absorption efficiency than CdTe at energies above 50 keV, at around 30 keV, it has the advantage that it is not as strongly affected by fluorescence.

However, as with many compound semiconductors, it is difficult to produce thick GaAs wafers with the combination of properties required for use in a high-performance pixel detector: high resistivity (so that when biased, there is a strong electric field throughout the sensor volume and low leakage current); long enough carrier lifetimes to ensure high charge collection efficiency; and high uniformity. One approach to producing detector-grade GaAs is to alter the properties of the material by doping. An effective way of doing this has been developed by Ayzenshtat et al. in 2001, where standard n-type GaAs wafers are doped with chromium by diffusion [16]. The Cr compensates the excess electrons in this material, thus achieving high-enough resistivity to allow it to be used as a photoconductor detector with ohmic contacts.

3.4.2.1 X-Ray Response

A LAMBDA module with a GaAs(Cr) sensor was produced as part of a German–Russian collaboration project, GALAPAD, between FEL (Functional Electronics Laboratory of Tomsk State University), JINR (Dubna), the University of Freiburg, the Karlsruhe Institute of Technology, and DESY [13]. The sensor was produced by FEL and consisted of a 500-μm-thick layer of GaAs(Cr) with 55-μm pixels. The sensor had an array of 768 × 512 pixels, with a sensitive area of 42 × 28 mm. This layout corresponds to 3 × 2 Medipix3 chips. The sensor was bonded to the readout chips by the University of Freiburg.

For testing, the GaAs sensor was connected to a bias voltage of –300 V and operated with the chip in electron-readout mode. Because the pixel size is small compared to the sensor thickness, most of the measured signal in each pixel will be due to the carriers drifting toward the pixelated side, rather than the back contact. This so-called "small pixel effect" is discussed in Ref. [17]. In a Cr-compensated GaAs, it has been shown that electrons have a higher mobility-lifetime product than holes, largely due to the inherently higher mobility of electrons in GaAs, and this means that electron readout will give higher collection efficiency [18].

Figure 3.6a [13] shows the detector response when uniformly illuminated with x-rays from a Mo target x-ray tube at 50 kV. The raw image shows a high level of non-uniformity, with a granular structure that is probably related to the original growth process of the wafer. The standard deviation of the pixel-to-pixel variation (excluding dark and noisy pixels) is 21.0%. However, by applying flat-field correction (i.e., using the flat-field response to correct subsequent images), the image quality can be greatly improved. Figure 3.6b shows a flat-field corrected x-ray image of a USB stick.

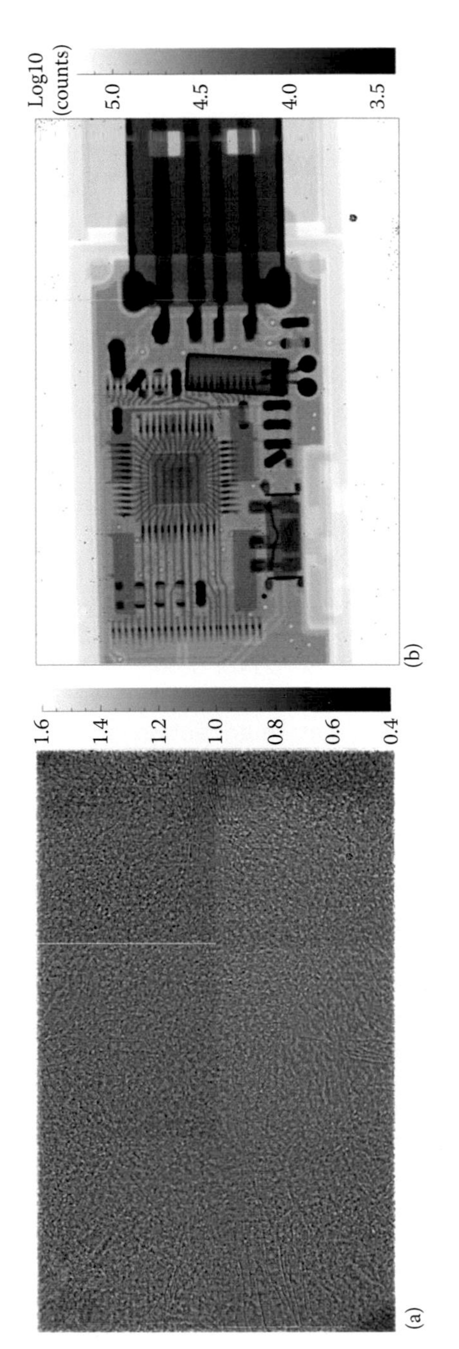

FIGURE 3.6
(a) Image of the response of the GaAs LAMBDA module to uniform illumination with x-rays from a Mo tube at 50 kV, using −300 V bias voltage and 10 keV threshold. (b) Flat-field corrected image of a USB stick taken with the GaAs LAMBDA module. The image is logarithmically scaled to show the large contrast range clearly. (Reproduced from Pennicard, D. et al., "The LAMBDA photon-counting pixel detector and high-Z sensor development," *Journal of Instrumentation* 9.12, C12026, 2014.)

The imaging conditions were the same as for the flat-field image, and the image is shown with a logarithmic color scaling. The image shows a good quality, with a high yield of functional pixels. Excluding a row of pixels that are nonfunctional due to a problem with the readout chip rather than the sensor, there are 640 dark pixels (0.16%) and 300 noisy pixels (0.08%), giving 99.76% working pixels. To test the effectiveness of the flat-field correction more quantitatively, a second set of flat-field images was acquired, and the first set of flat-field data was used to correct them. This reduced the standard deviation of the pixel-to-pixel variation (excluding dark and noisy pixels) from 21.0% to 0.15%.

It has also been shown elsewhere that the detector response is stable over time, which means that after the flat-field measurement is acquired, it can be consistently used to correct the detector response [19]. It should be noted that flat-field correction factors are dependent on photon energy, and when an object is imaged with polychromatic x-rays, the transmitted spectrum will differ from the direct tube spectrum, thus reducing the accuracy of flat-field correction. This, however, is not a problem for experiments with monochromatic beam at synchrotrons.

3.4.3 Ge

Germanium (Ge) is a well-known material for detection of particles, gamma rays, and x-rays [20]. It is widely used in high energy physics, nuclear physics, materials testing, and security. The main advantage of Ge over other high-Z materials is that it is available in large wafers and ingots with extremely high material quality. This in turn is due to Ge being an elemental semiconductor with a long history of development. The main drawback of Ge is that it has a narrow bandgap (0.67 eV), which means that it needs to be cooled to low temperatures to avoid excessive leakage current.

The possibility to produce very large volume Ge detectors with excellent performance has meant that Ge is commonly used for gamma ray detection, and development of Ge has primarily focused on large detectors based on high purity Ge (HPGe) [21]. However, there are also techniques for producing finely segmented detectors, such as photolithography, as described by Gutknecht [22]. Currently, Ge strip detectors with pitches down to 50 μm (single sided) or 200 μm (double sided) are commercially available.

While the high material quality of Ge makes it an appealing choice for hybrid pixel detectors, challenges in production and operation need to be overcome. Firstly, Ge is a relatively delicate material; its passivation layers are less robust than the typical SiO_2 used in Si detectors, and exposure to temperatures above 100°C can allow impurities to diffuse into Ge. So, the bump bonding process needs to be developed to avoid damaging the sensor. Secondly, as noted above, the detector needs to be cooled during operation. The first finely pixelated Ge hybrid pixel detectors were developed for use with the LAMBDA system in 2014 [23].

3.4.3.1 Ge Hybrid Pixel Production

The Ge sensors were produced on two 90-mm wafers of p-type high-purity Ge. The sensors have a photodiode structure, with the pixel contacts on the n-type implant side (electron collection). The high voltage (HV) side, or entrance window, is formed by a p-type implantation. A completed wafer is shown in Figure 3.7 [23]. Each wafer has 16 sensors with 55-μm pixel pitch and a layout of 256 × 256 pixels (65536 total), matching the Medipix3 chip. Around the pixel array, there are guard-ring structures with a total width of 750 μm. The thickness of the Ge sensor was 700 μm. This thickness was chosen as a compromise between performance and mechanical stability.

The sensors were bump-bonded to Medipix3 chips at Fraunhofer IZM, Berlin. The bonding was done using indium (In) bumps and a low-temperature thermocompression process. This ensured that the sensors were not damaged by temperature during the bonding process. Additionally, because In remains ductile even at low temperatures, this reduces the risk of bumps cracking during cooling due to the mismatch in thermal contraction between Ge and Si.

The sensors were cooled under vacuum during operation. Current–voltage tests showed that at a temperature of –100°C, the sensor's depletion voltage was 20 V. At –70°C, the depletion voltage increased to 40 V, and at warmer temperatures, the depletion voltage increased rapidly. Tests of 1-mm^2 diodes showed a bulk leakage current of 10 nA at –70°C, which is well below the 330 nA/mm^2 required for good performance with the Medipix3 chip.

FIGURE 3.7
90-mm high-purity germanium wafer with 16 Medipix3-compatible sensors, each with a 256 × 256 array of 55-μm pixels. The wafer is 700-μm-thick p-type material. (Reproduced from D. Pennicard et al., "A germanium hybrid pixel detector with 55μm pixel size and 65,000 channels," *JINST* 9, P12003, 2014.)

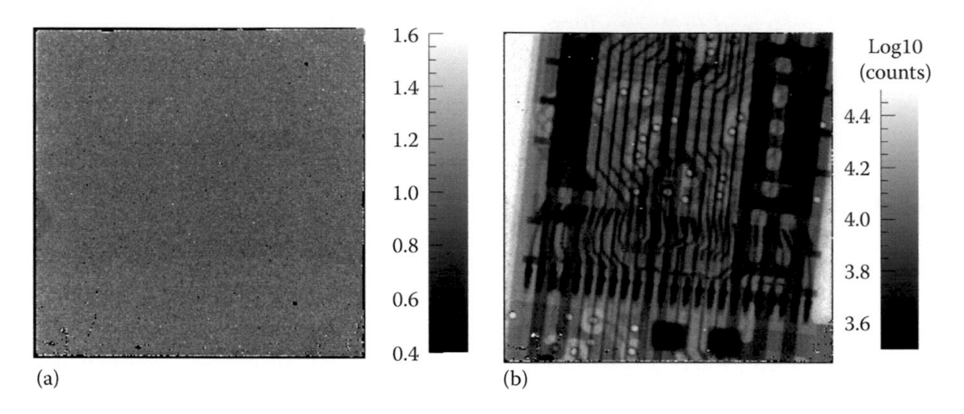

FIGURE 3.8
(a) Flat-field x-ray image taken with Ge sensor with 50-kV tube voltage, –100-V sensor bias, and 7.5-keV threshold. (b) X-ray image of a USB stick, taken with –200-V bias. The image is log-scaled but has not been flat-field corrected. (Reproduced from D. Pennicard et al., *JINST* 9, P12003, 2014.)

3.4.3.2 X-Ray Response

Figure 3.8a [23] shows a flat-field image taken with a Ge detector using an x-ray tube at 50 kV and 100-V bias. The image was taken with the sensor at –100°C and with a threshold setting of 7.5 keV. Pixels around the edge of the sensor are insensitive, due to excessive leakage current, but otherwise the sensor shows a good level of uniformity, without the systematic defects seen in the CdTe and GaAs sensors. The proportion of working pixels is 96.7%, or 99.77% in the central region ignoring the edges. The RMS spread in pixel intensities is 7.8%.

Due to the relatively good quality of the sensor, imaging tests show a reasonable image quality even without flat-field correction. Figure 3.8b shows an example of an x-ray image of a USB stick taken using the same x-ray source, without flat-field correction. The image is log-scaled.

3.4.3.3 Further Developments of Ge Sensors

A large area pixelated sensor is currently in preparation by DESY and Canberra. The sensor will have the size of 3 × 2 Medipix3 chips and cover an active area of approximately 45 × 30 mm.

3.5 Application Examples

Ever since the first prototypes of the LAMBDA camera have been developed, it has been used in a range of applications. The following examples have been chosen to demonstrate the capabilities of the system. LAMBDA has already found its way into routine operation at a few light sources, so the following examples highlight only a fraction of the many possible ways LAMBDA cameras can be used.

3.5.1 Conventional and Coherent Imaging

One of the most classic applications of an x-ray detector is imaging an object in transmission geometry. In this example, a tin of sardines was placed in front of the detector and illuminated by an x-ray tube more than a meter away. The resulting radiograph is shown in Figure 3.9, and it reveals both the structure of the can (e.g., scratches are seen as white vertical stripes) and the fine fish bones of the sardines with a very high resolution [24].

This example was one of the first demonstrations of the capability of the camera. It shows the high spatial resolution and sensitivity of the system. There are no gaps in the image; all photons that are interacting in the sensor are counted.

At synchrotron beamlines, objects can be imaged with higher spatial resolution using a range of "coherent imaging" techniques, which take advantage of x-ray diffraction, rather than just transmission. One common approach is ptychography. In this technique, a finely focused beam (hundreds of nanometers in size) is raster-scanned across a sample. At each point in the scan, a diffraction pattern is recorded by the detector. The step size is chosen so that overlapping regions are illuminated during the scan. From these diffraction data, it is possible to reconstruct an image of the sample on length scales significantly smaller than the beam (down to 10 nm). One particular appeal of this approach is that it can be combined with a range of other synchrotron techniques; for example, by measuring fluorescence emitted from the sample during the scan, it is possible both to image the sample and map its elemental composition. For these experiments, a detector with high sensitivity, high

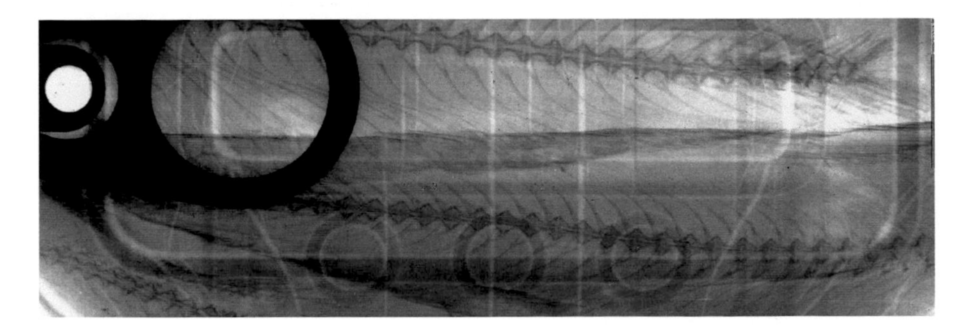

FIGURE 3.9
X-ray image of a sardine can, taken with a LAMBDA 750K camera with silicon sensor. The vertical white stripes are caused by scratches on the back of the tin [24]. (From M. Sarajlić et al., *JINST* 11, C02043. doi:10.1088/1748-0221/11/02/C02043, 2016.)

speed, and good spatial resolution is required to perform the scan in a reasonable length of time. LAMBDA has been used in ptychography experiments to obtain images with a spatial resolution of around 20 nm [25].

3.5.2 High Pressure Measurements in Diamond Anvil Cells

A high proportion of experiments at synchrotrons use x-ray diffraction to study a sample's structure on an atomic scale. Using hard x-rays for these experiments makes it possible to gain higher spatial resolution, to study large samples or samples made of heavier elements, and to probe samples while they are contained in sample environments. Extreme conditions experiments are a good example of the last example. By compressing a sample inside a diamond anvil cell while heating it with a laser, it is possible to recreate the extremely high pressures and temperatures found inside planets or in the outer layers of stars. If a highly focused x-ray beam is fired through the sample, the x-rays will be diffracted, and from the diffraction pattern, it is possible to infer the atomic-level structure. High-speed photon counting detectors like LAMBDA can then allow scientists to study the rapid changes that occur in a sample when the pressure and temperature change.

To investigate this possibility, the GaAs LAMBDA detector was tested at PETRA-III beamline P02.2, which is used for extreme conditions experiments [26]. A sample of a common test standard, CeO_2, was placed inside a diamond anvil cell in the x-ray beam. The x-ray beam had a photon energy of 42 keV; at this energy, the photoelectric absorption efficiency of 500-μm-thick GaAs is 74%, compared to 4.6% for the same thickness of silicon. The LAMBDA GaAs detector was placed at a distance of 35 cm, with a horizontal offset of 4 cm between the x-ray beam and the edge of the sensitive area. A series of images were then taken with different shutter times [13].

Figure 3.10a shows an image taken with an acquisition time of 1 ms. Because the sample is a coarse powder, the diffraction pattern consists of rings of diffraction spots, with each spot produced by scattering from a single grain of the material. The angle and intensity of these rings convey information about the crystalline structure of the material. Due to the short shutter time, the rings are faint, with most pixels only detecting a few photons. Nevertheless, due to the effectively noise-free behavior of the detector, when the photon hits in the 1-ms image are integrated as a function of scattering angle, a clear diffraction signal is obtained, with the rings clearly distinguished from the background level, as shown in Figure 3.10b. This demonstrates that the detector could be used to measure structural changes in a sample on a timescale of milliseconds.

More recently, two three-module (2 megapixel) GaAs systems have been constructed—one is shown on the right-hand side of Figure 3.2. These are now in routine use for diamond anvil cell experiments at 2-kHz frame rates.

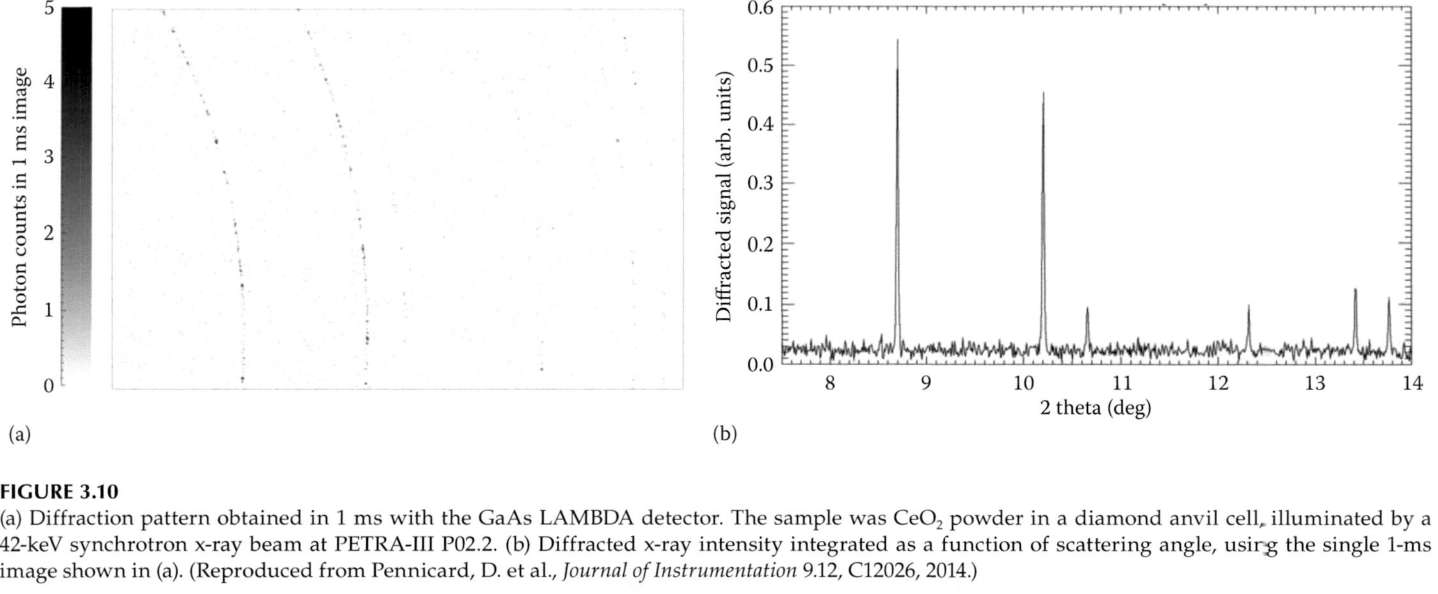

FIGURE 3.10
(a) Diffraction pattern obtained in 1 ms with the GaAs LAMBDA detector. The sample was CeO_2 powder in a diamond anvil cell, illuminated by a 42-keV synchrotron x-ray beam at PETRA-III P02.2. (b) Diffracted x-ray intensity integrated as a function of scattering angle, using the single 1-ms image shown in (a). (Reproduced from Pennicard, D. et al., *Journal of Instrumentation* 9.12, C12026, 2014.)

3.5.3 XPCS and Rheology in Colloids

X-ray photon correlation spectroscopy (XPCS) is a technique for studying dynamics in soft matter. If a beam of coherent x-rays is fired at a sample with an irregular structure, such as colloidal particles in a liquid, the resulting diffraction pattern will contain speckles whose positions depend on the particular arrangement of particles. As the particles move, the speckles will fluctuate, and by measuring these fluctuations, information about the particle dynamics can be obtained. This experimental technique requires a combination of high speed, high sensitivity, and small pixel size to successfully measure these fluctuations, which means that a detector like LAMBDA is particularly well-suited to XPCS.

A rheometer is an experimental setup for measuring the viscoelastic properties of material. A fluid sample is placed between a pair of plates, and the plates can then rotate or oscillate to produce different shear forces on the fluid and measure the corresponding deformation. By combining rheology with x-ray diffraction techniques such as XPCS, it becomes possible to relate these macroscopic viscoelastic properties to the microscopic behavior of the fluid. This has been demonstrated with LAMBDA [27]. For example, these experiments demonstrate that an effect called "shear thinning," where a colloidal sample becomes less viscous as the shear force on the colloid increases, occurs when the spacing of particles along the direction of flow becomes less regular. In these experiments, the x-ray beam passed vertically through the rheometer to reach the detector. This meant that the distance between the source and the detector was limited, and so the small pixel size of LAMBDA was a particularly important requirement.

3.6 Future Directions

The LAMBDA system is a versatile detector that has been successfully used for many synchrotron experiments. However, when tiling several modules together to cover larger areas, inactive gaps between the modules inevitably appear.

In the following, we will present two approaches to reduce the inactive gap, which, if implemented together, could possibly eliminate the gap almost completely.

3.6.1 Edgeless Sensors

The sensor of a hybrid detector normally has additional "guard ring" structures around its edges, beyond the pixel matrix, consisting of doped implants that can either be connected to constant voltages or left floating [28–30]. The structure of a conventional detector module with guard rings is shown on the left of Figure 3.11. The purpose of these guard structures is to

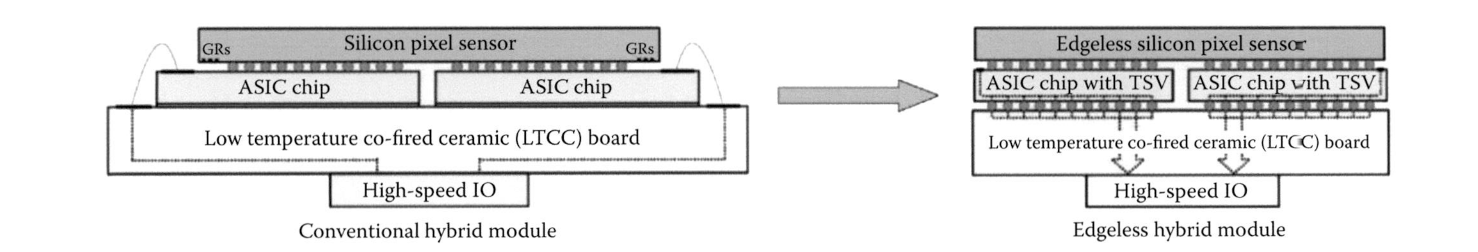

FIGURE 3.11
Illustration of a conventional x-ray detector module (left) and the concept of the detector with edgeless sensor and readout chip with TSV contacts (right). (Reproduced from J. Zhang et al., *JINST* 9, C12025. doi:10.1088/1748-0221/9/12/C12025, 2014.)

ensure a smooth and predictable drop in electric potential between the edge of the pixel array and the diced edge of the sensor. Without this guard ring structure, the trajectory of the generated charge in the edge region would be curved significantly and the resulting image would be distorted [29]. Other reasons to include a guard ring structure are to minimize the sensor leakage current, to prevent high-field regions from causing breakdown, and to prevent defects caused by dicing saw to introduce a current in the sensor region. An example of a typical sensor corner design is given in Figure 3.12 [31].

The presence of the guard rings introduces significant dead area in the case of multimodule detectors. To find a compromise and reduce the dead area without compromising the functionality, a new "edgeless" technology was developed, where special processing of the edge regions is used to achieve predictable behavior in these regions without large guard structures. This concept is depicted on the right of Figure 3.11. Special care must be taken in designing radiation-hard edgeless sensors; due to the short distance between the edge of the detector and the pixel array, electric fields within the detector are comparatively high, and insolation techniques must be designed to minimize high-field regions that could result in breakdown [29].

One pioneer was the VTT Institute in Finland, which started to develop edgeless sensors in 1995 [32]. In VTT's approach, the side of the sensor edge is implanted with either p+ or n+ dopants [33]. This implantation of the silicon has to be done at a particular angle and in multiple steps to ensure that all sides of the sensor receive defined amount of dopants. This is a very delicate process because the doping ions are coming to the edge surface at a certain angle that makes the process less controllable than standard ion implantation normal to the surface. Since all sensor chips are produced from a flat

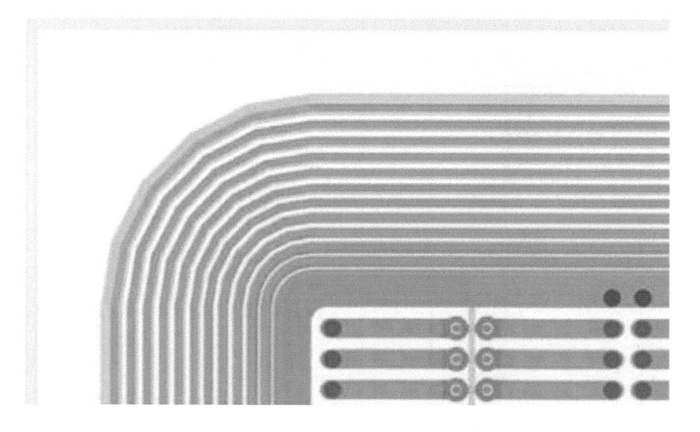

FIGURE 3.12
Example of current collection ring and guard ring structure. View on the sensor corner. (Reproduced from P. Weigell, "Investigation of Properties of Novel Silicon Pixel Assemblies Employing Thin n-in-p Sensors and 3D-Integration," PhD thesis, Technische Universität München, 2013.)

wafer, it is also necessary to first remove the part of the silicon material in between the chips to expose the sensor edge to the side implantation.

SINTEF, an independent research organization based in Norway, has been another major player for about a decade in the fabrication of edgeless sensors [34]. The key fabrication steps for edgeless sensors at SINTEF are given in Figure 3.13. The sensor wafer is bonded to a carrier wafer by fusion bonding to maintain the integrity of the sensor wafer during the subsequent processes. After that, narrow trenches are made through the sensor wafer by deep reactive ion etching (DRIE) to define the boundaries of each sensor. This is followed by the doping of the trenches by gas phase diffusion and the filling of trenches with polysilicon. Polysilicon filling is needed to "re-flatten" the sensor wafer so that the following photolithographic processes can be carried out easily without causing a significant yield loss. The remaining processing steps are the same steps involved in the fabrication of pixel sensors with guard ring design: ion implantation of pixels, metal contacts to pixels, etc. SINTEF cooperates with Fraunhofer IZM, Germany,

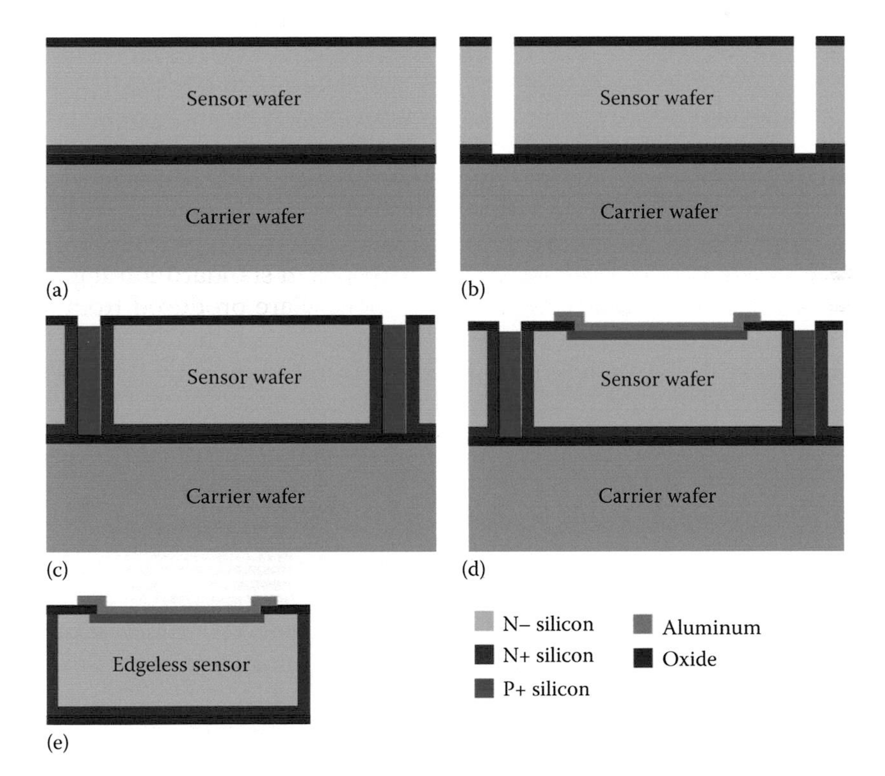

FIGURE 3.13
Key technology steps involved in fabrication of edgeless sensors at SINTEF. (a) Sensor wafer bonded to a carrier wafer. (b) Sensor edge trenches etched by DRIE. (c) Trenches doped and filled with polysilicon. (d) Standard planar sensor processing carried out. (e) Carrier wafer removed and edgeless sensor singulated. (Courtesy of SINTEF, Norway.)

to establish a process for removal of the carrier wafer and singulation of the sensors, which yet remains to be the most challenging process step in fabrication of edgeless sensors [35]. SINTEF is also currently investigating alternative processes not involving any carrier wafer in order to make edgeless sensors more commercially viable by reducing the complexity of fabrication and cost.

3.6.2 TSV Technology

In current hybrid pixel detectors, readout chips are connected to the readout electronics by wire bonding, and a significant part of the area on at least one side is consumed by the bond wires and bond pads. Figure 3.11 (left) illustrates this problem. "Through-silicon vias" (TSV) could eliminate this problem by replacing the wire bonds with metal-filled vias passing through the thickness of the readout chip, as shown on the right of Figure 3.11. In addition to reducing or eliminating the dead space due to wire bonds, this approach would provide greater flexibility in designing detector layouts; due to the space required for wire bonds, hybrid pixel detector modules usually have a "2 × N" chip layout, whereas TSV technology could make it possible, for example, to produce square modules with more than 2 × 2 chips.

The underlying technology for TSVs has been developed by the microelectronics industry to improve chip packaging, and to go beyond Moore's law and provide an increased density of integration, even when scaling of the components to the smaller sizes is not feasible anymore [36,37]. High-density integration enabled by TSV technology is called 3D integration. It enables microelectronic components to be stacked on top of each other and interconnected by vias between the chips. Although TSV technology existed more than 10 years ago, it has become increasingly popular in the last 10 years. Nowadays, many 3D integrated systems like DRAM, Flash memory, and FPGA units are commercially available [38].

3.6.2.1 Integrating TSVs into Medipix3 and LAMBDA

The Medipix3 chip was designed to be compatible with both wire bonds and TSV readout; in addition to wire bond pads for I/O on the surface of the chip, the chip has landing pads for TSVs in the lowest metal layer. One of the first successful applications of the TSV technology to x-ray detectors was achieved by CEA LETI [39] using the Medipix3 chip. Their processing gave an overall yield of 71% from one processed wafer, counting the number of the best chips after TSV processing and before. Measurements of noise in the pixels revealed only a minor increase in the values after TSV processing. Internal parameters, the so-called DAC values, were measured before and after TSV processing and showed no significant drift in respect to each other. Overall, no measurable degradation was observed, which means that this processing was very successful.

Development of LAMBDA systems with 3D integration is underway with Fraunhofer IZM, Berlin, Germany [40]. The Medipix3 chip and a via-last approach are used for this development. A shortened illustration of the technology steps needed for the TSV fabrication on Medipix3 chips is given in Figure 3.14. Medipix3 chips are fabricated in 130-nm technology on 200-mm (8-in.) wafers. The original thickness of the wafer is 725 μm. Wafers are then grinded to the thickness of 200 μm. This is needed to facilitate the process of via etching. The via is etched through the full grinded wafer thickness, stopping on the lowest metal layer (M1) of the backend-of-line metal layer stack. The vias have 60-μm diameter and 200-μm depth. These parameters represent an aspect ratio of 3:1, which is easily achievable with current technology. Next, an oxide passivation layer is deposited inside the via, followed by metal sputtering to form and barrier and a seed layer. A Cu liner filling process is then used for electrical interconnection to the M1 layer and backside redistribution layer (RDL). A cross section of TSVs processed from wafer backside is shown in Figure 3.15.

To form the connections between the vias and the PCB substrate, an RDL was designed. Figure 3.16 gives the proposed structure of the RDL. The contacts from a total of 114 vias are routed to 70 pads on the back side of the Medipix3. These pads will be connected using a ball grid array (BGA) to a PCB or LTCC (Ceramics) substrate. In the design of the RDL, it was taken care that the power and ground lines are routed with minimal electrical resistance and the LVDS signal lines are routed with negative and positive pairs in parallel. This is needed in order to keep the signal integrity of the LVDS lines.

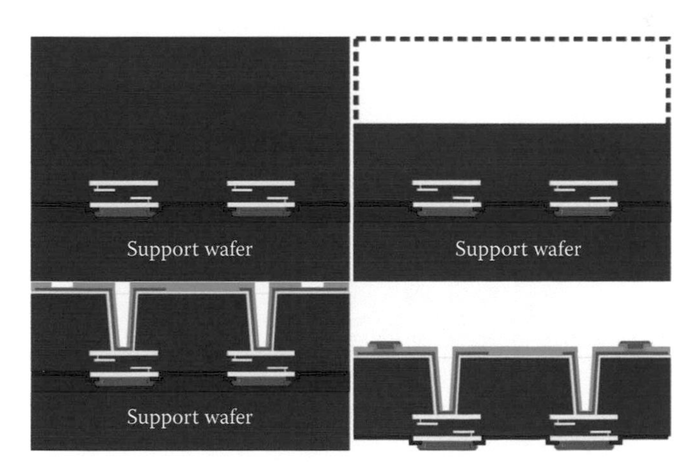

FIGURE 3.14
Technology steps involved in the production of the Medipix3 TSV module. Top left: Medipix3 wafer, thickness 725 μm, after UBM deposition on the front side and bonding to a support wafer. Top right: Medipix wafer after thinning. Bottom left: Wafer after TSV etch from the back side, passivation, TSV Cu filling and back side RDL. Bottom right: Wafer after UBM metallization on the back side and debonding of carrier wafer and cleaning of the front side. (Reproduced from M. Sarajlić et al., *JINST* 11, C02043. doi:10.1088/1748-0221/11/02/C02043, 2016.)

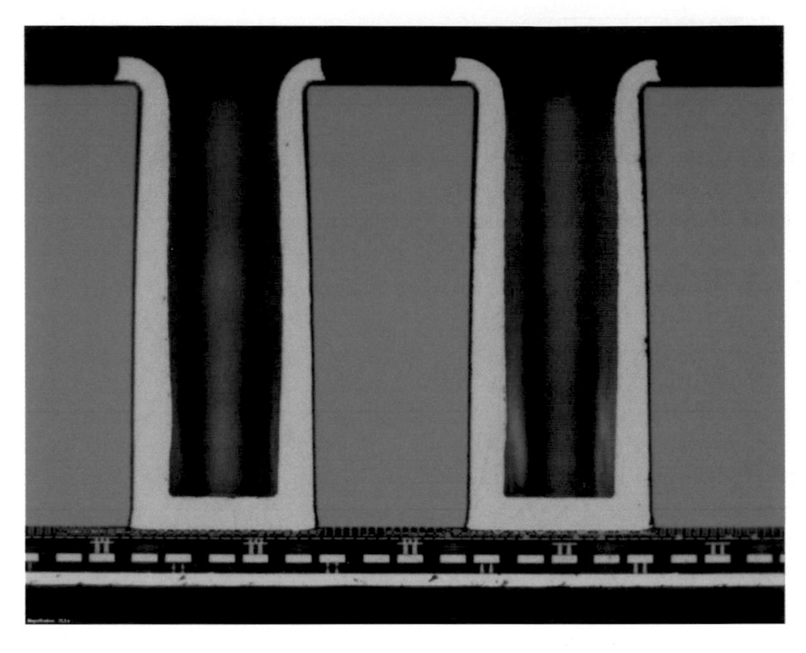

FIGURE 3.15
Cross section of Medipix3 chip after TSV last process from wafer backside. (Courtesy of Fraunhofer IZM.)

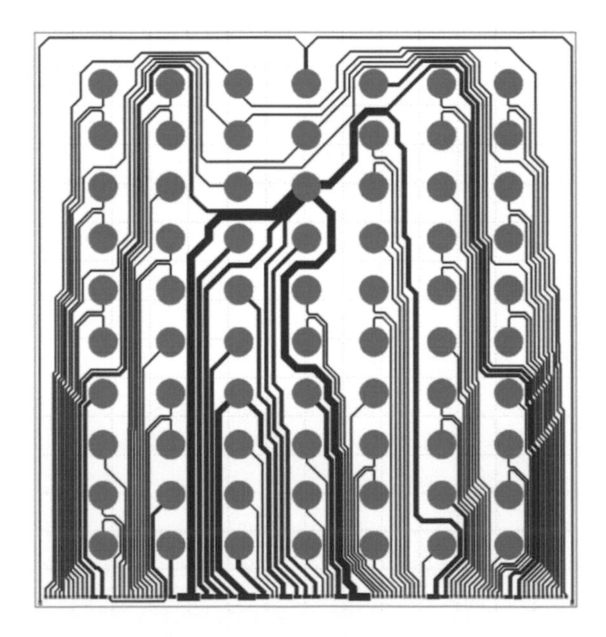

FIGURE 3.16
RDL as it is designed for the Medipix3 TSV chip. (Reproduced from M. Sarajlić et al., *JINST* 11, C02043. doi:10.1088/1748-0221/11/02/C02043, 2016.)

3.7 Summary

The rapidly increasing brilliance of synchrotron x-ray sources makes them powerful tools for research in a range of areas such as materials science and molecular biology. However, to make full use of this brilliance, new detectors are needed that can provide higher speed, sensitivity, and resolution.

The LAMBDA system is a photon counting pixel detector based on the Medipix3 chip. Its key features are high speed operation at 2000 frames per second (deadtime-free), a large tileable module design with small pixels, and single photon sensitivity. To achieve these features, both a custom ASIC design and a readout system with a high data bandwidth were needed.

A further feature of LAMBDA is its compatibility with high-Z materials for use in hard x-ray experiments. Systems using CdTe and GaAs have not yet demonstrated perfect sensor quality, but their quality is sufficiently good to allow practical use in synchrotron experiments. Prototype Ge sensors have demonstrated a good material quality, and development of larger Ge sensors is ongoing.

LAMBDA has been demonstrated in a range of synchrotron x-ray experiments such as coherent imaging and x-ray diffraction. In particular, it has been shown that LAMBDA systems with high-Z sensors can make it possible to detect millisecond-timescale changes in samples under extreme environments such as high pressure, thanks to their combination of high speed and sensitivity to hard x-rays.

Further developments to the LAMBDA system are aimed at greatly reducing dead areas between modules by using edgeless sensors and vertical integration with TSVs. As well as being used with LAMBDA, these techniques could play a more integral role in the design of new x-ray detectors.

Acknowledgments

The authors wish to thank all the current and former DESY staff involved in the development of the LAMBDA system, including Sergej Smoljanin, Sabine Lange, Helmut Hirsemann, Jiaguo Zhang, Bernd Struth, Igor Sheviakov, Qingqing Xia, Yuelong Yu, Andre Rothkirch, and FS-DS group leader Heinz Graafsma. We would also like to thank the staff involved in experiments at beamlines with LAMBDA; the experiments listed here were performed by Fabian Westermeier, Michael Sprung, Hanns-Peter Liermann, and Zuzana Konôpková.

Thanks to the Medipix3 designers Rafael Ballabriga Sune and Xavi Llopart Cudie, Jerome Alexandre Alozy, and CERN Microelectronics Group leader

Michael Campbell for their help on the details of Medipix3 technology and functionality.

Development of detectors with high-Z sensors was done in collaboration with the University of Freiburg (Michael Fiederle and Alex Fauler) and Tomsk State University (Anton Tyazhev and others).

We would like to express our gratitude to Fraunhofer IZM researchers: Thomas Fritzsch, Martin Wilke, Kai Zoschke, and Mario Rothermund for their help on 3D integration technology.

We wish to thank the researchers from SINTEF, Norway, Ozhan Koybasi, Angela Kok, and Anand Summanwar on their help in the field of edgeless sensors design and processing.

We are grateful to the engineers from Mirion Technologies (Canberra), France, Marie-Odile Lampert, Milan Zuvic, and Jeremy Flamanc, on their help on high purity and pixelated germanium sensors.

References

1. Light Sources. Available at http://www.lightsources.org/history, accessed 12/6/2016.
2. Shapiro, Stephen L. et al. "Silicon pin diode array hybrids for charged particle detection." *Nuclear Instruments and Methods in Physics Research Section A: Accelerators, Spectrometers, Detectors and Associated Equipment* 275.3 (1989): 580–586.
3. Ballabriga, R. et al. "The Medipix3 prototype, a pixel readout chip working in single photon counting mode with improved spectrometric performance." *2006 IEEE Nuclear Science Symposium Conference Record*. Vol. 6, IEEE, 2006.
4. Gimenez, Eva N. et al. "Medipix3RX: Characterizing the Medipix3 redesign with synchrotron radiation." *IEEE Transactions on Nuclear Science* 62.3 (2015): 1413–1421.
5. Ballabriga, Rafael et al. "The Medipix3RX: A high resolution, zero deadtime pixel detector readout chip allowing spectroscopic imaging." *Journal of Instrumentation* 8.02 (2013): C02016.
6. Pennicard, D., and H. Graafsma. "Simulated performance of high-Z detectors with Medipix3 readout." *Journal of Instrumentation* 6.06 (2011): P06007.
7. Pennicard, D. et al. "The LAMBDA photon-counting pixel detector," *Journal of Physics: Conference Series* 425 (2013): P062010.
8. Pennicard, D. et al. "LAMBDA—Large area Medipix3-based detector array." *Journal of Instrumentation* 7.11 (2012): C11009.
9. X-Spectrum website: http://www.x-spectrum.de.
10. Pennicard, D. et al. "High-speed readout of high-Z pixel detectors with the LAMBDA detector," *Journal of Instrumentation* 9.12 (2014): C12014.
11. Koenig, Thomas et al. "Charge summing in spectroscopic x-ray detectors with high-Z sensors." *IEEE Transactions on Nuclear Science* 60.6 (2013): 4713–4718.

12. Berger, M. J. et al. *NIST Standard Reference Database 8 (XGAM)*. Available at https://www.nist.gov/pml/xcom-photon-cross-sections-database.
13. Pennicard, D. et al. "The LAMBDA photon-counting pixel detector and high-Z sensor development," *Journal of Instrumentation* 9.12 (2014): C12026.
14. Funaki, M. et al. "Development of CdTe detectors in Acrorad," in *International Workshop on Semiconductor PET*, 2007.
15. Ruat, M., and C. Ponchut. "Defect signature, instabilities and polarization in CdTe x-ray sensors with quasi-ohmic contacts." *Journal of Instrumentation* 9.04 (2014): C04030.
16. Ayzenshtat, G. I. et al. "GaAs structures for x-ray imaging detectors." *Nuclear Instruments and Methods in Physics Research Section A: Accelerators, Spectrometers, Detectors and Associated Equipment* 466.1 (2001): 25–32.
17. Sellin, P. J. "Modelling of the small pixel effect in gallium arsenide x-ray imaging detectors." *Nuclear Instruments and Methods in Physics Research Section A: Accelerators, Spectrometers, Detectors and Associated Equipment* 434.1 (1999): 75–81.
18. Ayzenshtat, G. I. et al. "GaAs resistor structures for x-ray imaging detectors." *Nuclear Instruments and Methods in Physics Research Section A: Accelerators, Spectrometers, Detectors and Associated Equipment* 487.1 (2002): 96–101.
19. Hamann, E. et al. "Characterization of photon counting pixel detectors based on semi-insulating GaAs sensor material." *Journal of Physics: Conference Series*. Vol. 425. No. 6. IOP Publishing, 2013.
20. Canberra, Mirion technologies, http://www.canberra.com/default.asp.
21. ORTEC, http://www.ortec-online.com.
22. Gutknecht, D. "Photomask technique for fabricating high purity germanium strip detectors," *Nuclear Instruments and Methods A* 288 (1990): 13–18.
23. Pennicard, D. et al. "A germanium hybrid pixel detector with 55 μm pixel size and 65,000 channels," *JINST* 9 (2014): P12003.
24. Pennicard, D. et al. "LAMBDA—Large area Medipix3-based detector array," *Journal of Instrumentation* 7 (2012): C11009.
25. Wilke, R. N. et al. "High-flux ptychographic imaging using the new 55 μm-pixel detector Lambda based on the Medipix3 readout chip." *Acta Crystallographica Section A: Foundations and Advances* 70.6 (2014): 552–562.
26. Liermann, H.-P. et al. "The extreme conditions beamline P02.2 and the extreme conditions science infrastructure at PETRA III," *Journal of Synchrotron Radiation* 22 (2015): 908–924.
27. Westermeier, Fabian et al. "Connecting structure, dynamics and viscosity in sheared soft colloidal liquids: A medley of anisotropic fluctuations." *Soft Matter* 12.1 (2016): 171–180.
28. Maneuski, D. et al. "Edge pixel response studies of edgeless silicon sensor technology for pixellated imaging detectors," *JINST* 10 (2015): P03018. doi:10.1088/1748-0221/10/03/P03018.
29. Zhang, J. et al. "Optimization of radiation hardness and charge collection of edgeless silicon pixel sensors for photon science," *JINST* 9 (2014): C12025. doi:10.1088/1748-0221/9/12/C12025.
30. Peltola, T. et al. "Characterization of thin p-on-p radiation detectors with active edges," *Nuclear Instruments and Methods in Physics Research A* 813 (2016): 139–146. doi: 10.1016/j.nima.2016.01.016.

31. Weigell, P. "Investigation of Properties of Novel Silicon Pixel Assemblies Employing Thin n-in-p Sensors and 3D-Integration," PhD thesis, Technische Universität München, 2013.
32. Advacam. Available at http://www.advacam.com/en/technology/edgeless-sensor-technology.
33. Eranen, S. et al. "3D processing on 6 in. high resistive SOI wafers: Fabrication of edgeless strip and pixel detectors," *Nuclear Instruments and Methods in Physics Research A* 607 (2009): 85–88. doi:10.1016/j.nima.2009.03.243.
34. SINTEF, http://www.sintef.no/microsystems.
35. Fraunhofer, http://www.izm.fraunhofer.de/en.html.
36. Swaminathen, S. and Han, K.J., *Design and Modeling for 3D ICs and Interposers*, World Scientific, London, 2014, ISBN: 978-981-4508-59-9.
37. Lau, J. *Through-Silicon Vias for 3D Integration, The McGraw-Hill Companies*, 2012, ISBN: 978-0-07-178515-0.
38. Kumar, S. *3DIC & 2.5D TSV Interconnect for Advanced Packaging-Business Update Report 2016*, Yole Development, www.yole.fr.
39. Henry, D. et al. "TSV last for hybrid pixel detectors: Application to particle physics and imaging experiments," *Proceedings of IEEE Electronic Components & Technology Conference*, Las Vegas NV 28–31 May 2013.
40. Sarajlić, M. et al. "Development of edgeless TSV X-ray detectors," *JINST* 11 (2016): C02043. doi:10.1088/1748-0221/11/02/C02043.

4

Theoretical Consideration of the Response Function of a Two-Dimensional CdTe Detector

Hiroaki Hayashi

CONTENTS

4.1 What Is Response Function?

In this chapter, the response function of a CdTe detector is described. First of all, we have to take into consideration why the response function is an important parameter for photon-counting x-ray imaging. As described in the other chapters, one strong point for using a photon-counting detector is that scientists can analyze x-ray spectra [1,2]. The information of the x-ray spectrum is valuable in identifying different materials; therefore, it has the possibility to add a new function of the x-ray image, which can be used for medical and industrial applications [2–6]. For performing precise and accurate material identification, pure x-ray spectra should be measured. Unfortunately, x-ray spectra measured with a spectroscopic detector are not ideal, because the detector has insufficient efficiency. This insufficiency is caused by the interaction between incident x-rays and the detector's materials, which are reflected in the measured spectrum. In this chapter, this spectrum is known as the response function [7]. Using Figure 4.1, I will explain the concept of the response function.

In Figure 4.1, monoenergetic x-rays having energy E are incident to the detector, and via digital processing from the collected charges within the detector, the corresponding x-ray spectrum is obtained. The spectrum is reflected by the sum of all of the physics in the detector material and is equivalent to the probability distribution function (response function). This response function is considered to consist of two major components; one is concerned with the full-energy absorption event, which is energy E, and the other is affected by insufficient absorptions. What are the factors that affect the determination of a response function? The response functions reflect the interactions of the detector, and therefore characteristics of the detector, such as pixel size, detector thickness, material composition, and so on, are important parameters. That is why I will explain the relationship between response function and interaction in this section.

There is another important factor for the need to derive the response function of a multipixel-type detector. When researchers use a single-probe-type spectrometer, they can usually analyze x-ray spectra after the unfolding correction [8–10] is carried out. The concept of the unfolding correction is represented in Figure 4.2. Here, $G(E)$: black and $F(E)$: red represent the measured spectrum and unfolded spectrum (real x-ray spectrum), respectively. The response function is shown by $R(E',E)$: blue; here E' represents the energy of the incident monoenergetic x-ray. The mathematics of the unfolding procedure is described elsewhere, and there are many methods. A conventionally adopted method for continuous spectra (such as bremsstrahlung x-ray spectrum [9,10] and beta-ray spectrum [11]) is the stripping method, as shown in Figure 4.2. A feature of this method is ease

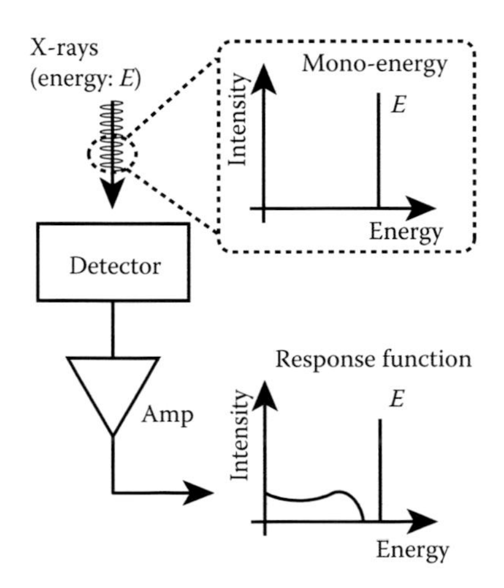

FIGURE 4.1
Concept of the response function.

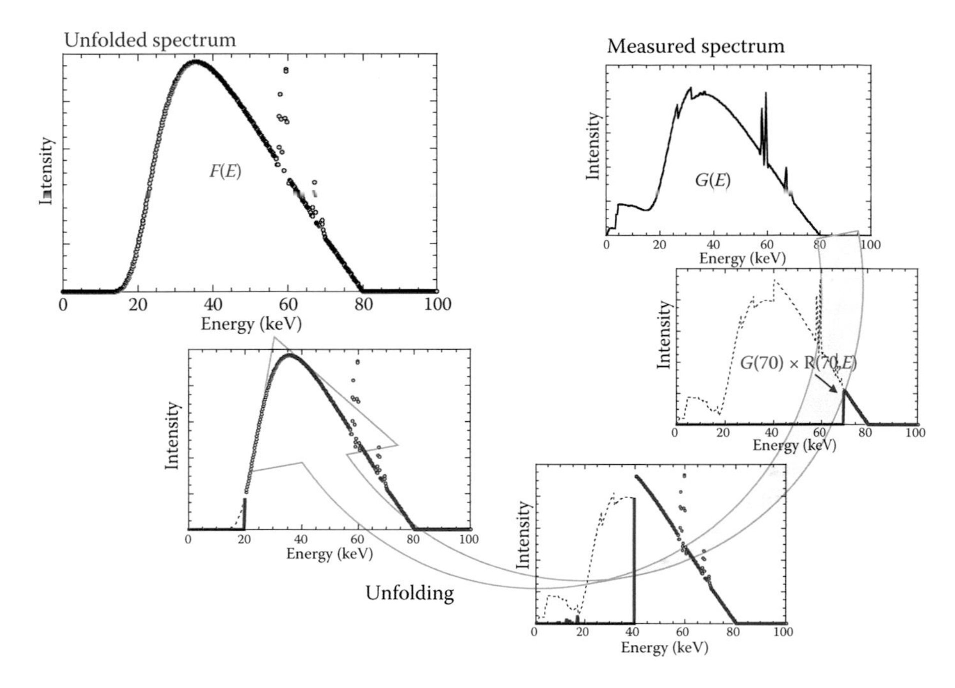

FIGURE 4.2
Concept of the unfolding procedure (stripping method). $G(E)$ and $F(E)$ represent measured and unfolded spectra, respectively. $R(E',E)$ indicates response function of the detector; E' shows incident monoenergetic x-ray. It should be noted that the unfolding correction can only be applied to the accumulated spectra.

of application. However, it should be noted that there are differences in the accuracies of energy regions; high accuracy is obtained for high-energy regions, and low accuracy is for low-energy regions. Fortunately, because the analysis of the x-ray spectrum and the beta-ray spectrum has focused attention on the higher-energy region rather than the lower-energy region, the stripping method works well.

Although the unfolding correction can derive the actual x-ray spectrum, it should be noted that the correction can only be applied to a "spectrum." Namely, a multichannel-analyzer is necessary for measuring the spectrum, and the unfolding correction can be applied to the accumulated spectrum after finishing the measurement. This is an essential problem in the development of a multipixel-type photon-counting detector. Most of the multipixel-type detectors produced using current technology can only perform under the following conditions: few energy bins with online (real-time) processing [12]. More easily said, it seems to be difficult to apply the conventional unfolding method to a measurement system using multipixel-type detectors. A solution will be proposed in the near future, but at present, most of the applications use measured spectrum with several bins. Therefore, we should take into consideration the importance of physics when using a multipixel-type detector.

Before describing the response function of a multipixel-type detector, I would like to explain the limitations of this chapter. In this chapter, I focus on the relationship between physics and measured spectrum without taking into consideration the charge transport processes of the detector and other phenomena based on a realistic apparatus. In other words, I will explain the creation of electron–hole pairs (charge cloud) in the detector material. I assume that the charge cloud is completely divided pixel by pixel. This situation is illustrated in Figure 4.3. Under this assumption, three main cases can be considered. In the left, the charge cloud is only created in each pixel, and it is completely contained in the same pixel. In the middle, the charge cloud is created in the middle area of the pixels, and only a part of the charge cloud is measured in each pixel. On the right, the first charge cloud is created in a certain pixel, but characteristic x-rays generate a *cross-talk event* between the neighboring pixels, and then secondary charge clouds are created within the pixels. Only in the first case (left), a full energy peak (FEP) is obtained. For the last two cases, portions of the energies are absorbed, and they do not form an FEP. However, these phenomena are close to ideal, and in the actual situation, charge transportations, which are called the *charge sharing effect*, should be taken into consideration as presented in Figure 4.4. There are many papers about charge sharing effects for test models of multipixel detectors [13–15]. This explanation goes into much greater detail than the description in this chapter. Therefore, I only present the situations in Figure 4.3.

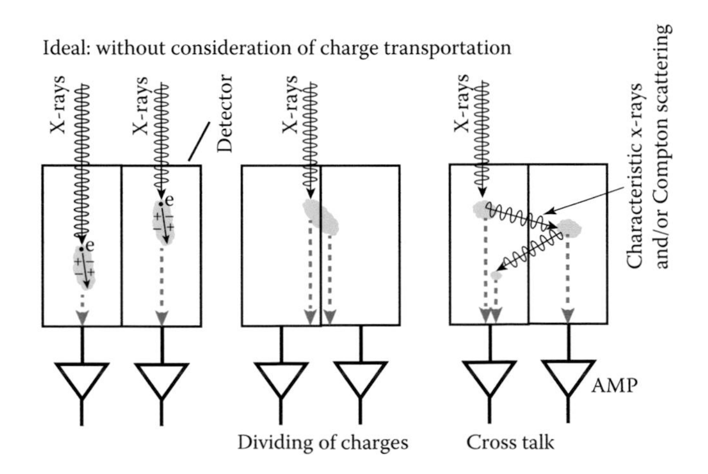

FIGURE 4.3
Relationship between incident and/or scattered x-rays and detector's responses. In this schematic drawing, the phenomena concerning the transportation of electrons are not considered.

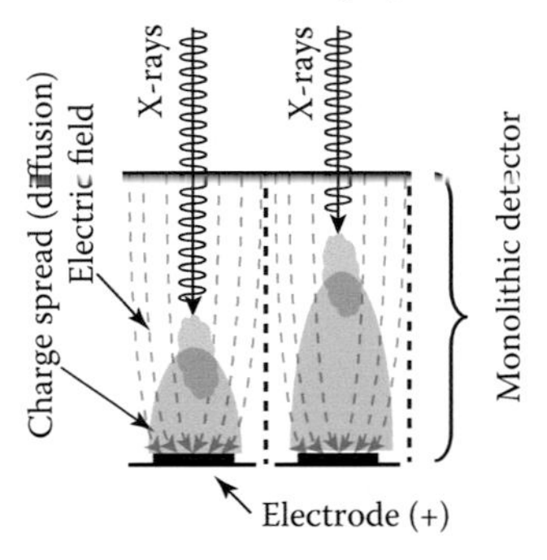

FIGURE 4.4
Schematic drawing of the actual situation in a multipixel-type detector. Electric fields and charge spread effects occur during electron transportation.

4.2 Interactions between Incident X-Rays and Detector Materials

4.2.1 Cross Sections

The probability of the interactions between incident x-rays and detector material can be described as *cross section* [16]. Currently, a CdTe detector is a detector expected to be used in the future; therefore, in this chapter, I exemplify the response function of the CdTe detector. Figure 4.5 shows cross sections of Cd and Te atoms as a function of energy. The main interactions are coherent

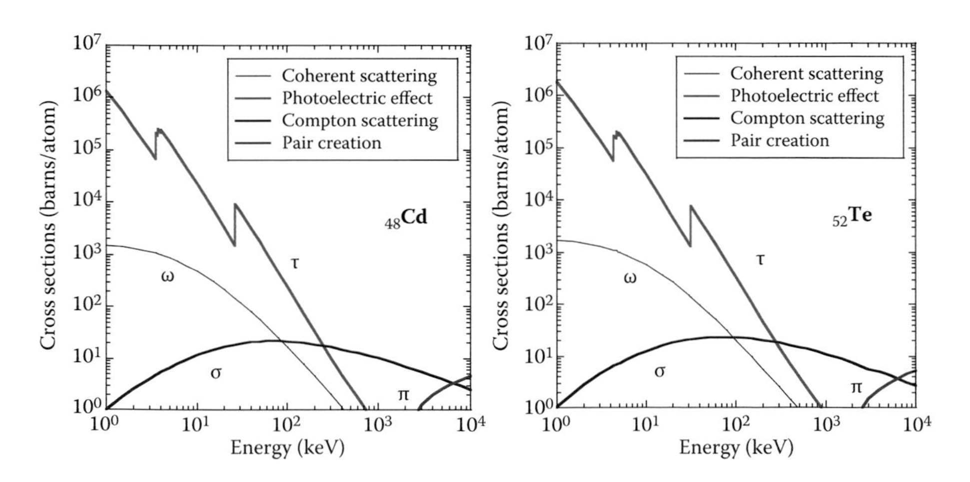

FIGURE 4.5
Photon cross sections of Cd and Te.

scattering, photoelectric effect, Compton scattering, and pair creation. Here, the cross sections resulting from these phenomena, are described as ω, τ, σ, and π for coherent scattering, photoelectric effect, Compton scattering, and pair creation, respectively. Then, the linear attenuation coefficient μ of a certain material is described as

$$\mu = \omega + \tau + \sigma + \pi. \tag{4.1}$$

Using μ, attenuation of x-rays can be calculated by the following equation:

$$\frac{dN}{dx} = \mu N, \tag{4.2}$$

where N and x represent the number of photons and distance from surface of the material. The equation indicates the quantitative rate of the interaction in minute range dx. Therefore, if we know μ, the interaction rate of the photons can be determined. This principle is used in the Monte Carlo simulation [17]. In the following sections, I will explain the phenomena.

4.2.2 Coherent Scattering

Coherent scattering is an elastic scattering in which photons are treated as a wave; namely, the direction of incident photon is changed, but the energy does not change. As shown in Figure 4.6, there are two types of coherent scatterings: one is Thomson scattering, which occurs from interaction with free electrons, and the other is Rayleigh scattering, which occurs from interaction with bound electrons. If coherent scattering occurs at a certain pixel in a multipixel-type detector, the photons energy is not absorbed in the pixel of interest. Therefore, in many cases, coherent scattering does not play an

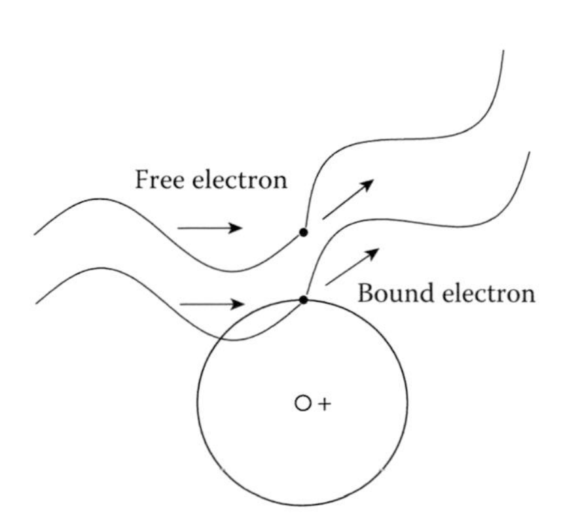

FIGURE 4.6
Schematic drawing of coherent scattering.

important role in the estimation of the response function of a multipixel-type detector.

4.2.3 Photoelectric Effect

A schematic drawing of the photoelectric effect is presented in Figure 4.7. This effect is the interaction between an incident photon and an orbital electron. The orbital electron is a situation in which electrons are strongly bound in the nucleus. In physics, conservation laws of energy and momentum should be obeyed; therefore, these formulas of conservations are considered for photons, electrons, and nuclide. This consideration leads to recoil of the nuclide, but fortunately, the energy of the recoiled nuclide is much smaller than those for other particles. As a result, the energy of the recoiled nuclide is treated as disregarded for general cases.

As shown in Figure 4.6, the photoelectric effect begins with the interaction of the incident photon having energy $E(= h\nu)$ with the orbital electron. Then, the energy E of the photon is transferred to the electron, which is bound with energy of $-W$. Then the electron is emitted from the atom with energy of $E = h\nu - W$, and is called a photoelectron. The electron range in the detector is very small, and the electron loses its energy in a small area. Then, a charge cloud is formed; the typical value is estimated to be from 1 μm (E = 10 keV) to 15 μm (E = 50 keV) [18]. What happens next? We should note that the atom is

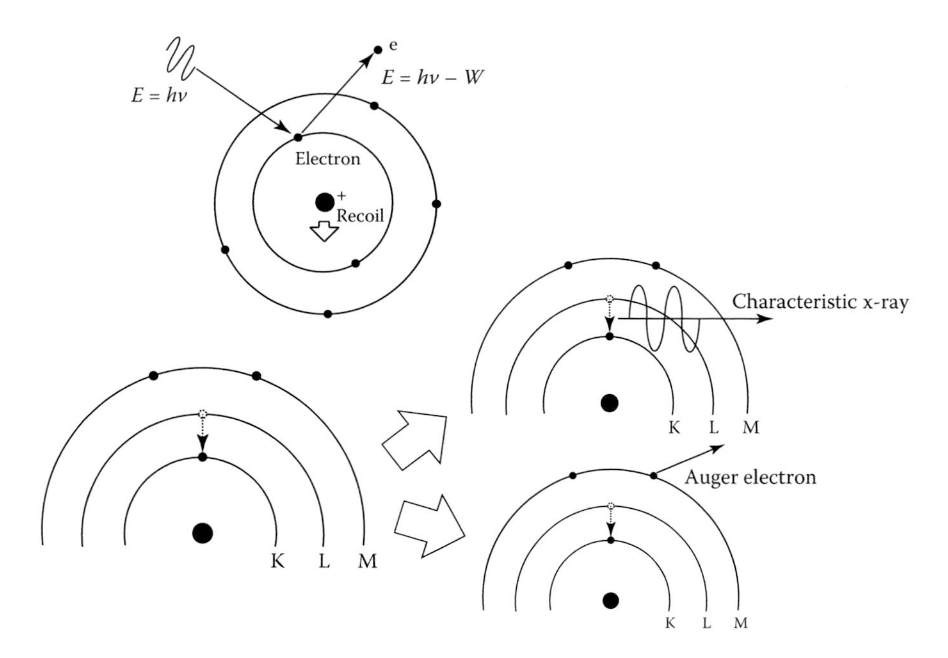

FIGURE 4.7
Schematic drawings of photoelectric effect (upper), and succeeding phenomena (lower).

left behind in an excited state after emitting the photoelectron, which results in a vacancy in the orbital. Therefore, as represented in the lower part of Figure 4.7, the remaining excitation energy is released via emission of a characteristic x-ray or Auger electron. If the Auger electron is emitted, its energy is absorbed readily in the detector; the resulting charge cloud cannot be distinguished in the cloud caused by the photoelectron.

The phenomena of characteristic x-ray emissions from Cd and Te are represented in Figure 4.8. The binding energies of Cd and Te are –26.7 and –31.8 keV [19]; therefore, the photons having energy above these values can occur in the photoelectric effect. In this chapter, we confine the discussion to K-shell interaction, because the characteristic x-rays caused by other shells are sufficiently low for photon counting in a realistic application. Therefore, it is important to know both the energies and intensities of K-x-rays. The lower graphs in Figure 4.8 show the K-x-ray values. In this graph, intensity is defined as emission probability when 100 vacancies appear in the K-shell orbital. It is clearly shown that the intensities of K_α (de-excitation between L shell to K shell) and K_β (de-excitation between M shell to K shell) are intense, and they are approximately 84–86% and 5–6%, respectively. Therefore, we should take into consideration the effect of characteristic x-ray emissions when we derive the response function for a multipixel-type detector.

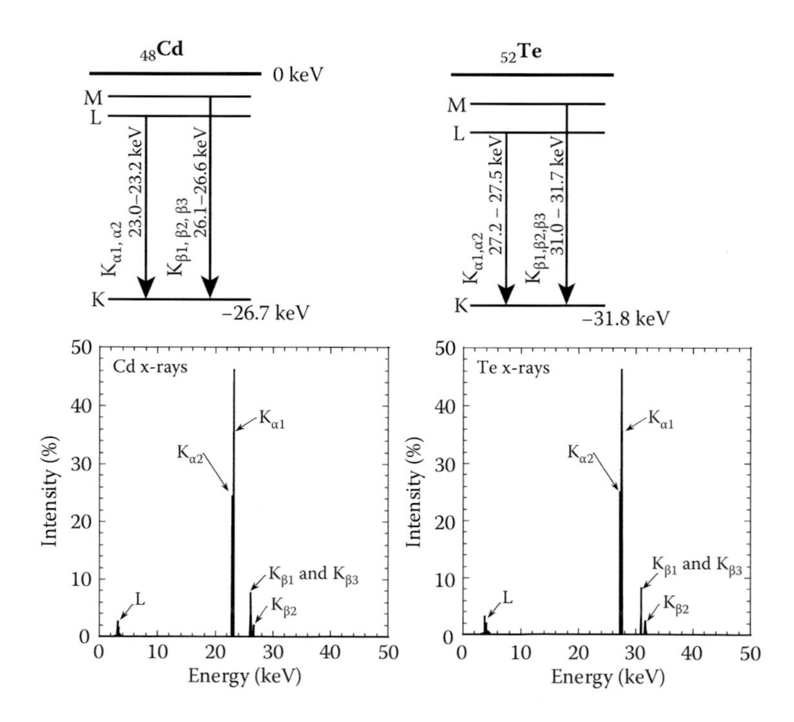

FIGURE 4.8
Characteristic x-rays emitted by Te and Cd. In these calculations, 100 vacancies are present in the K-shell.

4.2.4 Compton Scattering Effect

The Compton scattering effect is the interaction between photons and free electrons. In this situation, we should consider the conservation laws of energy and momentum between these two particles. A schematic drawing of the Compton scattering effect is shown in Figure 4.9. The equations for conservation laws can be solved analytically. The energy of scattered photons ($h\nu'$) is described as

$$h\nu' = h\nu \times \frac{1}{1 + \frac{h\nu}{m_e c^2}(1 - \cos\phi)}, \tag{4.3}$$

where $m_e c^2$ is the electron rest mass: e.g., 511 keV. Equation 4.3 demonstrates that the energy of the scattered photon is functions of incident photon energy $h\nu$ and scattering angle ϕ. The examples for 10 and 100 keV photons are presented in the upper graphs of Figure 4.10. In this graph, the characteristics of Compton scattering in the diagnostic x-ray region are clearly observed; the energy of scattered photons is similar to that of the incident photons. It demonstrates that most of the energies are carried off by scattered photons when the effect occurs in a multipixel-type detector. So, Compton scattering effect has an important role in response functions.

In Figure 4.5, the overall provability of Compton scattering has been presented. But when considering Compton scattering, the probability of angular dependence is very important. The mathematics is performed by the Klein–Nishina formula as

$$\frac{d\sigma}{d\Omega} = Zr_0^2 \left[\frac{1}{1+\alpha(1-\cos\phi)}\right]^2 \left[\frac{1+\cos^2\phi}{2}\right]\left[1 + \frac{\alpha^2(1-\cos\phi)^2}{(1+\cos^2\phi)[1+\alpha(1-\cos\phi)]}\right], \tag{4.4}$$

where α and r_0 represent $h\nu/m_e c^2$ and the classical electron radius, respectively. and α is a function of incident photon hν; therefore, this equation also contains

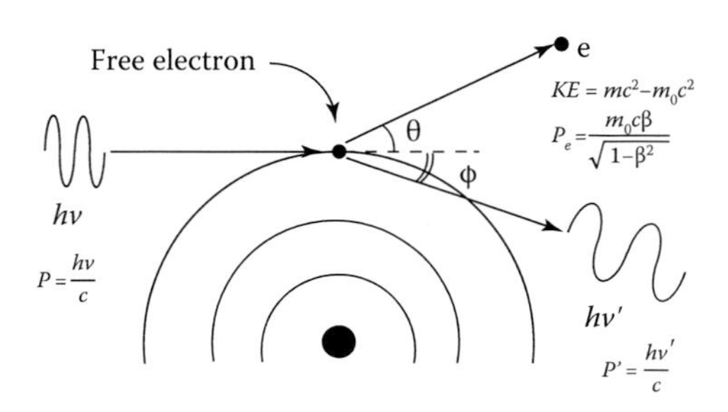

FIGURE 4.9
Schematic drawings of Compton scattering.

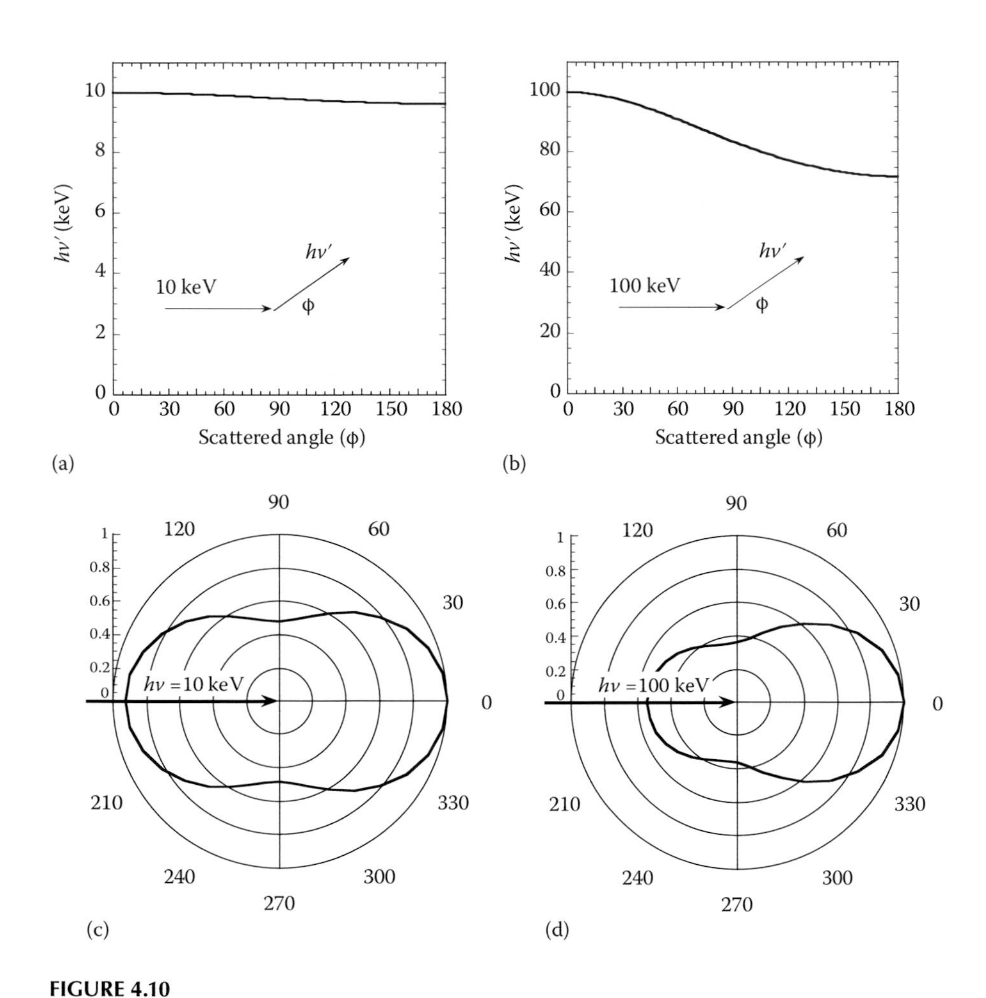

FIGURE 4.10
(a and b) Energies for scattered x-rays of incident x-rays at 10 and 100 keV, respectively. (c and d) Intensities for scattered x-rays of incident x-rays at 10 and 100 keV, respectively.

functions of incident photon energy $h\nu$ and scattering angle ϕ. The relative intensities for 10 and 100 keV photons as a function of angle are presented in Figure 4.10c and d. For the 10 keV photons, the intensities for forward (ϕ~0°) direction and backward (ϕ~180°) direction are much larger, and those of side directions (ϕ~90°, 270°) are relatively smaller. On the other hand, for the 100 keV photons, intensity for the forward direction is obviously high.

4.3 Response Function of Single-Probe-Type CdTe Detector

Let's begin to discuss the response function of a single-probe-type CdTe detector [9,10,20]. Figure 4.11 shows a schematic drawing of interactions of

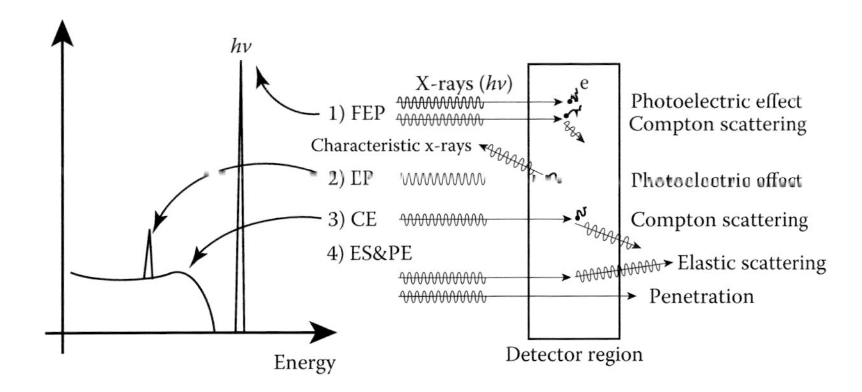

FIGURE 4.11
Schematic drawing of response function of a single-probe-type detector. Seen here are full energy peak (FEP), escape peak (EP), Compton escapes (CE), and elastic scattering and penetration (ES&PE). The corresponding phenomena are shown in the right.

incident x-rays in the single-probe-type CdTe detector. In the detector, differences between the full energy absorbing event and partial energy absorbing event are important. The main peak is formed by the full energy absorption of the incident photon by the detector; when all of the energies of the secondary produced particles are completely absorbed by the detector via the photoelectric effect and Compton scattering effect, the FEP appears. On the other hand, when the photoelectric effect occurs at the surface of the detector, there is the possibility that the succeeding characteristic x-rays cannot be absorbed by the detector. In this case, the absorbed energy becomes slightly smaller than the energy corresponding to the characteristic x-ray, and this event forms a peak, called an escape peak (EP). In reality, as described in the previous section (see Figure 4.8), the emission of the characteristic x-rays of Cd and Te is a little complex; therefore, there will be many EPs corresponding to the characteristic x-rays. When the Compton scattering effect occurs and the detector cannot absorb scattered photons, the corresponding events form continuous areas in the response function; these are called Compton escapes (CEs). The highest energy of the CE can be calculated by Equation 4.3; namely, the energy of incident photon (hν) minus energy of scattered photon at scattering angle of 180° ($h\nu'$). For the 10 and 100 keV photons, they become 0.4 and 28 keV, respectively. This estimation is based on single events, meaning that Compton scattering occurs only one time, but in an actual case, multiple events may occur. When the elastic scattering and penetration events (EE and PE) occurred, no events appeared in the response function. However, these events should not be ignored, because they lead to decreases in full energy events. It means that the FEP efficiency should be considered when the unfolding process is performed on the measured spectrum (see Figure 4.2).

A typical example of the response function of a single-probe-type CdTe detector is presented in Figure 4.12. The response functions are calculated using the Monte Carlo simulation code: EGS5 (electron-gamma-shower version 5) code [17]. The detector construction in the simulation is presented in the inset. The detector size is 3.0 mm width, 3.0 mm length, and 1.0 mm thickness; pencil and broad beams are applied. There are three figures corresponding to the incident x-ray energies of 30 keV, 60 keV, and 80 keV; in each figure, EPs and CE are clearly seen. As described previously, when 30 keV photons are incident to the detector, the photoelectric effect for only the Cd atom can occur. Then, in the response function for the 30 keV photons, the FEP and EP caused by characteristic x-ray escapes of Cd atoms can be seen. On the other hand, when energy of incident x-rays is over 32 keV, the photoelectric effect of both Cd and Te occurs. Therefore, in the response functions of 60 and 80 keV in Figure 4.12, EPs caused by Cd and Te are observed. The energies of EPs can be calculated theoretically, and they are summarized in Table 4.1. In an actual CdTe detector, the energy resolving power is not high enough to separate these seven EPs. Then, there may be three EPs: (1) E-Cd($K_{\alpha1}$) and E-Cd($K_{\alpha2}$) approximately 23 keV, (2) E-Cd($K_{\beta1}$, $K_{\beta3}$) approximately 26 keV, (3) E-Te($K_{\alpha1}$) and E-Te($K_{\alpha2}$) approximately 27 keV, and (4) E-Te($K_{\beta1}$, $K_{\beta3}$) and E-Te($K_{\beta2}$) approximately 31 keV. Figure 4.13 shows the ratio of contributions in the response function. The left and right figures indicate the results for pencil and broad beams, respectively. Focusing attention on the dependence of FEP, we can clearly find a significant strange trend; FEP is rapidly

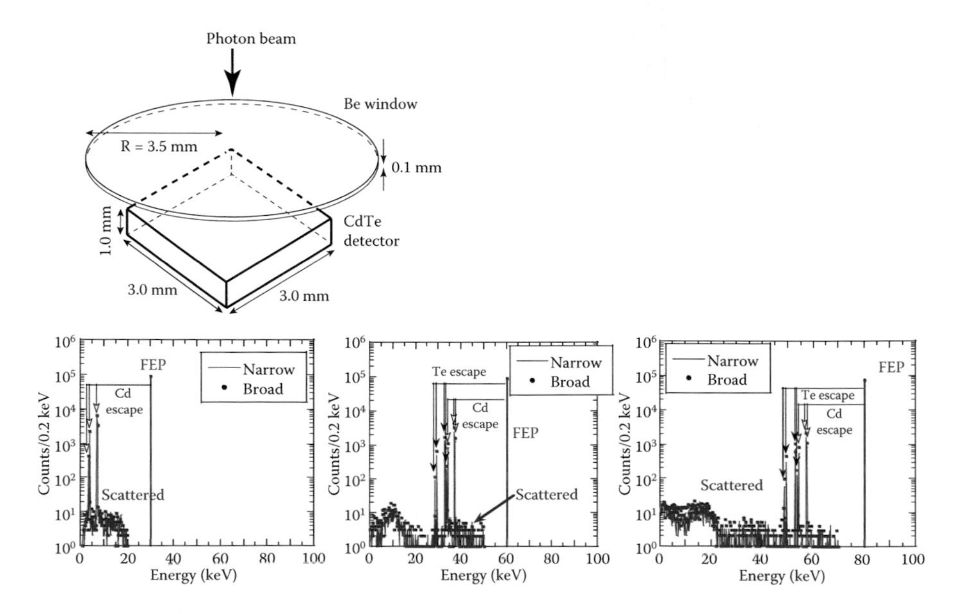

FIGURE 4.12
Typical example of the response function of a single-probe-type CdTe detector.

TABLE 4.1

Energies of the Escape Peaks in the Single-Probe-Type CdTe Detector

Phenomena	Description	Energy (keV)
Escape(Te)	E-Te($K_{\alpha1}$)	E-27.5
	E-Te($K_{\alpha2}$)	E-27.2
	E-Te($K_{\beta1}$, $K_{\beta3}$)	E-31.0
	E-Te($K_{\beta2}$)	E-31.7
Escape(Cd)	E-Cd($K_{\alpha1}$)	E-23.2
	E-Cd($K_{\alpha2}$)	E-23.0
	E-Cd($K_{\beta1}$, $K_{\beta3}$)	E-26.1

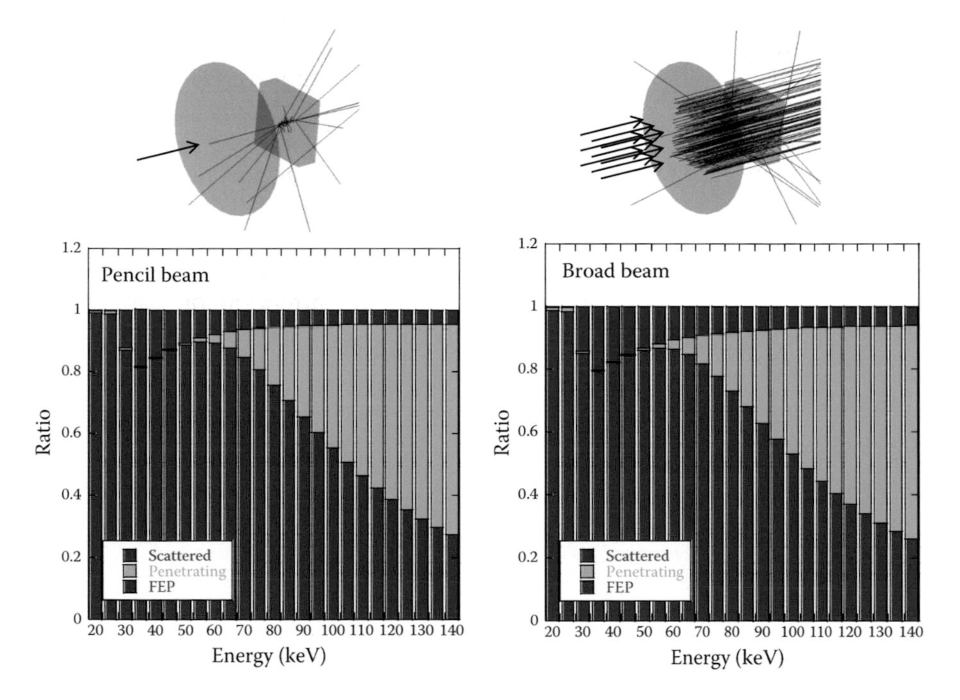

FIGURE 4.13
Ratio of phenomena for the response functions of a single-probe-type CdTe detector.

decreasing to 35 keV and gradually increasing to 55 keV, and then rapidly decreasing. These phenomena are mainly caused by the photoelectric effects of Cd and Te. For energies below 60 keV, which is just around the effective energy in medical diagnosis, the events caused by the scattered photons are relatively high.

The CdTe detector is convenient to use in x-ray spectroscopy because it can be operated at room temperature and high-energy resolution can be obtained. In addition, detection efficiency is not bad, because of the use of high-atomic-number materials ($_{48}$Cd and $_{52}$Te), but leads to the appearance of EPs for lower-energy x-rays. Therefore, x-ray spectra measured with a single-probe-type CdTe detector need to have the unfolding correction applied as described in Figure 4.2.

4.4 Response Function of Multipixel-Type CdTe Detector

In this section, I describe the response function of a multipixel-type CdTe detector. Here, I define the response function as a spectrum concerning just one pixel in the two-dimensional detector array. First of all, an illustration is presented in Figure 4.14. The left and right figures show response functions for 1 pixel (200 μm) in a monolithic CdTe detector with different irradiation areas (side length = L) of 200 and 400 μm, respectively. We can observe obvious differences; the left graph is similar to that of the response function of the single-probe-type CdTe detector as preliminary represented in Figure 4.12, but the right graph is different. What happens to these response functions? I will explain this using physics. Remember that the binding energies of Cd and Te are −26.7 and −31.8 keV, respectively, and K-x-rays of Te are 27.2–27.5 keV; namely, the energy of K-x-rays of Te are a little bit higher than the binding energy of Cd. It means that the K-x-rays of Te can interact with

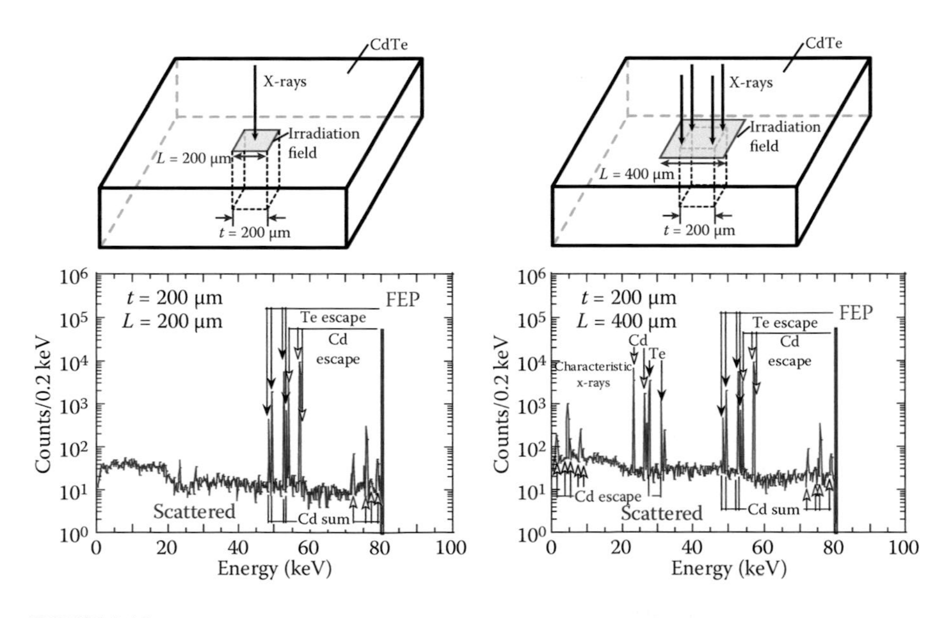

FIGURE 4.14
Differences of response functions between a very small irradiation field (left: L = 200 μm) and larger one (right: L = 400 μm).

Cd in the detector. When this phenomenon occurs in the same pixel, it leads to no particular effect. On the other hand, when this phenomenon occurs in different pixels, especially neighboring pixels, an additional consideration is needed. Figure 4.15 shows a schematic drawing of the interactions between the K-x-rays of Te and Cd. When the photons incident to a certain pixel interact with the Te and the photoelectric effect occurs, K-x-rays of Te are emitted. Then, the x-rays interact with the Cd in the neighboring pixels; the K-x-rays of Cd are also generated. In this case, the K-x-rays of Cd come back to the pixel of interest, and they are absorbed. The total absorbed energy is estimated to be E-K(Te) + K(Cd), where E, K(Te), and K(Cd) represent the energy of the incident photon, energy of K-x-rays of Te, and energy of K-x-rays of Cd, respectively. More precisely, all of the interactions and corresponding absorbed energies are listed in Table 4.2. In this table, "Escape(Te)" and "Escape(Cd)" represent EPs of the characteristic x-ray escapes. The same phenomena have been discussed in the previous section (see Figure 4.12 and Table 4.1). The phenomena in the third row show the interactions in Figure 4.15; note that the energies of K-x-rays ($K_{\alpha1}$, $K_{\alpha2}$, $K_{\beta1}$, $K_{\beta2}$, $K_{\beta3}$) of Cd are superimposed to the EPs of the K-x-rays [E-Te($K_{\alpha1}$), E-Te($K_{\alpha2}$), E-Te($K_{\beta1}$, $K_{\beta3}$), and E-Te($K_{\beta2}$)] of Te, and all of the combinations appear. In addition, the characteristic x-ray peaks of Cd and Te are also observed in the response functions; when a larger irradiation area is set, the characteristic x-rays are generated in the neighboring pixels. We should note that the EPs from the characteristic x-rays of Te are also shown; the phenomena appear in the neighboring pixels as represented in Figure 4.15.

The aforementioned consideration means that the response function strongly depends on the size of the x-ray irradiation field. In order to estimate the dependency, I will present the results of the Monte Carlo simulation [17]. Figure 4.16 shows the simulation condition and typical result. We selected the parameters of a monolithic CdTe detector sized 10 mm in width, 10 mm in length and 1.5 mm in thickness. As shown in the inset, pixel sizes

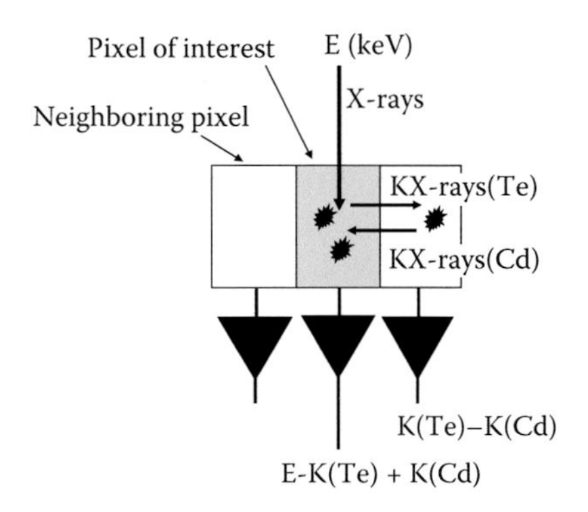

FIGURE 4.15
Schematic drawing of interactions of K-x-rays of Te in the neighboring pixels. Because the energy of K-x-rays of Te is higher than the binding energy of Cd, these x-rays can cause the photoelectric effect in Cd. Then, K-x-rays of Cd are emitted from the Cd atom.

TABLE 4.2

Phenomena and Corresponding Energies for the Interactions for Escape Peaks in the Multipixel-Type CdTe Detector

Phenomena	Description	Energy (keV)
Escape(Te)	E-Te($K_{\alpha1}$)	E-27.5
	E-Te($K_{\alpha2}$)	E-27.2
	E-Te($K_{\beta1}$, $K_{\beta3}$)	E-31.0
	E-Te($K_{\beta2}$)	E-31.7
Escape(Cd)	E-Cd($K_{\alpha1}$)	E-23.2
	E-Cd($K_{\alpha2}$)	E-23.0
	E-Cd($K_{\beta1}$, $K_{\beta3}$)	E-26.1
	E-Cd($K_{\beta2}$)	E-26.6
Escape(Te)+ Characteristic x-rays(Cd)	E-Te($K_{\alpha1}$) + Cd($K_{\alpha1}$)	E-27.5 + 23.2 = E-4.3
	E-Te($K_{\alpha1}$) + Cd($K_{\alpha2}$)	E-27.5 + 23.0 = E-4.5
	E-Te($K_{\alpha1}$) + Cd($K_{\beta1}$, $K_{\beta3}$)	E-27.5 + 26.1 = E-1.4
	E-Te($K_{\alpha1}$) + Cd($K_{\beta2}$)	E-27.5 + 26.6 = E-0.9
	E-Te($K_{\alpha2}$) + Cd($K_{\alpha1}$)	E-27.2 + 23.2 = E-4.0
	E-Te($K_{\alpha2}$) + Cd($K_{\alpha2}$)	E-27.2 + 23.0 = E-4.2
	E-Te($K_{\alpha2}$) + Cd($K_{\beta1}$, $K_{\beta3}$)	E-27.2 + 26.1 = E-1.1
	E-Te($K_{\alpha2}$) + Cd($K_{\beta2}$)	E-27.2 + 26.6 = E-0.6
	E-Te($K_{\beta1}$, $K_{\beta3}$) + Cd($K_{\alpha1}$)	E-31.0 + 23.2 = E-7.8
	E-Te(K_{b1}, K_{b3}) + Cd($K_{\alpha2}$)	E-31.0 + 23.0 = E-8.0
	E-Te($K_{\beta1}$, $K_{\beta3}$) + Cd($K_{\beta1}$, $K_{\beta3}$)	E-31.0 + 26.1 = E-4.9
	E-Te($K_{\beta1}$, $K_{\beta3}$) + Cd($K_{\beta2}$)	E-31.0 + 26.6 = E-4.4
	E-Te($K_{\beta2}$) + Cd($K_{\alpha1}$)	E-31.7 + 23.2 = E-8.5
	E-Te($K_{\beta2}$) + Cd($K_{\alpha2}$)	E-31.7 + 23.0 = E-8.7
	E-Te($K_{\beta2}$) + Cd($K_{\beta1}$, $K_{\beta3}$)	E-31.7 + 26.1 = E-5.6
	E-Te($K_{\beta2}$) + Cd($K_{\beta2}$)	E-31.7 + 26.6 = E-5.1
Characteristic x-rays	Te($K_{\alpha1}$), Te($K_{\alpha2}$)	27.5, 27.2
	Te($K_{\beta1}$, $K_{\beta3}$), Te($K_{\beta2}$)	31.0, 31.7
	Cd($K_{\alpha1}$), Cd($K_{\alpha2}$)	23.2, 23.0
	Cd($K_{\beta1}$, $K_{\beta3}$), Cd($K_{\beta2}$)	26.1, 26.6
Escape of x-rays(Cd) from characteristic x-rays(Te)	Te($K_{\alpha1}$)-Cd($K_{\alpha1}$)	27.5 – 23.2 = 4.3
	Te($K_{\alpha1}$)-Cd($K_{\alpha2}$)	27.5 – 23.0 = 4.5
	Te($K_{\alpha1}$)-Cd($K_{\beta1}$, $K_{\beta3}$)	27.5 – 26.1 = 1.4
	Te($K_{\alpha1}$)-Cd($K_{\beta2}$)	27.5 – 26.6 = 0.9
	Te($K_{\alpha2}$)-Cd($K_{\alpha1}$)	27.2 – 23.2 = 4.0
	Te($K_{\alpha2}$)-Cd($K_{\alpha2}$)	27.2 – 23.0 = 4.2
	Te($K_{\alpha2}$)-Cd($K_{\beta1}$, $K_{\beta3}$)	27.2 – 26.1 = 1.1
	Te($K_{\alpha2}$)-Cd($K_{\beta2}$)	27.2 – 26.6 = 1.1
	Te($K_{\beta1}$, $K_{\beta3}$)-Cd($K_{\alpha1}$)	31.0 – 23.2 = 7.8
	Te($K_{\beta1}$, $K_{\beta3}$)-Cd($K_{\alpha2}$)	31.0 – 23.0 = 8.0

(Continued)

TABLE 4.2 (CONTINUED)

Phenomena and Corresponding Energies for the Interactions for Escape Peaks in the Multipixel-Type CdTe Detector

Phenomena	Description	Energy (keV)
	Te($K_{\beta1}$, $K_{\beta3}$)-Cd($K_{\beta1}$, $K_{\beta3}$)	31.0 – 26.1 = 4.9
	Te($K_{\beta1}$, $K_{\beta3}$)-Cd($K_{\beta2}$)	31.0 – 26.6 = 4.4
	Te($K_{\beta2}$)-Cd($K_{\alpha1}$)	31.7 – 23.2 = 8.5
	Te($K_{\beta2}$)-Cd($K_{\alpha2}$)	31.7 – 23.0 = 8.7
	Te($K_{\beta2}$)-Cd($K_{\beta1}$, $K_{\beta3}$)	31.7 – 26.1 = 5.6
	Te($K_{\beta2}$)-Cd($K_{\beta2}$)	31.7 – 26.6 = 5.1

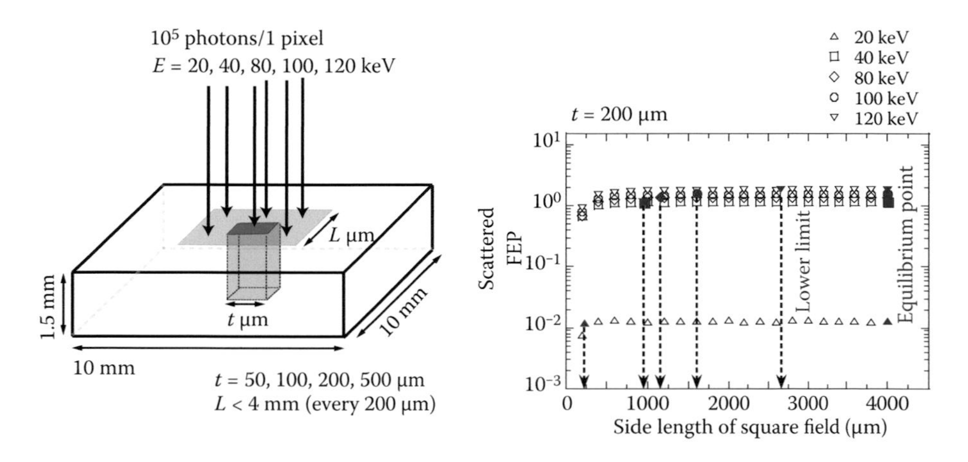

FIGURE 4.16
Demonstration to estimate proper size of an irradiation field.

(side length: *t*) varied from 50, 100, 200, to 500 μm. The side lengths of the irradiation fields (side length: *L*) continuously changed (side length *L* is limited up to 4 mm). The typical results for a *t* = 200 μm pixel detector are presented in Figure 4.16; the vertical axis shows the ratio of the scattered area (blue color in Figure 4.15) to FEP (red color in Figure 4.15). We can clearly find that the small irradiation field did not provide equilibrium, and the irradiation field having a side length of 4 mm is considered to be sufficient. In the following description, I will use the response function under the irradiation field.

Figure 4.17 shows a comparison of response functions for different pixel sizes: the results for 50, 100, 200, and 500 μm are presented. There are many peaks in addition to the FEP, but they can be easily identified using Table 4.2. The response function for a 50 μm pixel has a relatively small FEP and large scattered areas (EPs and CEs) compared with those for larger pixels. In recent medical diagnoses, the detector having a pixel size of approximately 200 μm is used [21]; therefore, in the following description, I will show the results for 200 μm. Figure 4.18 shows a comparison of response functions for different

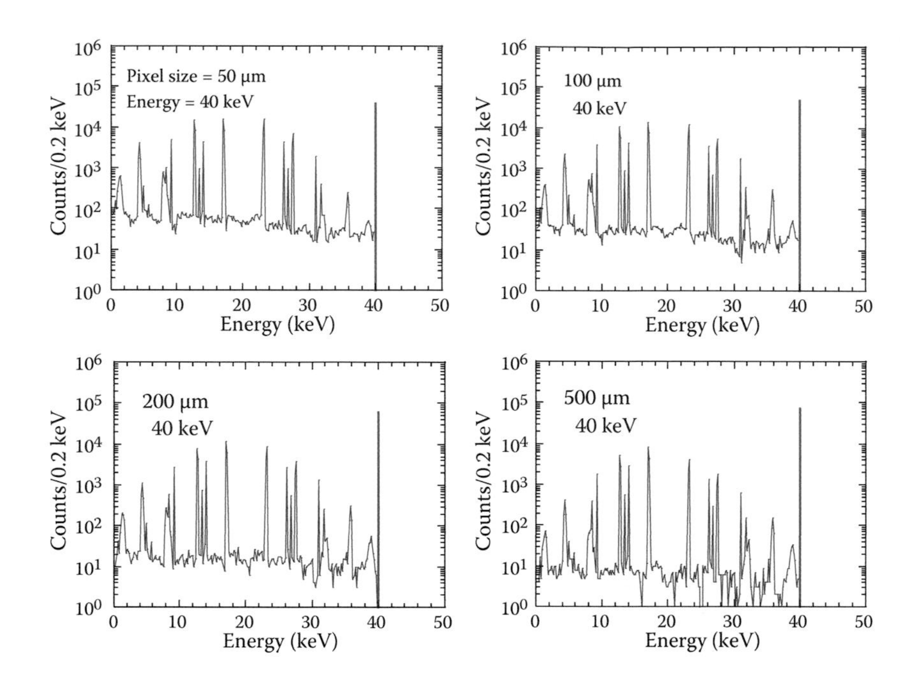

FIGURE 4.17
Comparison of response functions for different pixel sizes.

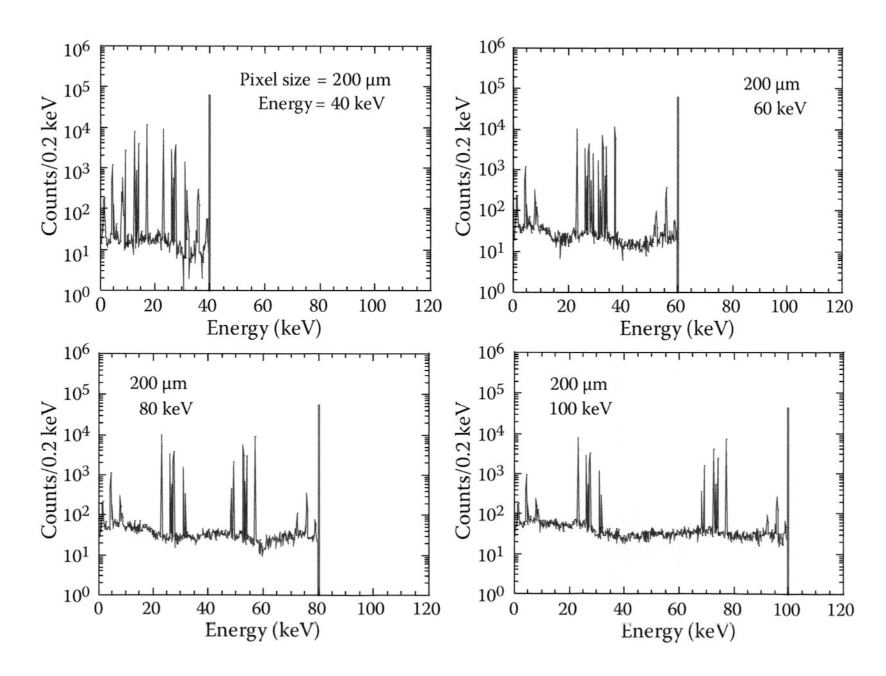

FIGURE 4.18
Comparison of response functions for different incident energies.

incident energies; incident energies of 40, 60, 80, and 100 keV are presented. In the plots, characteristic x-rays of Cd and Te are observed at the energy regions of between 20 and 30 keV, while the EPs take variable values depending on the incident energies.

Next, I will discuss quantitative analysis results for each component of the response function. Figure 4.19 shows a comparison of the ratio of each component in the response functions for different pixel sizes. The ratio is normalized by the incident photons (irradiated photons) for the pixel of interest. Compared with the result for the single-probe-type CdTe detector (see Figure 4.13), the most important thing is that the sum of the ratios has a value of over 1.0. These phenomena are caused by the contamination from scattered x-rays, which are generated in the neighboring pixels. Therefore, contamination rapidly decreases as pixel size becomes larger.

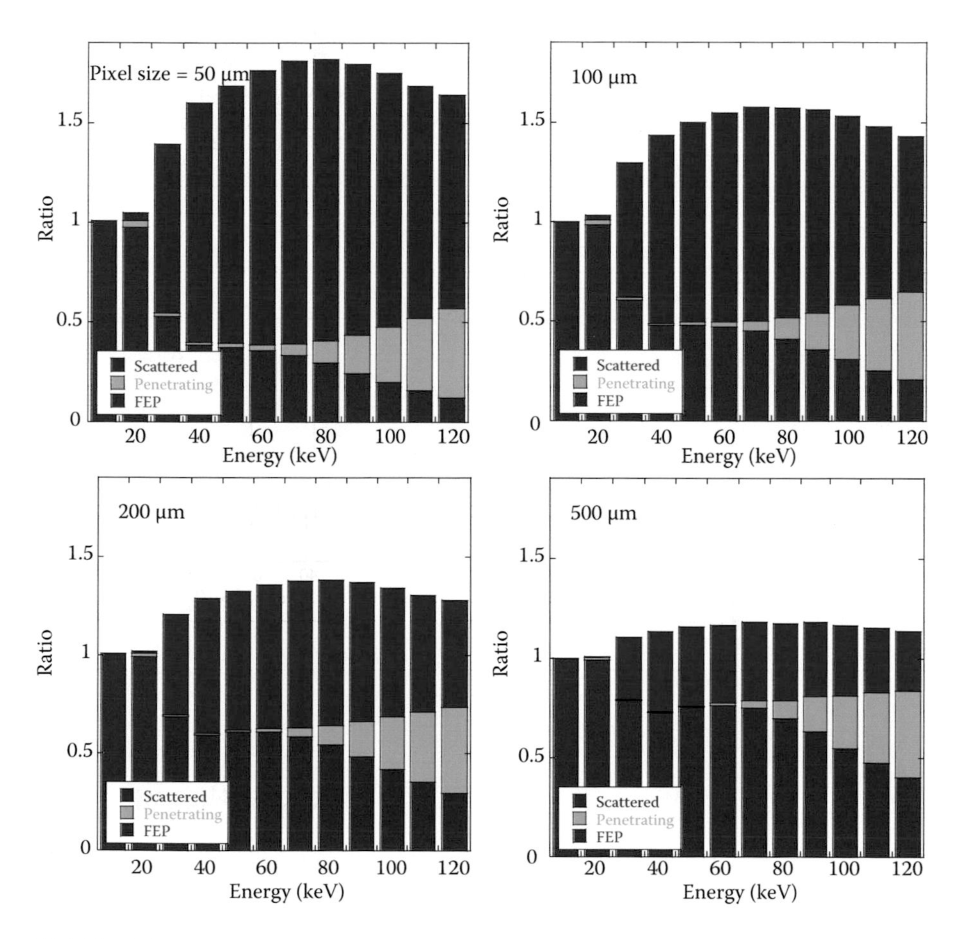

FIGURE 4.19
Comparison of ratio of each component in the response functions for different pixel sizes.

4.5 Effect of Response Functions on the Measured X-Ray Spectra

In this section, I will try to determine spectra measured with a multipixel-type detector for medical diagnosis using x-rays. A diagram to explain the spectra is represented in Figure 4.20. The original x-ray spectrum, which can be determined by the semiempirical formula [22–24], is shown in the upper graph. The middle graph represents an x-ray spectrum measured with one certain pixel of a multipixel-type detector, and the right lower plot indicates the x-ray spectrum measured with the single-probe-type detector. Comparing these two measured spectra, the difference between the former spectrum and original x-ray spectrum is larger than that of the latter. This can be explained by the response function. Currently, a high-speed processing technique has been developed and is applied to the analysis of charge sharing correction for a multipixel-type detector. In this technique, absorbed energies are analyzed event by event using a coincidence summing module [13–15]. This correction is considered to be effective, but we should note that the corrected

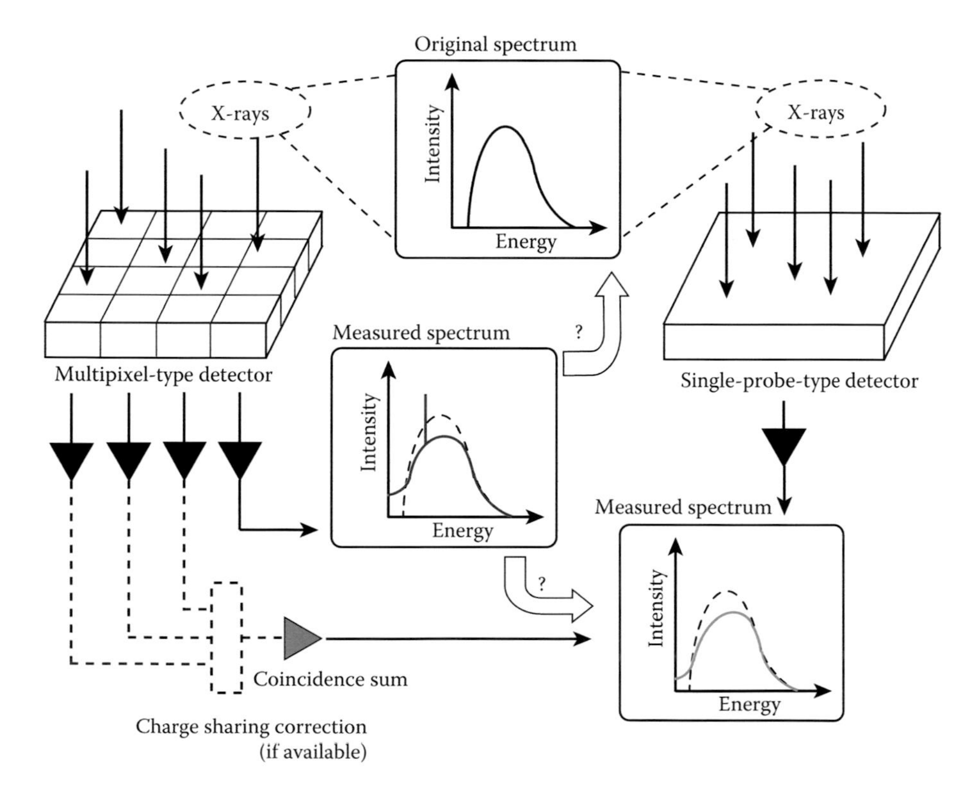

FIGURE 4.20
Diagram to explain original and measured x-ray spectra.

spectrum is the same as the spectrum measured with the single-probe-type detector. Because the x-ray spectrum measured with the single-probe-type detector needs correction (see Figure 4.2), the x-ray spectrum corrected by the charge sharing effect is still incomplete. If proper methods for correcting the measured spectrum (for the multipixel-type CdTe detector) to the original spectrum are developed with/without information on the response function, the method will be valuable. I anticipate that new research will lead to new findings and complete correction can be carried out.

Finally, I present several examples of ways to measure x-ray spectra. In these spectra, the effect of response function is clearly observed. Figure 4.21 shows a comparison of predicted x-ray spectra for tube voltages of 40, 60, and 80 kV. Black data show incident x-ray spectrum. Blue and red spectra are folded spectra for a single-probe-type CdTe detector and multipixel-type CdTe detector (for pixel size of 200 μm), respectively. The upper, middle, and lower graphs indicate spectra without an absorber, penetrating soft tissue 20 mm and penetrating aluminum 10 mm, respectively. As described previously, the

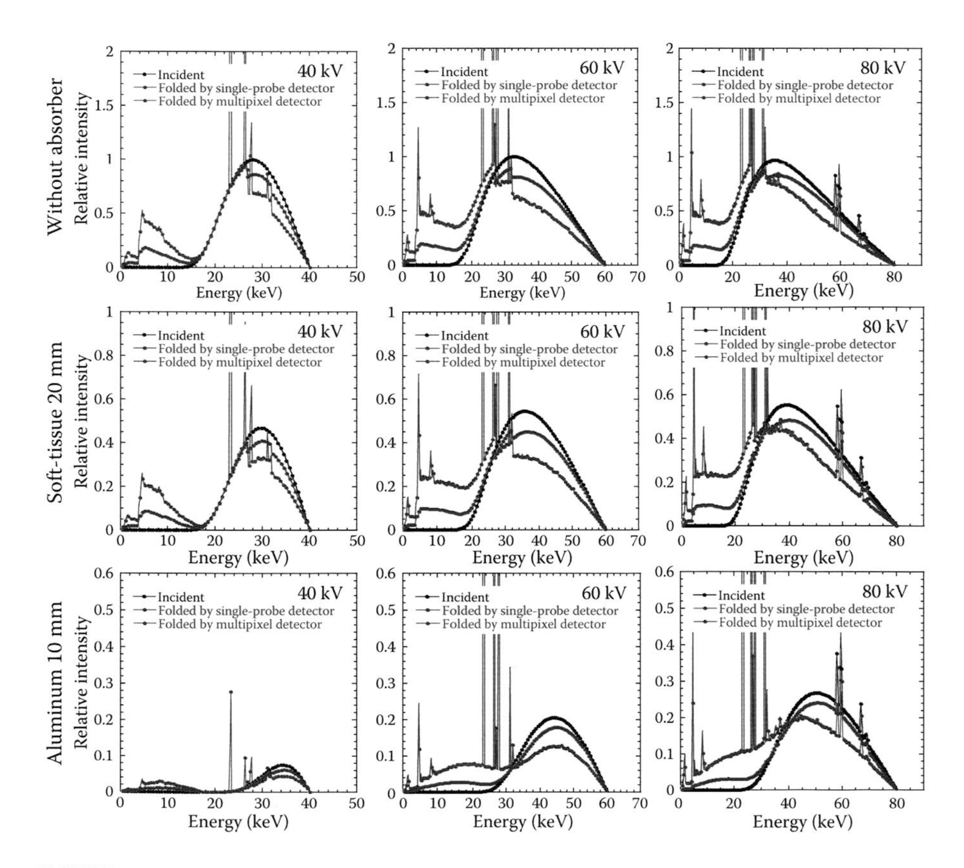

FIGURE 4.21
Comparison of predicted x-ray spectra.

black (incident) and blue (single-probe-type) spectra are similar, but unfolding correction should be applied to analyze the spectra measured with the CdTe detector. On the other hand, the red spectrum (multipixel type) indicates specific trends, especially for the 20–30 keV region. The characteristic x-rays of Cd and Te are clearly observed. In the response functions, there are many peaks, but most peaks depend on incident energy (see Table 4.2). When the x-ray spectra (continuous x-ray spectra composed of various *E*) are measured, the corresponding EPs have different values. On the other hand, the characteristic x-rays are commonly observed not depending on incident x-ray energies. Therefore, the peaks of characteristic x-rays are enhanced. In the near future, for the development of a new medical diagnostic apparatus, the effect of response function of a multipixel-type CdTe detector on the measured x-ray spectra should be considered carefully.

4.6 Summary of This Chapter

In this chapter, the response function of a multipixel-type CdTe detector was described. In order to explain the response function from the point of view based on the physics, the interaction between detector material (Cd and Te) and photon was initially described. The photoelectric effect plays an important role. Next, the response function of a single-probe-type CdTe detector was described; the point was to demonstrate the existences of EPs. FEP efficiency is strongly related to the cross section of the photoelectric effect and also related to the EPs. Then, the description was expanded to the response function of a multipixel-type CdTe detector, and I demonstrated that there are many peaks in the response functions. Finally, as a demonstration of the application, I predicted x-ray spectra measured with a multipixel-type CdTe detector. I hope that the description in this chapter is useful to the application of using a multipixel-type CdTe detector.

Acknowledgments

The descriptions in this chapter were partially supported by a collaborative research between Tokushima University and Job Corporation. The author wishes to express gratitude to Mr. Tsutomu Yamakawa and Dr. Syuichiro Yamamoto for their valuable advice. All of the presented data were originally simulated. I wish to thank graduate school students Ms. Natsumi Kimoto, Mr. Takashi Asahara, and Mr. Hiroki Okino, for getting many of the results.

References

1. K. Taguchi et al., Vision 20/20: Single photon counting x-ray detectors in medical imaging, *Medical Physics* 40, 100901-1-19, doi: 10.1118/1.4820371, 2013.
2. N. Kimoto et al., "Precise material identification method based on a photon counting technique with correction of the beam hardening effect in X-ray spectra," *Applied Radiation and Isotopes* 124, 16–26, doi: 10.1016/j.apradiso.2017.01.049, 2017.
3. N. Kimoto et al., "Development of a novel method based on a photon counting technique with the aim of precise material identification in clinical X-ray diagnosis," *Proceedings of SPIE* 10132, 1013239-1-11, doi: 10.1117/12.2253564, 2017.
4. X. Wang et al., Material separation in x-ray CT with energy resolved photon-counting detectors, *Medical Physics* 38, 1534–1546, doi: 10.1118/1.3553401, 2011.
5. J. Rinkel et al., Experimental evaluation of material identification methods with CdTe x-ray spectrometric detector, *IEEE Transactions on Nuclear Science* 58(5), 2371–2377, doi: 10.1109/TNS.2011.2164266, 2011.
6. H. Hayashi et al., A fundamental experiment for novel material identification method based on a photon counting technique: Using conventional x-ray equipment, *IEEE Nuclear Science Symposium & Medical Imaging Conference*, Conference Record, 2015.
7. G. F. Knoll, *Radiation Detection and Measurement*, John Wiley & Sons, Inc., 2010.
8. N. Tsoulfanidis, *Measurement and Detection of Radiation*, Hemisphere Publication, 1983.
9. K. Maeda et al., Compton-scattering measurement of diagnostic x-ray spectrum using high-resolution Schottky CdTe detector, *Medical Physics* 32, 1542–1547, doi: 10.1118/1.1921647, 2005.
10. S. Miyajima et al., CdZnTe detector in diagnostic x-ray spectroscopy, *Medical Physics* 29, 1421–1429, doi: 10.1118/1.1485975, 2002.
11. Y. Kojima et al., A precise method of Q_β determination with small HPGe detector in an energy range of 1–9 MeV, *Nuclear Instruments and Methods in Physics Research A* 458, 656–669, doi: 10.1016/S0168-9002(00)00899-8, 2001.
12. R. Ballabriga et al., Medipix3: A 64 k pixel detector readout chip working in single photon counting mode with improved spectrometric performance, *Nuclear Instruments and Methods in Physics Research A* 633, S15–S18, doi: 10.1016/j.nima.2010.06.108, 2011.
13. C. Ullberg et al., Measurements of a dual-energy fast photon counting detector with integrated charge sharing correction, *Proceedings of SPIE* 8668, 86680P-1, doi: 10.1117/12.2007892, 2013.
14. A. Brambilla et al., CdTe linear pixel x-ray detector with enhanced spectrometric performance for high flux x-ray imaging, *IEEE Nuclear Science Conference Record*, R18-5, 4825–4828, 2011.
15. G. Pellegrini et al., "Performance limits of a 55 μm pixel CdTe detector," *IEEE Transactions in nuclear Science* 53, doi: 10.1109/NSSMIC.2004.1462678, 2004.
16. E. Storm et al., Photon cross sections from 1 keV to 100 MeV for elements $Z = 1$ to $Z = 100$, *Nuclear Data Tables* A7, 565–681, 1970.
17. H. Hirayama et al., The EGS5 code system, *KEK Report* 2005-8, 1–418, 2005.

18. T. E. Everhart et al., Determination of kilovolt electron energy dissipation vs penetration distance in solid materials, *Journal of Applied Physics* 42, 5837, doi: 10.1063/1.1660019, 1971.
19. R. B. Firestone et al., *Table of Isotopes*, 8th edition, John Wiley and Sons, Inc., ISBN 0471-14918-7, 1998.
20. H. Hayashi et al., Experimental evaluation of response functions of a CdTe detector in the diagnostic region with the aim of carrying out a basic experiment concerning a next generation photon counting system, *European Congress of Radiology*, EPOS (C-0006) 1–28, doi: 10.1594/ecr2016/C-0006, 2016.
21. R. R. Carlton et al., *Principles of Radiographic Imaging*, 5th edition, Delmar Cengage Learning,. ISBN-13: 978-1-4390-5872-5, 2003.
22. R. Birch et al., Computation of Bremsstrahlung x-ray spectra and comparison with spectra measured with a Ge(Li) detector, *Physics in Medicine and Biology* 24(3), 505–517, 1979.
23. D. M. Tucker et al., Semiempirical model for generating tungsten target x-ray spectra, *Medical Physics* 18(2), 211–218, 1991.
24. J. P. Bissonnette et al., A comparison of semiempirical models for generating tungsten target x-ray spectra, *Medical Physics* 19, 579–582, doi: 10.1118/1.596848, 1992.

5

Monte Carlo Modeling of Solid State X-Ray Detectors for Medical Imaging

Yuan Fang and Aldo Badano

CONTENTS

5.1 Introduction

Semiconductor x-ray detectors are important components of medical imaging systems and are used in a wide range of modalities including general radiography, full-field digital mammography (FFDM) [1], and computed tomography (CT) [2]. Characterization of semiconductor x-ray detectors provides insight into performance limitations and guides the development and optimization of imaging systems. By employing semiconductive materials to convert x-rays directly into electric signal, these detectors allow for good energy resolution, high efficiency, and high carrier yield [3].

Monte Carlo simulation is a statistical numerical technique that relies on random numbers to sample models that describe stochastic physical processes in radiation transport. This technique provides the ability to manage complex models using well-known atomic interaction models. Using Monte Carlo methods, the mean value of the outcome of a stochastic process can be estimated by integrating the results of many random trials. Due to these inherent advantages, Monte Carlo methods can be used to study charge generation and transport in semiconductor materials, as well as for design validation and optimization of imaging systems.

In this chapter, direct and indirect detection methods using semiconductor and scintillator materials are presented to provide a background on x-ray detection materials used in medical imaging applications (Section 1.2). A summary of modeling approaches using Monte Carlo (MC) methods are covered (Section 1.3). The theory and implementation of a detailed MC model for direct x-ray detectors is then described (Section 1.4). Practical applications of the model and simulation results are presented (Section 1.5).

5.2 X-Ray Detector Technologies

5.2.1 Scintillator-Based Detectors

In scintillator-based detectors, a phosphor scintillator converts an incident x-ray photon into multiple optical light photons, detected as electric signals in a photodiode or a photomultiplier tube, hence an indirect conversion process [4]. Figure 5.1a shows the structure of a scintillator detector.

One of the advantages of the indirect detection method is high absorption efficiency of the scintillator material (e.g., CsI) capable of absorbing a large fraction of incident x-ray photons in the medical imaging energy range. However, the major disadvantage of the indirect method is the loss of spatial resolution due to isotropic generation of optical photons, shown in Figure 5.1a. Thinner scintillator material can be used to limit the effect of optical spreading, at a cost of reduced absorption efficiency. Scintillators with columnar phosphor structures are also used to confine and reduce the spreading of optical photons.

5.2.2 Semiconductor-Based Detectors

In semiconductor-based detectors, x-ray photons are absorbed in the photoconductor material and converted directly into electron–hole pairs (EHPs).

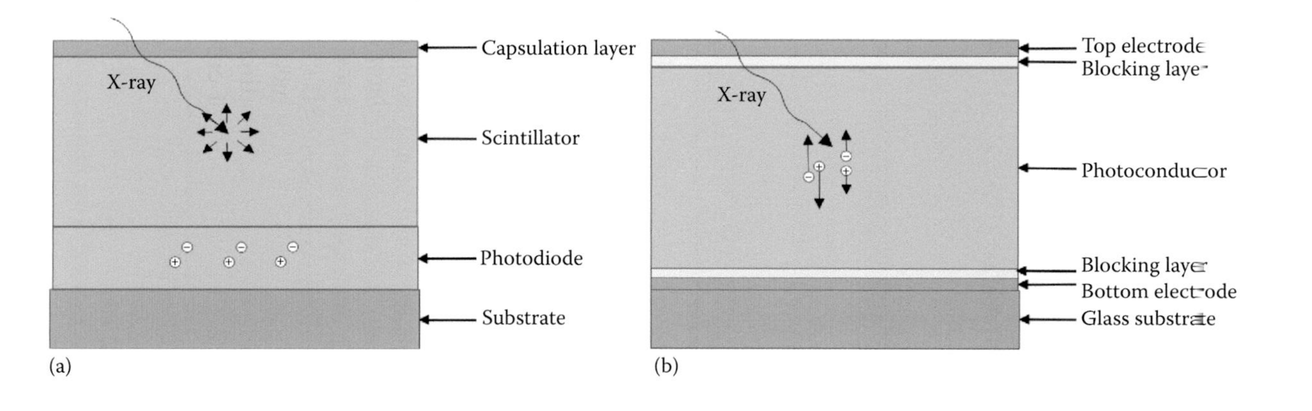

FIGURE 5.1
(a) Scintillator-based indirect digital x-ray imaging detector. (b) Semiconductor-based direct digital x-ray imaging detector. (From Fang, Y. et al., *Radiation Detector for Medical Imaging*, pp. 27–46, 2015. CRC Press, Taylor & Francis.)

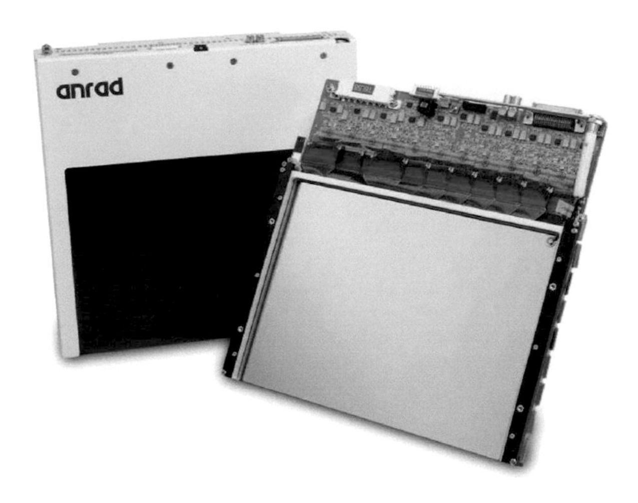

FIGURE 5.2
Commercial a-Se semiconductor direct conversion x-ray detector for FFDM. (Courtesy of ANRAD Corp. From Fang, Y. et al., *Radiation Detector for Medical Imaging*, pp. 27–46, 2015. CRC Press, Taylor & Francis.)

Figure 5.1b is an illustration of the direct conversion method. As x-ray photons are absorbed in the photoconductive material, a large number of EHPs are generated near the region of interaction, which are eventually collected at the electrodes.

With direct conversion, resolution of semiconductor x-ray detectors depends only on spreading of incident x-ray photons, secondary high-energy electrons, and EHPs. In general, photoconductive materials used in semiconductor detectors have lower atomic numbers compared to scintillators, and thus require a thicker detector to absorb the same amount of incident x-rays. A bias voltage is applied to collect the EHPs and blocking or (electron/hole) transport layers are used to prevent leakage of the charge carriers. Figure 5.2 shows a typical, commercially available amorphous selenium (a-Se) direct conversion x-ray detector manufactured by ANRAD Corporation.

5.3 Modeling Approaches

The complete signal formation process in semiconductor x-ray detectors from incident x-rays to electric signal can be divided into four subprocesses: incident x-ray interactions, secondary electron interactions, EHP generation, and charge transport. This sequence is illustrated in Figure 5.3. Incident x-ray photons can interact within the semiconductor material through Rayleigh scattering, Compton scattering, and photoelectric effect.

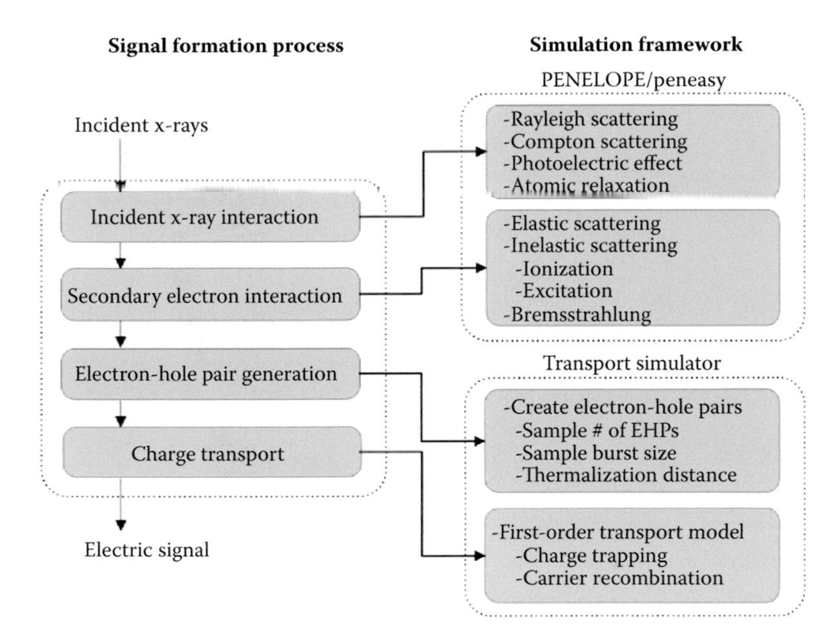

FIGURE 5.3
Block diagram of the signal formation process in semiconductor x-ray detectors including incident x-ray interaction, secondary electron interaction, EHP generation, and charge transport [5]. (From Fang, Y. et al., *Radiation Detector for Medical Imaging*, pp. 27–46, 2015. CRC Press, Taylor & Francis.)

High-energy electrons can be generated from photoelectric absorption and Compton scattering events, where the electron's kinetic energy is deposited in the semiconductor via inelastic scattering events. Models for the generation of EHPs include sampling algorithms, and spatial distribution of EHPs through calculation of burst and thermalization distances. The charge transport models include carrier recombination and trapping. Photon–electron and EHP interactions can be simulated separately or coupled together in MC models. This section briefly describes some existing methods available for modeling direct x-ray imaging detectors.

5.3.1 Photon–Electron Interactions

Incident x-ray photon and secondary electron interactions can be modeled with a number of available MC simulators: PENELOPE, EGSnrc, GEANT4, MCNP, ETRAN, ITS3, and FLUKA [6–12]. Figure 5.4a and b shows the photon and electron interaction cross sections in selenium, generated with PENELOPE 2006 [8]. The interaction cross sections are functions of particle energy and material properties. For x-ray energy range of medical applications, the primary interaction mechanisms include Rayleigh scattering, Compton scattering, and photoelectric absorption. For electrons, the main

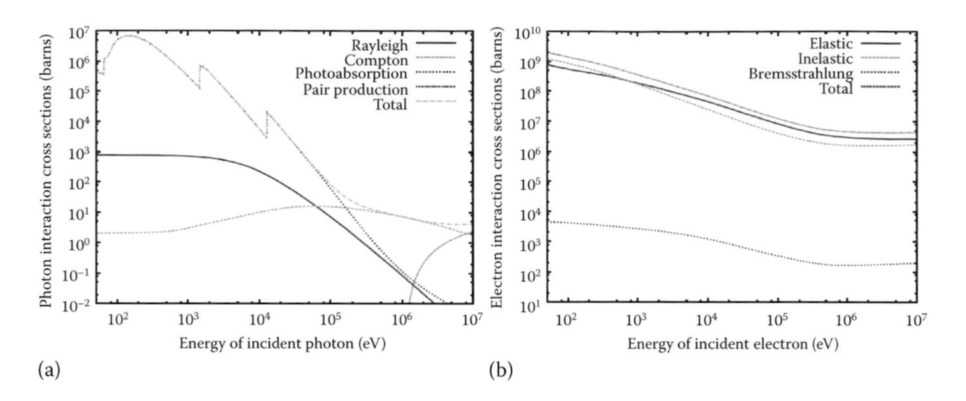

FIGURE 5.4
(a) PENELOPE photon interaction cross sections for selenium from 100 eV to 10 MeV. (b) PENELOPE electron interaction cross sections for selenium from 100 eV to 10 MeV. Note: 1 barn = 10^{-24} cm^2 [5]. (From Fang, Y. et al., *Radiation Detector for Medical Imaging*, pp. 27–46, 2015. CRC Press, Taylor & Francis.)

mechanisms are elastic and inelastic scattering and Bremsstrahlung radiation. The total interaction cross section can be computed as the sum of the cross sections for all possible interaction mechanisms and is used to compute the mean free path required for sampling the location of the random scattering events in the material.

Existing MC models offer advanced geometric packages that allow for simulation of complex detector geometries and experimental setups. The interaction models and cross sections have already been benchmarked and validated with established databases and offer an accurate model for the simulation of various interactions. MC models require simulation of many histories in order to reach high accuracy and low estimate variance. Some MC codes allow for different modes of simulation such as condensed, detailed, and mixed. For instance, in condensed simulation of electrons, many soft interactions that do not change significantly the direction and energy of the particle are reproduced into a single interaction, using multiple scattering theories to reduce simulation time. In a case where interaction locations and small energy changes are needed, the detailed simulation mode can be used to track all electron interactions. A mixed simulation mode can be used to optimize simulation detail and runtime. One limitation of many MC simulators for modeling of semiconductor detectors is lack of simulation libraries and models for generation and transport of EHPs.

5.3.2 EHP Transport

Some custom MC simulators have been developed to focus on modeling of EHP interactions specifically [13,14]. The effect of trapping and recombination of EHPs on the sensitivity reduction, ghosting [15], and time-of-flight

(TOF) simulations of EHPs to determine the density of state in a-Se [14] have been previously studied. The sensitivity reduction in a-Se detectors was found to be dependent on different detector operating conditions, such as applied electric field, x-ray spectrum, and photoconductor layer thickness, that affect recombination and trapping in the detector. TOF simulations take into account carrier drift and trapping.

Custom EHP simulators allow for focused studies of the carrier transport in semiconductor x-ray detectors, and offer significant flexibility in implementation of complex recombination and trapping models. Compared to detailed photon–electron simulations, in some cases, exponential attenuation models for x-ray photons are used assuming complete absorption of incident energy for carrier generation and one-dimensional model for EHP transport. Such an approach ignores charge spreading due to high-energy photoelectric and Compton electrons, lateral spreading of EHPs due to diffusion, and noise in the detector response from Compton scattering and fluorescent x-rays.

5.3.3 Coupled MC Simulation

The coupled simulation method takes advantages of the previous two methods by combining the simulation of x-ray radiation with EHP transport. For example, energy deposition events in the semiconductor or scintillator material can be simulated with an available MC simulator in combination with a custom simulator for EHPs. The MANTIS package is an efficient and flexible simulation tool for research and development of scintillator-based indirect radiation systems. The package consists of PENELOPE for photon–electron interactions, including x-ray scattering and DETECTII routines for simulation of optical photons, and allows for detailed study of 3D optical blur with realistic columnar model.

Combined simulation can be used to simulate the complete signal formation process in x-ray detectors by utilizing existing validated MC simulators for photon–electron interactions and allows for significant customization of EHP transport models. However, in-depth knowledge and modifications to the existing simulation packages are often required to efficiently interface the codes. A high number of simulation histories are required to achieve low variance for studies, such as the point response function, needed for modulation transfer function (MTF) and detective quantum efficiency (DQE) calculations. These limitations drive the need to further improve simulation efficiency, including code parallelization and utilization of graphics processing units [16].

5.3.4 Analytical Methods

It is important to note that even though this chapter focuses on MC methods, analytical methods have also been widely used for modeling of x-ray detectors. For example, the small pixel effect for minimization of trapping of slow carriers on the electric signal has been previously studied by Barrett et al. [17]. This work assumes a homogeneous slab of semiconductor material, where the current induced on each pixel electrode is calculated via Shockley–Ramo theorem, and reductions in low energy tails of the pulse-height spectra are validated with experimental results. Compared to MC, analytical methods do not require long simulation time and are efficient at solving radiation transport problems with simple electric field distributions that can be mathematically represented. However, analytical methods can be limited in modeling three-dimensional charge clouds of secondary carriers inside the detector material, and taking into account the stochastic events that affect radiation transport, such as trapping and recombination inside a non-uniform electric field.

5.4 MC Simulation of Radiation Transport

5.4.1 Theory

5.4.1.1 Charge Generation

For optical photon detection, only one EHP is generated and it is assumed that the carriers lost their initial kinetic energy and are separated by a finite distance r_o due to thermalization. This distance can be estimated for the given photon energy hv and applied electric field E_{app} using the Knight–Davis equation [18], where D is the diffusion coefficient, E_{gap} is the band gap of the semiconductor material, ε is the dielectric constant, and e is the elementary charge:

$$\frac{r_o^2}{D} = \frac{(\text{hv} - E_{\text{gap}}) + \dfrac{e^2}{4\pi\varepsilon r_o} + \text{eE}_{\text{app}} r_o}{\text{hv}_p^2}. \tag{5.1}$$

Compared to optical detectors, the charge generation models for radiation detectors are more complex due to generation of many EHPs by a single incident photon. Photoelectric absorption is the dominant x-ray interaction mechanism in the energy range of interest. It creates a secondary photoelectron, with most of the energy of the initial x-ray, capable of further ionizing the material and producing a significant number of EHPs. X-ray photons that are Compton-scattered can also produce energetic electrons capable of creating many EHPs; however, the particle's kinetic energy is lower compared to the photoelectron. As the high energy ionizing electron travels through the detector material, it gradually loses its energy through inelastic scattering, and the energy lost, E_d, is deposited in the semiconductor material, leading to the generation of EHPs. The mean number of EHPs generated, $\bar{N}_{\text{EHP}}$, can be estimated via Poisson sampling from the energy deposited and the material ionization energy, W_o:

$$\bar{N}_{\text{EHP}} = \frac{E_d}{W_o}. \tag{5.2}$$

The ionization energy for semiconductors was originally derived by Klein [19] and is given by the following equation:

$$W_o \approx K * E_{\text{gap}} + r\text{hv}_p, \tag{5.3}$$

where the phonon energy is hv_p, and r is a uniform random number between 0 and 1, representing the ionization and photon emission components. The dimensionless constant K is found to be 2.8 for crystalline materials through semi-empirical calculations and 2.2 for amorphous materials [20].

Several models have been developed to model the carrier generation process in silicon. Some models assume all the EHPs generated in a sphere following either Gaussian or uniform distribution [21–23], while others use MC simulations of a large number of electron tracks to estimate the center-of-mass and uniformly distribute portions of the photon energy into a bubble and a line [24]. In silicon, W_o is not field-dependent and the dominant effect of charge sharing is diffusion of carriers. In a-Se, however, carrier drift also plays a major role due to the field dependence of carrier generation and transport. The concept of EHP bursts is proposed for modeling carrier generation in a-Se. A burst is defined as the cloud (spatiotemporal distribution) of electrons and holes generated after a local deposition of energy [25]. Energy deposited in electron inelastic collisions with outer-shell electrons can lead to excitation of plasma waves and generate multiple EHPs [26]. These pairs constitute a burst, and the burst size is dependent on the energy of the incident particle and the material plasma frequency. According to the Bohr adiabatic criterion [27], the burst size, r_b, can be approximated using the following expression:

$$r_b = \frac{\upsilon}{\omega_{\text{pe}}}, \tag{5.4}$$

where υ is the velocity of the incident particle, and ω_{pe} is the plasma frequency, dependent on the electron mass of the material and density. The concept of a burst is introduced in conjunction with thermalization of carriers in order to provide a three-dimensional distribution model for EHP generation.

5.4.1.2 Recombination and Trapping

There exist two models to study recombination of carriers in a-Se: geminate and columnar recombination. Geminate recombination is used by Onsager to model EHP recombination due to optical photons and assumes that carriers can only recombine with their original geminate pair. Columnar recombination occurs when a high-energy electron produces EHPs continuously in a column surrounding its track, and carriers from different interactions recombine in a columnar fashion. Our model takes into account both processes by considering both geminate and columnar recombination in bursts. Recombination can occur between any electron and hole traveling toward each other, and trapping can occur when an electron or a hole reaches a lower energy state due to material impurities. The drift component takes into account the applied electric field, E_{app}, and the Coulomb field due to other charge carriers. For the ith charge carrier, the resulting electric field acting on it is given by

$$E_i = E_{\text{app}} + \sum_{j \neq i} \frac{1}{4\pi\varepsilon} \frac{e_j}{r_{ij}^2} \hat{r}_{ij}, \tag{5.5}$$

where r_{ij} is the separation distance between charge carrier i and j, and $\hat{r}_{ij}$ is the field direction vector. In turn, the displacement in x, y, and z direction due to drift can be found:

$$\Delta x_{\text{drift}} = uE_x \Delta t, \tag{5.6}$$

where μ is the carrier mobility, E_x is the x component of the electric field, and Δt is the simulation time step. The components of the y and z directions can be found similarly. The diffusion component can be found by sampling the polar and azimuthal angles from a uniform distribution, where the diffusion distance is given by $\sqrt{(6D\Delta t)}$ [28], and the total displacement in each direction is a sum of the drift and diffusion components:

$$\Delta x = \Delta x_{\text{drift}} + \Delta x_{\text{diffusion}}. \tag{5.7}$$

During transport, both drift and diffusion of carriers are calculated at each time step. The drift component depends on the carrier mobility, the electric field acting on the carrier, and the simulation time step. The diffusion component depends on the diffusion coefficient and the time step as shown in Equations 5.5 through 5.7. Depending on the material properties, the carrier mobilities may differ, and the drift and diffusion components are affected, causing the carrier to travel faster/slower in the semiconductor material.

Many trapping effects have been modeled previously in one dimension (thickness) for a-Se detectors [15]. These include deep trap, shallow trap, trap releasing, trap filling, and trap center generation due to incident x-rays. Deep and shallow trapping differs in the trapping time of carriers. Deep traps have longer trapping times on the order of seconds to minutes, while shallow traps may release carriers in fractions of a microsecond or less. For simulation purposes, when a carrier is trapped in a deep trap, it is considered lost. However, when a carrier is trapped in a shallow trap, the release of this trapped carrier (perhaps in subsequent exposures) can contribute to the detected signal as well. As EHPs start to move in the material and get trapped, the number of available traps decreases as a function of time, x-ray exposure, and carrier concentration. At the same time, a competing process of trap center creation is occurring due to x-ray bombardment of the semiconductor material. The current implementation of trapping uses a simple model that only considers deep trapping. The probability of trapping, P_{trapping}, can be calculated as [15]

$$P_{\text{trapping}} = 1 - e^{-\frac{\Delta t}{\tau_{\text{trapping}}}}, \tag{5.8}$$

where τ_{trapping} is the trapping time. Constant trapping times are used for electrons and holes to give an estimate of the average carrier lifetime and the effect of applied electric field on carrier trapping probabilities in the semiconductor material. The probability of carrier trapping depends on the total carrier transit time, and applied electric field can be used to collect carriers from the interaction site to the appropriate electrodes.

5.4.2 Implementation

An MC transport code, ARTEMIS (pArticle transport, Recombination, and Trapping in sEMiconductor Imaging Simulation), was developed for the purpose of simulation of the signal formation process in direct x-ray detectors [29]. Various functions are implemented to model the physics outlined in the theory section. The flow diagram for the implemented simulation framework is shown in Figure 5.5.

X-ray photon and secondary electron interactions in the presence of an external electric field are modeled by PENELOPE [8], and the locations of inelastic electron interactions with energy deposition are coupled with the transport routines for EHP simulations. Figure 5.6 shows the photon and electron particle tracks of 100 keV mono-energetic x-rays. Figure 5.6a depicts the absorption of a pencil beam of x-ray photons perpendicularly incident on the a-Se detector (in green). Most photons are absorbed in the center of the detectors, and the off-center photons are due to Compton scattering and fluorescence. Figure 5.6b is a close-up showing the secondary electrons move in random walk and deposit energy at random locations in the photoconductor (in red).

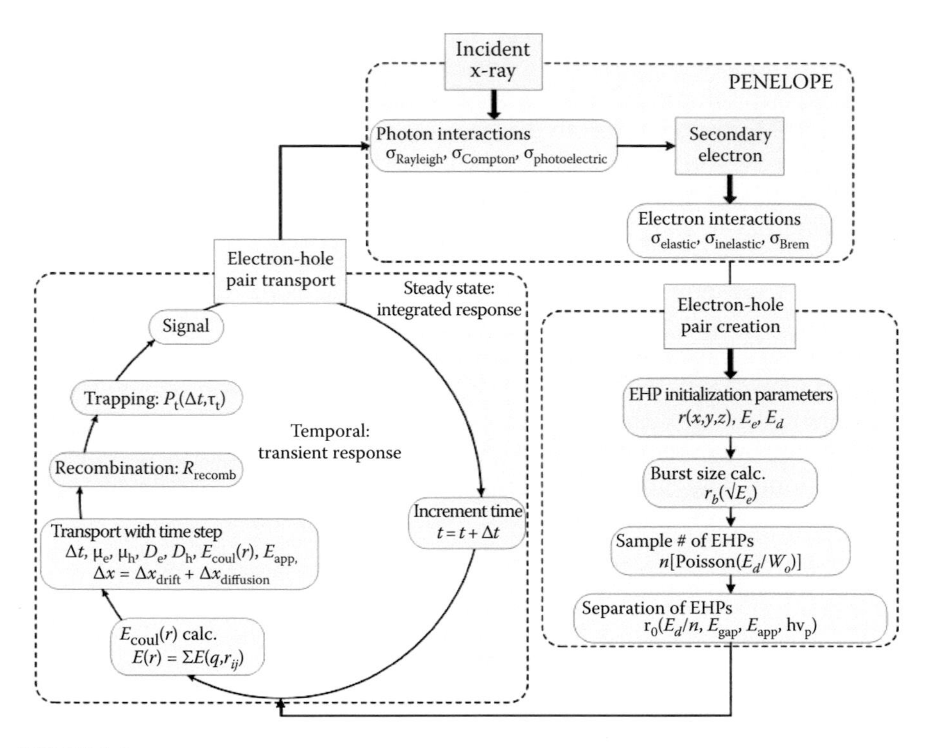

FIGURE 5.5
Flow chart for the simulation of the signal formation process in semiconductor x-ray detectors. Simulation of photon and secondary electron with PENELOPE is coupled with transport for detailed spatiotemporal simulation of EHPs [29]. (From Fang, Y. et al., *Radiation Detector for Medical Imaging*, pp. 27–46, 2015. CRC Press, Taylor & Francis.)

To further show the energy deposition events as high-energy electron loses kinetic energy in the semiconductor material, Figure 5.7a and b depicts the single high-energy electron track produced by a 40 and 140 keV x-ray photon. The bubble size is largest at the beginning of the track where the high-energy electron is created and gradually decreases as energy is deposited in the semiconductor material. PENELOPE has been modified to take into account the effect of electric field for high-energy electron interactions.

Figure 5.8a illustrates the generation of bursts of EHPs from sites of energy deposition. Once the EHPs are generated, the applied electric field pulls the holes and electrons to the opposing electrodes. However, these charge carriers could be lost as they travel within the photoconductor (illustrated in Figure 5.8b) by two processes: recombination and trapping [5]. Currently, due to the large number of EHPs, each burst is simulated separately for the transport including recombination and trapping considerations.

Recombination of carriers is checked at each simulation step and occurs when an electron and a hole are getting sufficiently close together, making

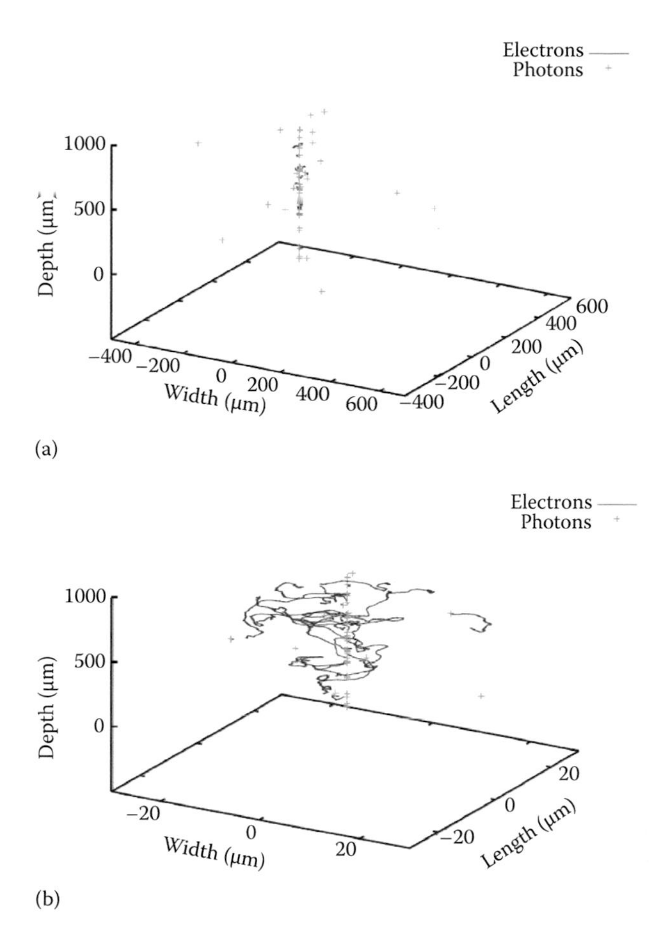

FIGURE 5.6
(a) Particle track of 100 keV incident photons (100 histories) in selenium. (b) Close-up of (a) [5]. (From Fang, Y. et al., *Radiation Detector for Medical Imaging*, pp. 27–46, 2015. CRC Press, Taylor & Francis.)

the Coulomb attraction so strong that they cannot escape each other. As carriers approach each other due to Coulomb attraction, their drift component increases inversely proportional to the separation distance squared. Thus, as the separation distance is reduced, the simulation time step also should be reduced in order to accurately capture the movement of the carriers as they come close to each other. However, this comes at the expense of simulation time. To solve this problem, a recombination distance was used by Bartczak et al [30]. in their study of ion recombination in irradiated nonpolar liquids. Figure 5.9a and b shows the sample transport tracks of three EHPs in electric field taking into account drift alone, and drift and diffusion.

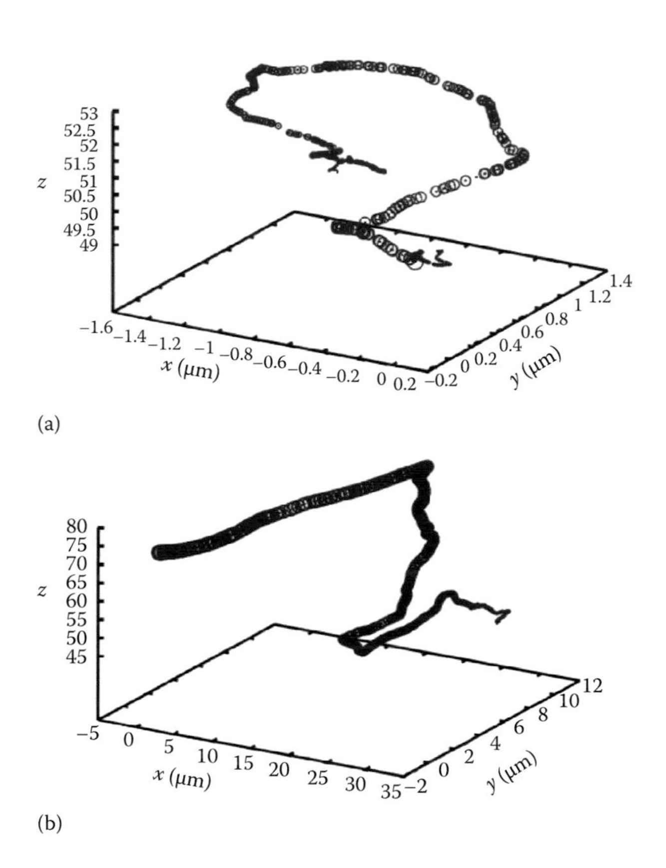

FIGURE 5.7
3D bubble plot of energy deposition events by secondary electrons in PENELOPE by (a) 40 keV photon and (b) 140 keV photon. (From Fang, Y. et al., *Radiation Detector for Medical Imaging*, pp. 27–46, 2015. CRC Press, Taylor & Francis.)

5.5 Applications

5.5.1 Charge Sharing in Photon-Counting Mode Detectors

Photon-counting detectors with energy discrimination capabilities can be used for medical x-ray imaging applications [31,32]. Photon-counting detectors with the ability to estimate the energy of transmitted photons at each pixel location can enable improved material decomposition, higher spatial resolution, and implementation of the beam hardening corrections. In addition, the use of photon-counting detectors with energy discrimination abilities can potentially improve signal-to-noise ratio (SNR) performance in existing modalities and allow for the development of novel applications [31,34–38], including spectral CT [39–45].

Charge sharing is a major challenge for photon-counting detectors. This phenomenon is most significant near pixel boundaries, where the charge

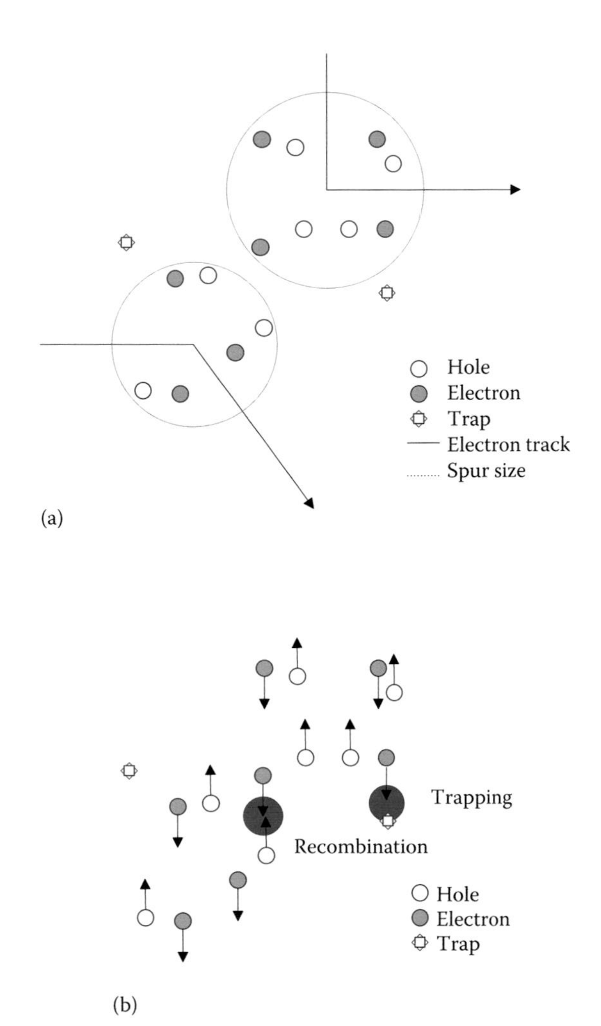

FIGURE 5.8
(a) Generation of EHPs from inelastic electron interactions, with varying burst size and thermalization distance. (b) Transport of EHPs; charged carriers can be lost due to recombination and trapping [5]. (From Fang, Y. et al., *Radiation Detector for Medical Imaging*, pp. 27–46, 2015. CRC Press, Taylor & Francis.)

cloud created by the x-ray quantum may be divided and detected simultaneously by multiple pixels recording energies lower than the photon energy. The redistribution of the energy carried by the x-ray quantum among multiple pixels can cause distortions in the spectral response. The significance of this effect depends on the detection material properties such as charge carrier mobility, detector design features such as the pixel size and applied bias, and the location of x-ray absorption with respect to the pixel boundary and depth of interaction within the active detector layer.

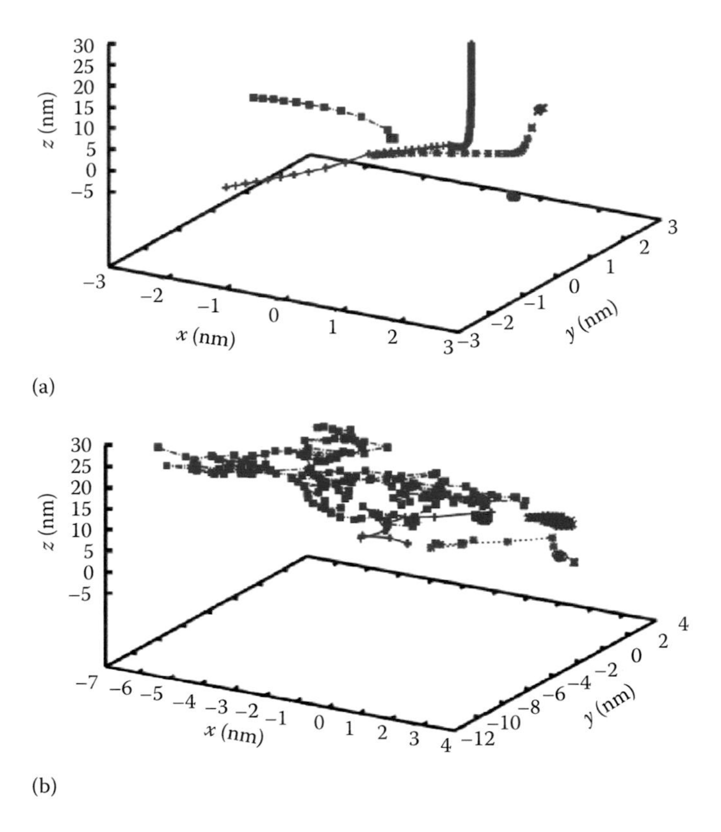

FIGURE 5.9
(a) Sample transport simulation track of three EHPs in electric field taking into account drift. (b) Sample transport simulation track of three EHPs in electric field taking into account drift and diffusion. Blue and red dots represent hole and electron tracks, respectively. (From Fang, Y. et al., *Radiation Detector for Medical Imaging*, pp. 27–46, 2015. CRC Press, Taylor & Francis.)

Figure 5.10 illustrates the silicon strip detector geometry and the setup used in this study. Specifically, the pixel size is 50 μm wide, with a strip length of 1 cm and thickness of 500 μm. The simulation results include nine x-ray absorption locations to cover a range relative to the interpixel boundary and the electrodes. This includes three incident x-ray locations 5, 15, and 25 μm away from the pixel boundary and three x-ray absorption depths of 50, 250, and 450 μm from the pixel electrode.

Pulse-height simulation results for 25 keV mono-energetic x-ray photons are presented in Figure 5.11. The *x*-axis is the number of holes collected from each interacted x-ray photon, and the *y*-axis is the number of primary x-ray absorption events normalized to unity at the peak.

For the 25 keV case where the incident x-ray is absorbed close (5 μm) to the pixel boundary and 50 μm from the pixel electrode, pulse-height spectrum (PHS) contains a spectral tail due to diffusion of charge carriers. The PHS further degrades in terms of total count of carriers detected, and the spectral

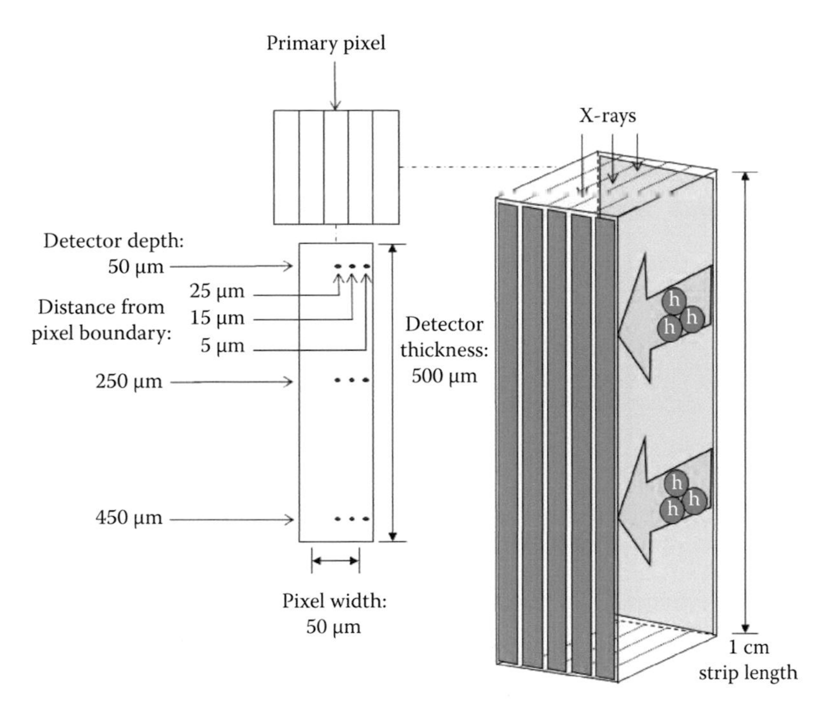

FIGURE 5.10
Silicon strip detector in the on-edge geometry for simulation purposes. The pixel size is 50 μm and the incident x-ray locations are 5, 15, and 25 μm away from the pixel boundary. The detector thickness is 500 μm, and the absorption depths are 50, 250, and 450 μm.

width increases as the absorption depth increases from the pixel electrode (250 and 450 μm). This is due to the more significant diffusion effects as a result of the increase in the distance the charge carrier must travel to reach the pixel electrode and due to the random walk by the high-energy electron leading to a charge cloud of EHPs created. As the incident x-ray photon interaction location moves further away (15 and 25 μm) from the pixel boundary, the PHS shape is recovered and the effects of charge carrier transport become less pronounced.

For the 75 keV case shown in Figure 5.12, similar trends are observed in PHS as a function of x-ray interaction location from pixel boundary and absorption depth in the detector. The main difference between the 25 and 75 keV cases is the increase in counts at lower energies due to Compton scattering.

5.5.2 Detective Quantum Efficiency

One of the most important performance metrics of a pixelated (or pixelized) solid-state x-ray detector is the DQE. The DQE is used to evaluate detector performance because it describes how effectively the x-ray detector can produce

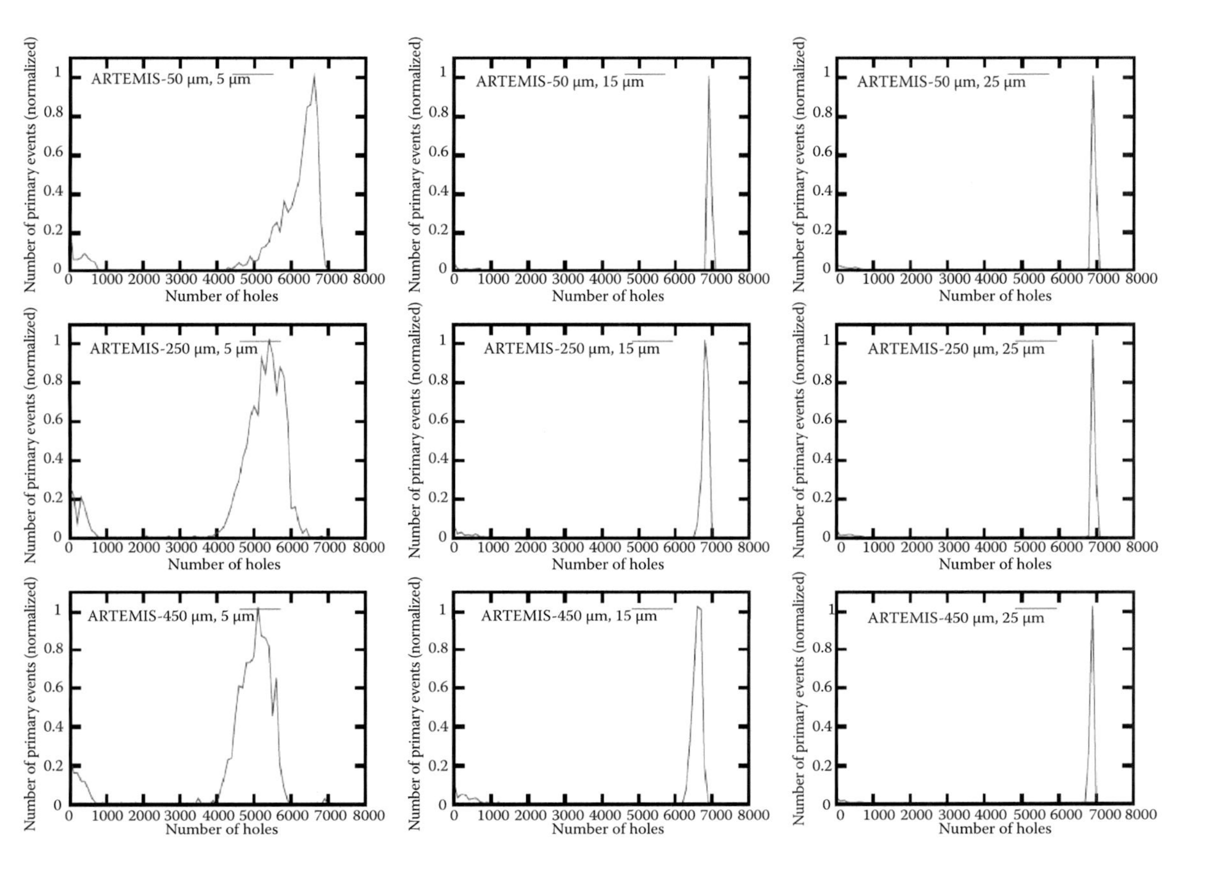

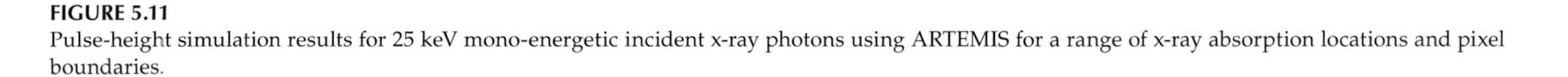
FIGURE 5.11
Pulse-height simulation results for 25 keV mono-energetic incident x-ray photons using ARTEMIS for a range of x-ray absorption locations and pixel boundaries.

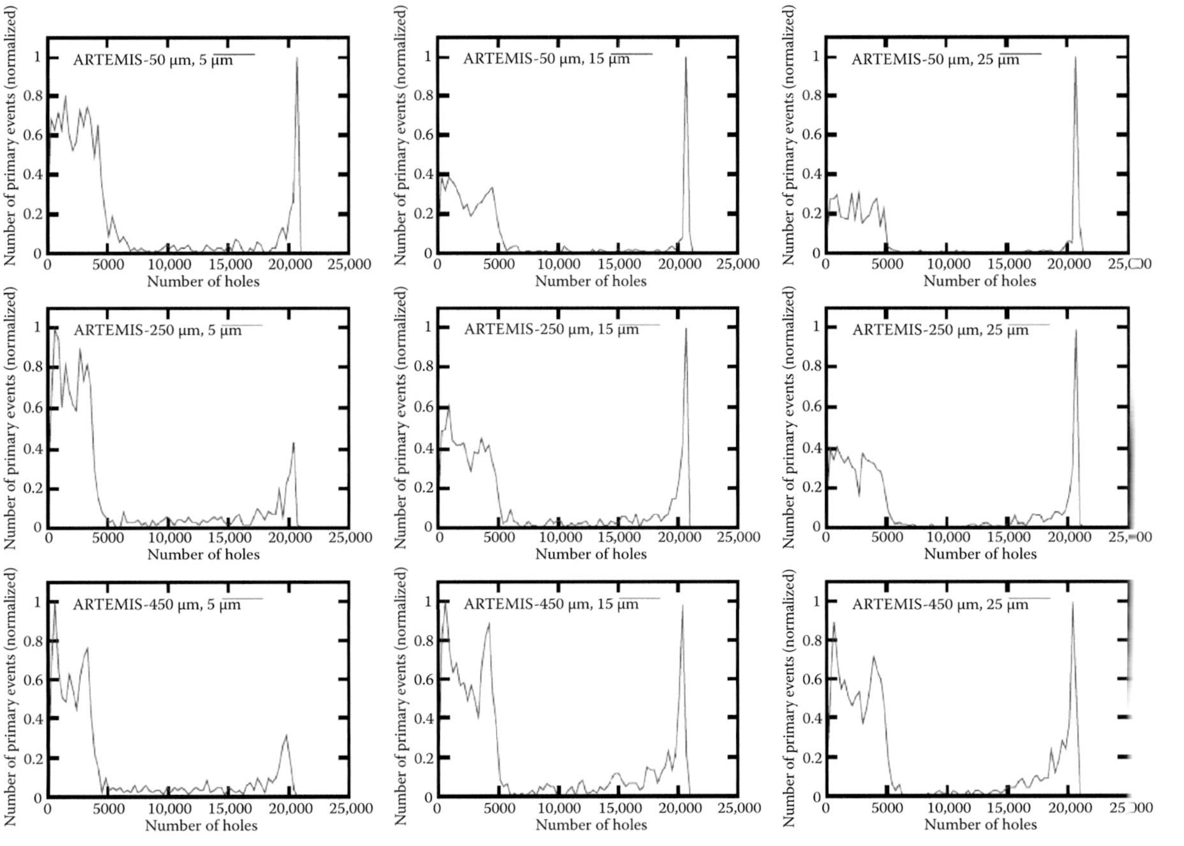
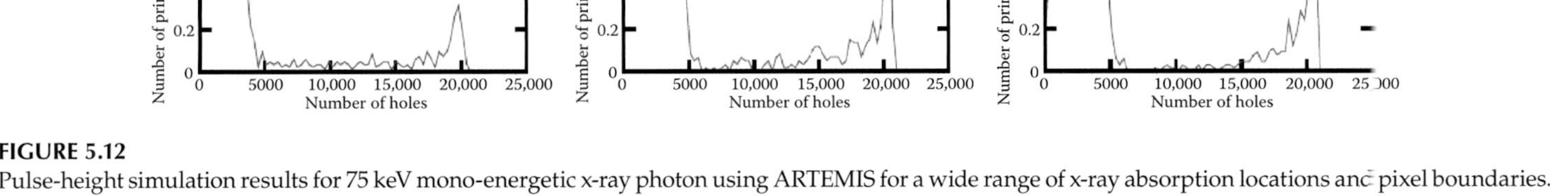

FIGURE 5.12
Pulse-height simulation results for 75 keV mono-energetic x-ray photon using ARTEMIS for a wide range of x-ray absorption locations and pixel boundaries.

an image with a high SNR relative to an ideal detector. IEC 62220-1-3:2008 was developed to standardize the testing procedures for measuring DQE for solid-state x-ray detectors. In this section, we show some simulated DQE results for an a-Se x-ray detector for some typical clinical beam qualities.

The Fujita–Lubberts–Swank method [46] was used for DQE simulations, which take into account the detector MTF, noise power spectrum (NPS), and Swank factor, with details of the DQE simulation methods described. [47] In order to compute the NPS, the simulated point spread (PSF) function is summed along one direction to yield a PSF projection that is Fourier transformed and squared. The shape of the 1D NPS is obtained by averaging the simulation results over a number of runs. The DQE expression is given as:

$$\mathrm{DQE}(f) = \mathrm{DQE}(0) \times \frac{\mathrm{MTF}(f)^2}{\mathrm{NPS}(f)}, \tag{5.9}$$

where DQE(0) is the zero-frequency DQE defined by

$$\mathrm{DQE}(0) = \mathrm{QE} \times I. \tag{5.10}$$

The interaction efficiency of detection material is given by QE, and I is the Swank factor defined as the statistical variation in the detected signal per primary x-ray quantum [48,49]. RQA beam qualities are used as the x-ray input in this work. The simulated x-ray spectra are generated with the method described by Boone et al. [50] shown in Figure 5.13, and the simulated and

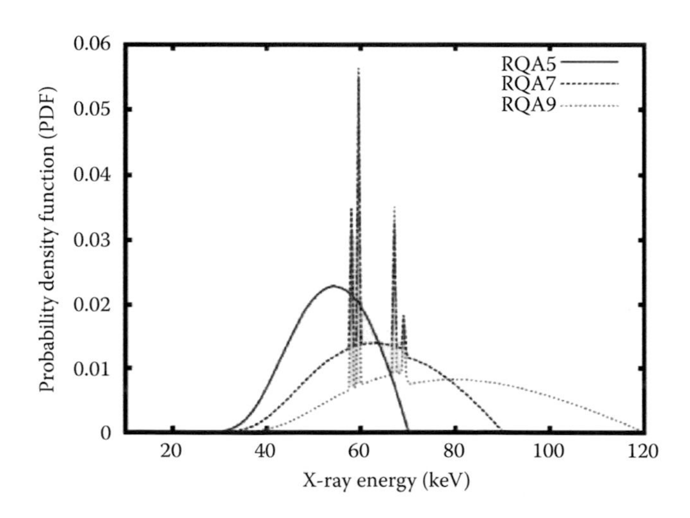

FIGURE 5.13
RQA x-ray beam qualities used in the simulation study.

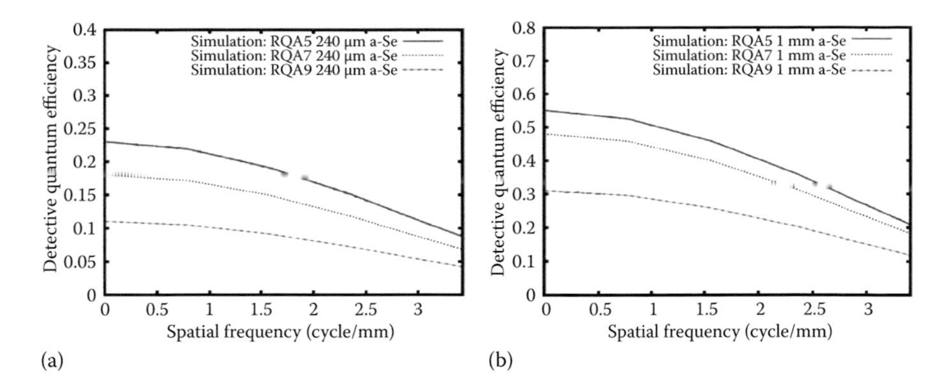

FIGURE 5.14
Simulated DQE curves for a-Se detector with (a) 240 μm and (b) 1 mm thickness and 10 V/μm applied electric field.

experimental DQE for two detectors with thicknesses of 240 μm and 1 mm are shown in Figure 5.14.

5.6 Summary

Semiconductor x-ray detectors are important components of x-ray medical imaging systems and can be used in a wide range of modalities and applications. MC methods can be used for modeling solid-state x-ray detectors and provide insight into the fundamental physics and theoretical performance limitations of imaging detectors. There are still many areas that need improvement in modeling the complete signal formation process in semiconductor detectors for x-ray imaging applications. As the high-energy photoelectron or Compton electron deposits energy in the detection material, each burst of EHPs is simulated separately due to the large number of carriers to be considered at one time. When the electron energy is high, the mean free path is larger and deposition events occur far away and apart from each other. However, with lower electron energy, energy deposition events occur more locally, and a need to consider multiple bursts may arise to more realistically model charge generation and recombination processes. In the application section, simulation results are shown for a-Se detectors in the conventional geometry and Si detectors in edge on geometry operating in photon-counting mode.

Detector thickness and carrier mobility can greatly affect the transport properties and hence the detector performance. As the detector thickness increases, the carriers require more time to travel to the electrodes, thus increasing the probability of recombination and trapping.

Disclaimer

The mention of commercial products herein is not to be construed as either an actual or implied endorsement of such products by the Department of Health and Human Services.

References

1. R. Schulz-Wendtland, K. Hermann, E. Wenkel, B. Adamietz, M. Lell, K. Anders and M. Uder, "First experiments for the detection of simulated mammographic lesions: Digital full field mammography with new detector with double plate of pure selenium," *Radiologe*, 51, 130–134 (2004).
2. A.L. Goertzen, V. Nagarkar, R.A. Street, M.J. Paulus, J.M. Boone and S.R. Cherry, "A comparison of x-ray detectors for mouse CT imaging," *Phys. Med. Biol.*, 49, 5251–5265 (2004).
3. G. Knoll, *Radiation Detection and Measurement.* New Jersey: Wiley Interscience/ John Wiley & Sons, Inc., pp. 265–366 (2010).
4. S.O. Kasap and J.A. Rowlands, "Direct-conversion flat-panel x-ray image detectors," *IEE Proc. CDS*, 149, 85 (2002).
5. Y. Fang, A. Badal, N. Allec, K.S. Karim and A. Badano, "Monte Carlo simulation of amorphous selenium imaging detectors," *Proc. SPIE* 7622, 762214 (2010).
6. M.J. Berger and S.M. Seltzer, Chapters 7, 8, and 9 in *Monte Carlo Transport of Electrons and Photons*, eds. Jenkins, T.M., Nelson, W.R. and Rindi, A., New York: Plenum (1988).
7. J.A. Halbleib, P.P. Kensek, T.A. Mehlhorn, G.D. Valdez, S.M. Seltzer and M.J. Berger, "ITS version 3.0: The integrated TIGER series of coupled electron/ photon Monte Carlo transport codes," *Report SAND*91-1634 (1992).
8. F. Salvat, J.M. Fernandez-Varea and J. Sempau, "PENELOPE-2006: A code system for Monte Carlo Simulation of electron and photon transport," Issy-les-Moulineaux, France: OECD/NEA Data Bank (2006).
9. I. Kawrakow and D.W.O. Rogers, "The EGSnrc code system: Monte Carlo simulation of electron and photon transport," *Report PIRS*-701 (2001).
10. X-5 Monte Carlo Team "MCNP—A general Monte Carlo N-particle transport xode, Version 5," Report No. LA-UR-03-1987 (2003).
11. S. Agostinelli et al., "Geant4—A simulation toolkit," *Nucl. Instrum. Meth. A*, 506 (2003).
12. A. Ferrari, P.R. Sala, A. Fassμo and J. Ranft, "FLUKA: A multi-particle transport code (Program version 2005)," CERN-2005-10, INFN/TC-05/11, SLAC-R-773 (2005).
13. E. Fourkal, M. Lachaine and B.G. Fallone, "Signal formation in amorphous-Se-based x-ray detectors," *Phys. Rev. B.*, 63, 195204 (2001).
14. M. Yunus, M.Z. Kabir and S.O. Kasap, "Sensitivity reduction mechanisms in amorphous selenium photoconductive x-ray image detectors," *Appl. Phys. Lett.*, 85, 6430–6432 (2004).

15. K. Koughia, Z. Shakoor, S.O. Kasap and J.M. Marshall, "Density of localized electronic states in a-Se from electron time-of-flight photocurrent measurement," *J. Appl. Phys.* 97, 033706-1-033706-11 (2005).
16. D. Sharma, A. Badal and A. Badano, "hybridMANTIS: A CPU-GPU Monte Carlo method for modeling indirect x-ray detector with columnar scintillators," *Phys. Med. Biol.* 57, 2357 (2012).
17. H.H. Barrett, J.D. Eskin and H.B. Barber, "Charge transport in arrays of semiconductor gamma-ray detectors," *Phys. Rev. Lett.* 75, 156–159 (1995).
18. J. Knight and E. Davis, "Photogeneration of charge carrier in amorphous selenium," *J. Phys. Chem. Solids* 35, 543–554 (1975).
19. C.A. Klein, "Bandgap dependence and related features of radiation ionization energies in semiconductors," *J. Appl. Phys.* 39, 2029 (1968).
20. W. Que and J. Rowlands, "X-ray photogeneration in amorphous selenium: Geminate versus columnar recombination," *Phys. Rev. B* 51, 10500–10507 (1995).
21. J.R. Janesik, *Scientific Charge-Coupled Devices.* Bellingham, WA: SPIE Press (2001).
22. L.K. Townsley, P.S. Broos, G. Chartas, E. Moskalenko, J.A. Nousek, G.G. Pavlov, "Simulating CCDs for the Chandra advanced CCD imaging spectrometer," *Nucl. Instrum. Methods A* 486, 716 (2002).
23. O. Godet, P. Sizun, D. Barret, P. Mandrou, B. Cordier, S. Schanne and N. Remoue, "Monte-Carlo simulations of the background of the coded-mask camera for X-and gamma-rays on-board the Chinese-French GRB mission SVOM," *Nucl. Instr. Methods A* 494, 775 (2009).
24. C. Xu, M. Danielsson and H. Bornefalk, "Validity of spherical approximation of initial charge cloud shape in silicon detectors," *Nucl. Instrum. Methods A* 648, S190 (2011).
25. M. Lachaine et al., "Calculation of inelastic cross-sections for the interaction of electrons with amorphous selenium," *J. Phys. D Appl. Phys* (2000).
26. D. Sharma, Y. Fang, F. Zafar, K. S. Karim and A. Badano, "Recombination models for spatio-temporal Monte Carlo transport of interacting carriers in semiconductors," *Appl. Phys. Lett.* 98, 242111 (2011).
27. N. Bohr, "The penetration of atomic particles through matter," *K. Dan. Vidensk. Selsk. Mat. Fys. Medd.* 18, 8 (1948).
28. H.H. Barrett and K. Myers, *Foundation of Imaging Science* New Jersey: Wiley Interscience/John Wiley & Sons, Inc., pp. 748–763 (2004).
29. Y. Fang, A. Badal, N. Allec, K.S. Karim and A. Badano, "Spatiotemporal Monte Carlo transport methods in x-ray semiconductor detectors: Application to pulse-height spectroscopy in a-Se," *Med. Phys.* 39, 308–319 (2012).
30. W.M. Bartczak, M.P. DeHaas and A. Hummel, "Computer simulation of the recombination of the ions in tracks of high-energy electrons in nonpolar liquids," *Radiat. Phys. Chem.* 37, 401–406 (1991).
31. K. Taguchi and J.S. Iwanczyk, Vision 20/20: "Single photon counting x-ray detectors in medical imaging," *Med. Phys.* 40(10), 100901 (2013).
32. E. Fredenberg, M. Hemmendor, B. Cederstrom, M. Aslund and M. Danielsson, "Contrast-enhanced spectral mammography with a photon-counting detector," *Med. Phys.* 37(5), 2017–2029 (2010).
33. M. Overdick, C. Baumer, K.J. Engel, J. Fink, C. Hermann, H. Kruger, M. Simon, R. Steadman and G. Zeitler, "Towards direct conversion detectors for medical imaging with x-rays," *IEEE Nucl. Sci. Conf.* R. NSS 1527 (2008).

34. J.S. Iwanczyk, E. Nygard, O. Meirav, J. Arenson, W.C. Barber, N.E. Hartsough, N. Malakhov and J.C. Wessel, "Photon counting energy dispersive detector arrays for x-ray imaging," *IEEE Trans. Nucl. Sci.* 56(3), 535–542 (2009).
35. W.C. Barber, E. Nygard, J.S. Iwanczyk, M. Zhang, E.C. Frey, B.M. Tsui, J.C. Wessel, N. Malakhov, G. Wawrzyniak, N.E. Hartsough and T. Gandhi, "Charge transport in arrays of semiconductor gamma-ray detectors," *SPIE Med. Imaging Int. Soc. Optics Photon.* 725824 (2009).
36. S. Kappler, F. Glasser, S. Janssen, E. Kraft and M. Reinwand, "A research prototype system for quantum-counting clinical CT," *SPIE Med. Imaging Int. Soc. Optics Photon.* 76221Z (2010).
37. S. Kappler, T. Hannemann, E. Kraft, B. Kreisler, D. Niederloehner, K. Stierstorfer and T. Flohr, "First results from a hybrid prototype CT scanner for exploring benefits of quantum-counting in clinical CT," *SPIE Med. Imaging Int. Soc. Optics Photon.* 83130X (2012).
38. R.J. Acciavatti and A.D. Maidment, "A comparative analysis of OTF, NPS, and DQE in energy integrating and photon counting digital x-ray detectors," *Med. Phys.* 37(12), 6480–6495 (2010).
39. H. Chen, B. Cederström, C. Xu, M. Persson, S. Karlsson and M. Danielsson, "A photon-counting silicon-strip detector for digital mammography with an ultrafast 0.18-μm CMOS ASIC," *Nucl. Instrum. Methods Phys. Res. A* 749, 1–6 (2014).
40. H. Chen, B. Cederström, C. Xu, M. Persson, S. Karlsson and M. Danielsson, "Photon-counting spectral computed tomography using silicon strip detectors: A feasibility study," *Phys. Med. Biol.* 55, 1999–2022 (2010).
41. M. Persson, B. Huber, S. Karlsson, X. Liu, H. Chen, C. Xu, M. Yveborg, H. Bornefalk and M. Danielsson, "Energy-resolved $_{540}$ ct imaging with a photon-counting silicon-strip detector," *Phys. Med. Biol.* 59, 6709–6727 (2014).
42. E. Roessl and R. Proksa, "K-edge imaging in x-ray computed tomography using multi-bin photon-counting detectors," *Phys. Med. Biol.* 52, 4679 (2007).
43. J.P. Ronaldson, R. Zinon, N.J.A. Scott, S.P. Gieseg, A.P. Butler, P.H. Butler and N.G. Anderson, "Toward quantifying the composition of soft tissues by spectral CT with Medipix3," *Med. Phys.* 39, 6847–6857 (2012).
44. J. Schlomka, E. Roessl, R. Dorscheid, S. Dill, G. Martens, T. Istel, C. Baumer, C. Herrmann, R. Steadman, G. Zeitler and A. Livne, "Experimentally feasibility of multi-energy photon counting k-edge imaging in pre-clinical computed tomography," *Phys. Med. Biol.* 53, 4031 (2008).
45 P.M. Shikhaliev, "Energy-resolved computed tomography: First experimental results," *Phys. Med. Biol.* 53, 5595 (2008).
46. J. Star-Lack, M. Sun, A. Meyer, D. Morf, D. Constantin, R. Fahrig and E. Abel, "Molybdenum, rhodium, and tungsten anode spectral models using interpolating polynomials with application to mammography," *Med. Phys.* 41, 031916 (2014).
47. Y. Fang and A. Badano, "DQE simulation of a-Se x-ray detectors using ARTEMIS," *Proc. SPIE*, 978314, (2016).
48. R. Swank, "Absorption and noise in x-ray phosphors," *J. App. Phys.* 44, 4199 (1973).
49. A. Ginzburg and C. Dick, "Image information transfer properties of x-ray intensifying screens in the energy range from 17 to 320 keV," *Med. Phys.* 20, 1013 (1993).
50. J.M. Boone and R. Jennings, "Molybdenum, rhodium, and tungsten anode spectral models using interpolating polynomials with application to mammography," *Med. Phys.* 24, 1863 (1997).

6

Synthesizable Inverter-Based Analog Processor for Radiation Detection Read-Out Front Ends

Lampros Mountrichas, Thomas Noulis, and Stylianos Siskos

CONTENTS

6.1 Introduction

Radiation detection is a fundamental technique in various applications like radioactivity control, high-energy physics, space science, and medical applications. The use of application-specific integrated circuit (ASIC) readout systems is gaining importance as the implementation of readout electronics and semiconductor detectors onto the same chip offers enhanced detection sensitivity thanks to improved noise performance [1–5]. Placing the very first stage of the front-end circuit close to the detector electrode allows the noise optimization capacitive matching criterion [6,7] to be satisfied more effectively, unlike the use of relative high-gate-capacitance discrete transistors. The preamplifier semi-Gaussian shaper (S-G shaper) structure is commonly adopted in the design of the aforementioned systems. Radiation events are detected by an inverse biased diode, generating electron–hole pairs proportional to the absorbed energies. A low-noise charge-sensitive preamplifier (CSA) is widely used at the front end due to its low-noise and gain insensitivity to the detector capacitance variations. The CSA output is fed to a pulse-shaping amplifier to optimize the S/N system ratio. The resulting output is a narrow pulse, which is typically sampled by an analog-to-digital converter (ADC) to accommodate further processing.

The majority of the shapers and, in general, of the CSA–shaper readout processing channels that were proposed so far are based on operational amplifiers (OpAmps) and operational transconductance amplifiers (OTAs) [8–14]. In this paper, an advanced and detailed analysis concerning inverter-based shapers for front-end systems is carried out, and novel shaper designs are proposed. Additionally, a complete chain of CSA, shaper, and ADC based on inverter cells is proposed. Inverters as digital blocks are easily described in hardware description language (HDL), significantly reducing the design time, while the whole chain presents low power consumption, high performance, and low silicon real estate. Specifically, a novel design of an inverter-based shaper and an improved low area design of a 10-bit SAR ADC are proposed. Advanced and detailed analyses and extended simulations are performed validating the proposed design.

6.2 Inverter-Based Shaper

Pulse-shaping filters are used to measure the energy of charge particles [15]. The purpose of such filters is to provide a voltage pulse with a height proportional to the energy of the detected particle. A well-known technique is to use a capacitor-resistor (CR) high-pass filter (HPF) and a number (n) of resistor-capacitor (RC) low-pass filters (LPF) to create a CR-RC^n filter (semi-Gaussian [S-G] shaper) [15]. The HPF sets the duration of the pulse by introducing a decay time constant and n LPFs increase the rise time in order to limit the noise bandwidth.

Typically, the shaping amplifier transforms a narrow sensor pulse into a broader pulse with a gradually rounded maximum at the peaking time. Pulse shaping has two conflicting objectives. The first one is to restrict the bandwidth to match the measurement time. A large bandwidth will increase the noise without increasing the signal. The second objective is to constrain the pulse width so that successive signal pulses can be measured without undershoot or overlap (pileup). Reducing the pulse duration increases the allowable signal rate, but at the expense of electronic noise. While designing the shaper, it is necessary to balance these conflicting goals. Many different considerations lead to the compromise that optimum shaping depends on the application [2].

There already are inverter-based CSAs in the literature [16]. The main advantage of an inverter-based amplifier is that it operates in class AB and has a large bandwidth with high slew rate. The main drawbacks are the increased sensitivity to power supply noise and the limited open loop gain, especially when short channel transistors are used. Following that trend, a novel inverter-based synthesizable shaper is analyzed in the following sections. A synthesizable shaper can be described in HDL offering fast implementation, low power consumption, high performance, and low silicon footprint. Unlike the shaper, the CSA cannot be described in HDL because the strict performance characteristics force the use of nonminimum transistor length.

6.2.1 The Complementary Metal-Oxide Semiconductor Inverter as an Amplifier

The use of a complementary metal-oxide semiconductor (CMOS) inverter as an amplifier can be advantageous as it works at low supply voltage and can be automatically synthesized using digital tools. Figure 6.1 shows the direct current (DC) voltage characteristic of the inverter.

By biasing the input around $V_{dd}/2$, when V_{dd} is the supply voltage, the inverter can operate as an amplifier. The operating point $V_{OUT} = V_{dd}/2$ when $V_{IN} = V_{dd}/2$ can approximately be obtained by setting $W_p/W_n \cong \mu_n/\mu_p$ with $L_n = L_p$, where μ_n and μ_p are the respective N-channel Metal-Oxide Semiconductor (NMOS) and P-channel Metal-Oxide Semiconductor (PMOS) mobility, W_n and W_p represent the transistors' channel widths, and L_n and L_p the channel lengths.

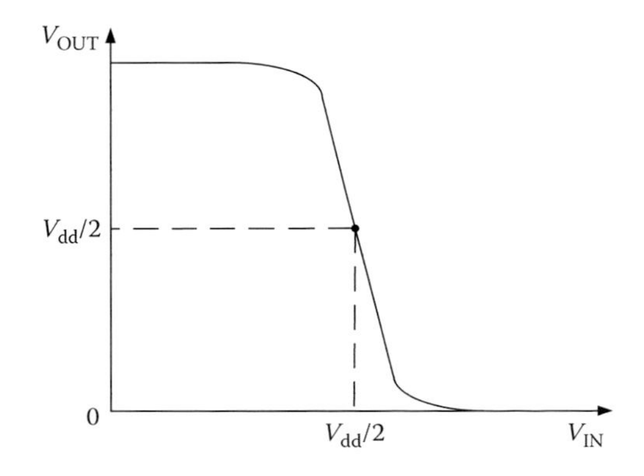

FIGURE 6.1
CMOS inverter input–output.

6.2.2 Second-Order Inverter-Based Shaper

A basic implementation of an inverter-based CR-RC^2 shaper is illustrated in Figure 6.2. It consists of an HPF formed by C_{HPF} and R_{HPF}, two LPFs formed by R_{LPF} and C_{LPF}, and three inverter-based amplifiers.

The first inverter is biased at the middle of its operation region through R_{HPF} so that it works as an amplifier. Based on Figure 6.1, the output of the inverter will be at the middle of the operation region, and thus, the second inverter will also be biased correctly. The same is true for any number of inverters. This biasing scheme is not that robust as mismatch variations slightly alter the biasing point of each inverter, forcing the outputs in either the positive or negative rail.

6.2.3 Robust Second-Order Inverter-Based Shaper over Mismatch and Process Variations

To further enhance the stability over mismatch variations without reducing the shaper gain, the self-biased inverter-based CR-RC^2 shaper illustrated in Figure 6.3 is proposed. The shaper consists of an HPF formed by C_{HPF} and the R_{HPF}, two LPFs formed by R_{LPF} and C_{LPF}, and three inverter-based amplifiers. A low-frequency negative feedback loop is closed around the shaper, creating an appropriate bias voltage. The HPF elements act as an LPF loop from the output to the input, allowing only the low-frequency component to reach the input and bias the inverter chain.

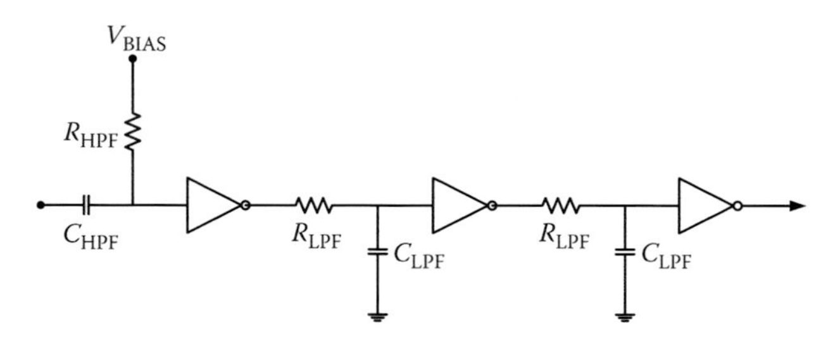

FIGURE 6.2
CR-RC^2 inverter-based shaper.

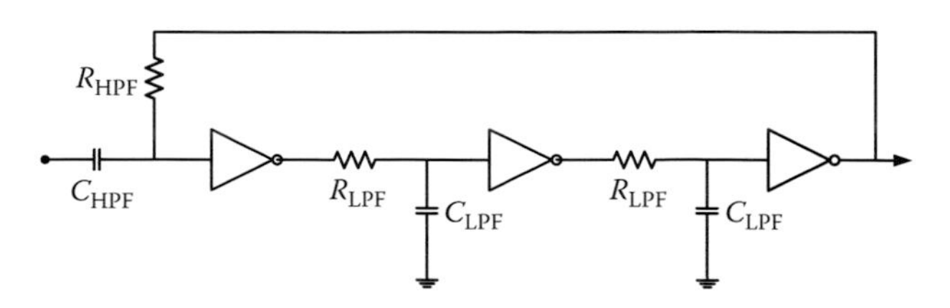

FIGURE 6.3
CR-RC^2 self-biased inverter-based shaper.

TABLE 6.1

Shaper Performance over Process and Mismatch Variations

	Process Variation			Mismatch Variation		
Biasing Type	**Mean Output**	**Sigma Variation**	**Sigma Variation (%)**	**Mean Output**	**Sigma Variation**	**Sigma Variation (%)**
Single bias (Figure 6.2)	120.10 mV	32.35 mV	26.9	2.2 mV	11.1 mV	–
Self-biased (Figure 6.3)	110.10 mV	28.50 mV	25.9	114.8 mV	5.6 mV	5.1

This type of shaper is much more robust over mismatch and process variations and in addition doesn't suffer the reduced gain. In case more amplification is needed, additional inverters can be added. These inverters must be added in pairs of two in order to maintain the negative feedback of the biasing loop. Depending on the needed amplification, the shaper output can be taken from either of these inverters.

Table 6.1 summarizes the performance of the presented shapers across process and mismatch variations. The differentiator time constant is set at about 160 μs (C_{HPF} = 100 pF and R_{HPF} = 1.6 MΩ) and the integrator time constant at 800 ns (C_{LPF} = 5 pF and R_{LPF} = 160 kΩ). Depending on the application, these can change. The bandwidth (BW) of the shaper is about 200 kHz. The Table 6.1 results are for a 24 fC input charge, translating to about 4 mV at the shaper input (6 pF CSA feedback capacitance). Both biasing schemes perform similarly over process variations, having approximately the same variation, but only the self-biased shaper (Figure 6.3) performs well over mismatch variations. Due to the small transistor size, mismatch variations are large, affecting the bias point of each inverter significantly. For this reason, the single-bias shaper (Figure 6.2) is inoperable under mismatch variations, and the outputs of the inverters settle in either the positive or negative power rail. Contrarily, the self-biased (Figure 6.3) shaper auto-regulates the operating point providing robust operation across both process and mismatch variations.

6.2.3.1 Shaper Dynamic Range Optimization

In terms of the dynamic range performance, the shaper gain can be reduced by introducing a voltage divider to the LPF. In Figure 6.4, a voltage divider by 2 is introduced after each LPF. The total gain is reduced by 4, increasing the input dynamic range. To maintain the same LPF RC constant, the values of the resistors are doubled, as in the small signal equivalent circuit, the additional resistors are in parallel with the filter resistance. The common node of the additional resistors is tied to a shorted biasing inverter.

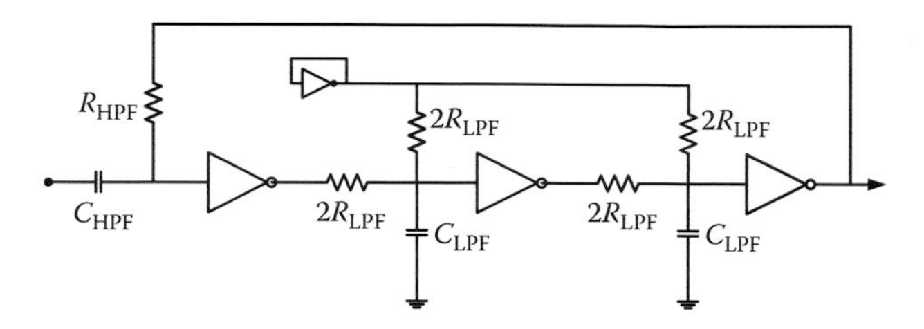

FIGURE 6.4
CR-RC^2 shelf-biased shaper with improved dynamic range.

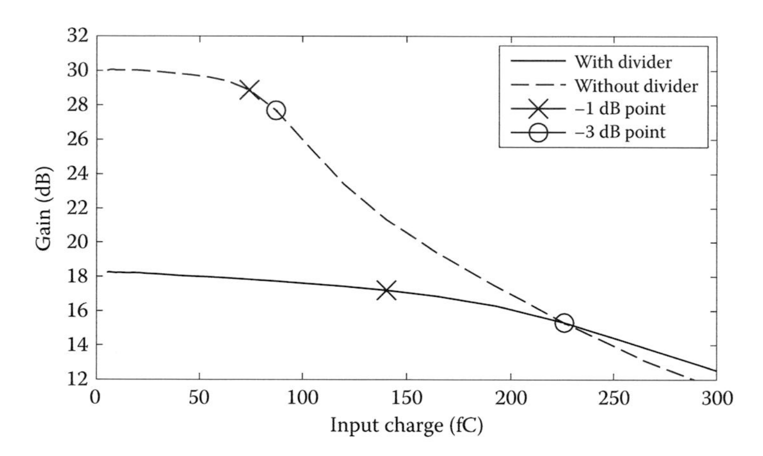

FIGURE 6.5
Improved dynamic range of CR-RC^2 shelf-biased shaper (2 pF detector capacitance).

Figure 6.5 illustrates the linearity improvement with the addition of the voltage divider. The gain is reduced 12 dB. The –1 dB compression point when no divider is used is at about 75 fC (for a 6 pF CSA feedback capacitance). The inclusion of the divider enhances the linearity of the shaper, resulting in a –1 dB point at 140 fC.

6.2.3.2 Shaper Noise Optimization

If a slight overshoot of the output voltage can be tolerated, which is the case when just the peak of the pulse is sampled, it is possible to improve the shaper noise and reduce the power consumption at the same time. Instead of tying the divider resistors to a biasing inverter, the inverter can be removed, and the resistors can be tied together forming a loop around the second inverter (Figure 6.6). The voltage at the resistors' common terminal remains relatively stable with a mean value at about the second inverter bias voltage, but due to a small disturbance when the event pulse arrives (~2 mV), the output exhibits some overshoot. The removal of the additional inverter offers a

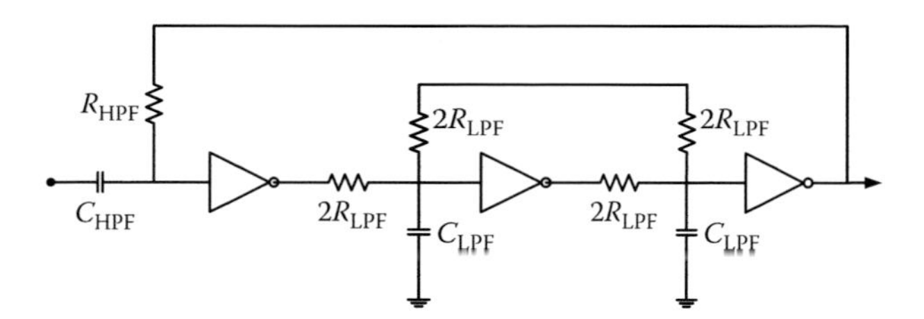

FIGURE 6.6
Shaper with improved noise characteristics.

25% improvement on the current consumption. Furthermore, the noise of the shaper is reduced due to the negative feedback. For a bandwidth of 1 kHz to 200 kHz, the rms noise of the shaper of Figure 6.4 is 103.7 μV, whereas the root mean square (rms) noise of the shaper of Figure 6.6 is 49.67 μV. Figure 6.7 illustrates the effect the removal of the extra bias inverter has on the output waveform.

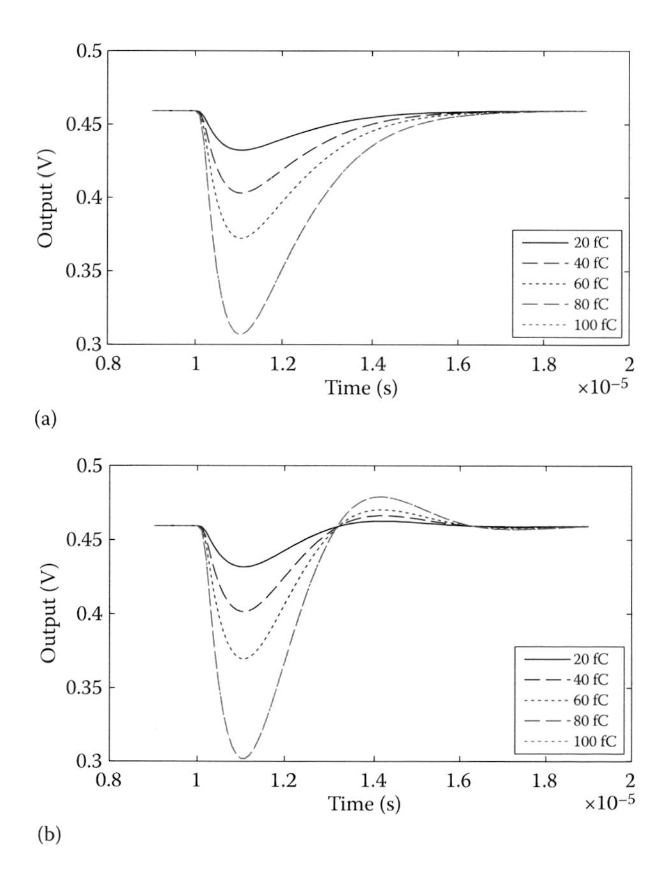

FIGURE 6.7
Shaper output (a) with inverter and (b) without inverter.

6.2.4 Resistor Implementation

To avoid using a large-footprint passive resistor, a metal-oxide-semiconductor field-effect transistor (MOSFET)–based resistor can be used. Typically, an active resistor would be implemented as a transistor, either p-channel or n-channel, with a controlled gate voltage defining the equivalent resistance (Figure 6.8a).

An alternative resistor implementation consists of two parallel diode-connected transistors of either n or p type. Figure 6.8b illustrates such a resistor using p-type MOSFETs. Since the current through those transistors is minimal, the transistors are biased in the subthreshold region, offering large resistance at the expense of large area, compared to the gate-driven (GD) variant. These resistors can be calibrated by manipulating the bulk voltage of the transistors.

While at first glance, the GD resistors seem the appropriate choice for the application since smaller transistors can be used for the same resistance, detailed investigation reveals numerous advantages for the bulk-driven (BD) circuit. Figure 6.9 illustrates the resulting equivalent resistance over different

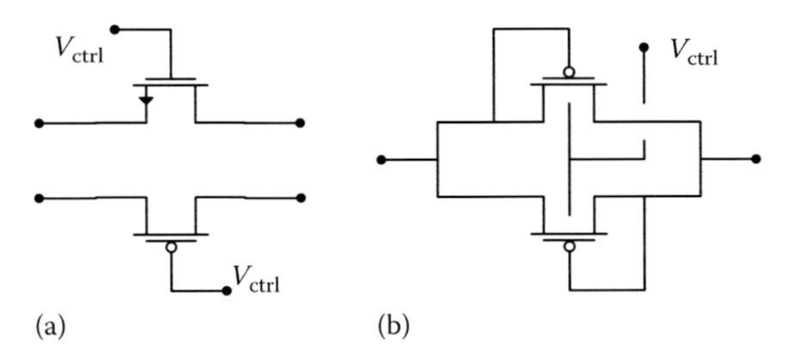

FIGURE 6.8
MOS resistors. (a) Gate-driven. (b) Bulk-driven.

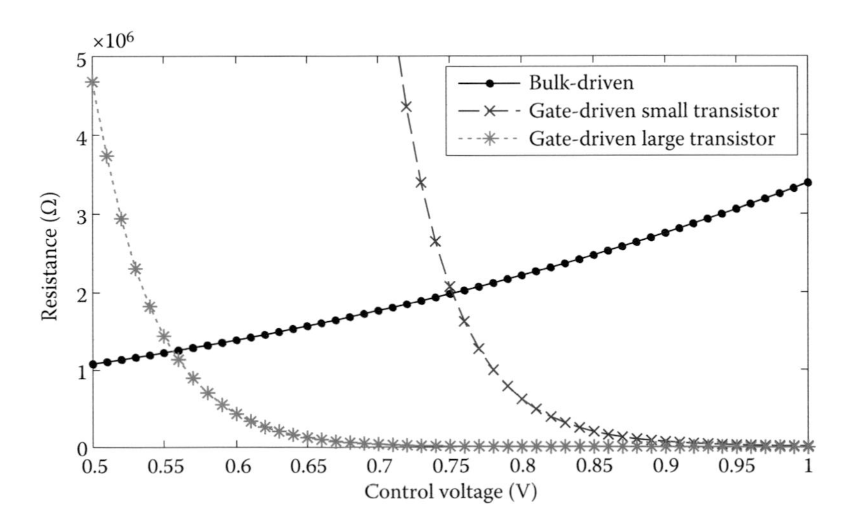

FIGURE 6.9
Resistance versus control voltage for a 500 mV input voltage.

control voltages for a DC input of 500 mV. Two GD plots are given, one for small transistors and the other for large transistors with equal footprint as the BD resistor of the same resistance. The BD circuit provides better control of the resistance, while the GD circuit is oversensitive in controlling voltage variations. This results in resistance variation due to noise.

Furthermore, as depicted in Figure 6.10, the BD circuit provides a relatively constant resistance for different input voltage levels, for comparable resistances. This is crucial if fairly constant RC is to be maintained even in the presence of various height pulses. In addition, the size advantage of the GD resistor is diminished in light of the overly large mismatch variations, as seen in the mismatch variation data provided in Table 6.2 for the BD and

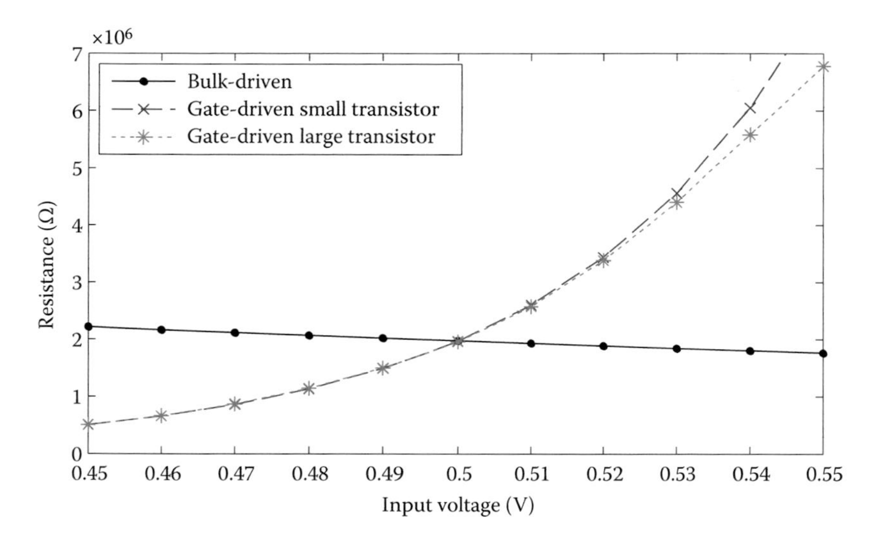

FIGURE 6.10
Resistance versus input voltage.

TABLE 6.2

Mismatch Variation per Active Resistor Type

Type	Resistance (Ω)	Area (μm²)	Sigma Variation (Ω)	Sigma Variation (%)
Passive polysilicon resistor P+POLY without salicide	160 k	28.8	205	0.13
BD PMOS	160 k	48	1.281 k	0.8
Small GD NMOS	160 k	0.12	24.86 k	15.5
Large GD NMOS	160 k	48	1.36 k	0.85
Passive resistor P+POLY without salicide	1.6 M	288	690	0.04
BD PMOS	1.6 M	4.8	37.4 k	2,3

Note: P+POLY, polysilicon resistor made by p+ source/drain ion implantation.

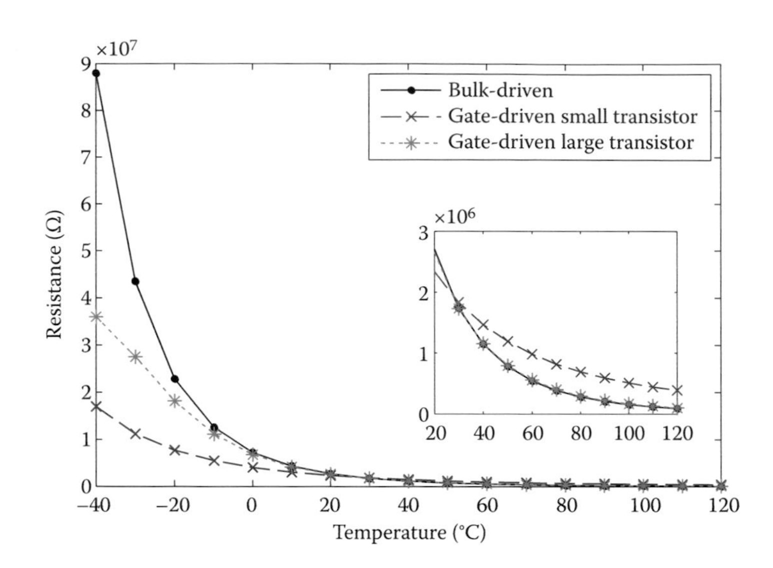

FIGURE 6.11
Resistance versus temperature.

GD resistors. A 160 kΩ resistor using a small-footprint GD resistor exhibits excessive variation. On the other hand, a GD and a BD resistor of equal footprint have similar, relatively low variation. The passive resistor occupies a smaller area and exhibits the smallest variation. For the large feedback resistor, the passive resistor occupies an excessive area, making the BD resistor more appropriate.

An advantage of the GD resistor is its temperature stability over the BD circuit (Figure 6.11), but even this is negligible in the 0°C to 120°C temperature range, especially when large-footprint transistors are used.

Finally, Figure 6.12 illustrates the shaper output for various input pulses for BD (Figure 6.12a), large-area GD (Figure 6.12b), and passive resistors (Figure 6.12c). Both BD and GD resistor-based shapers exhibit varying RC constant as active resistors cannot replace the stability over pulse height that passive resistors offer. Using GD resistors, the RC constant increases with increasing input height, and the output pulse compresses, while using BD resistors, the RC constant decreases, and although the output pulse maintains good shape, the linearity is impaired. For the designed system, passive resistors are used for the LPF, and an active BD resistor is used for the large feedback resistor and the loop around the second inverter as well.

6.2.5 High-Speed Power-Down System

A high-speed power down/power up system is included in this design. That way, power pulsing mode can be used to limit the current consumption [17]. Powering down the shaper of Figure 6.6 is achieved by switching the

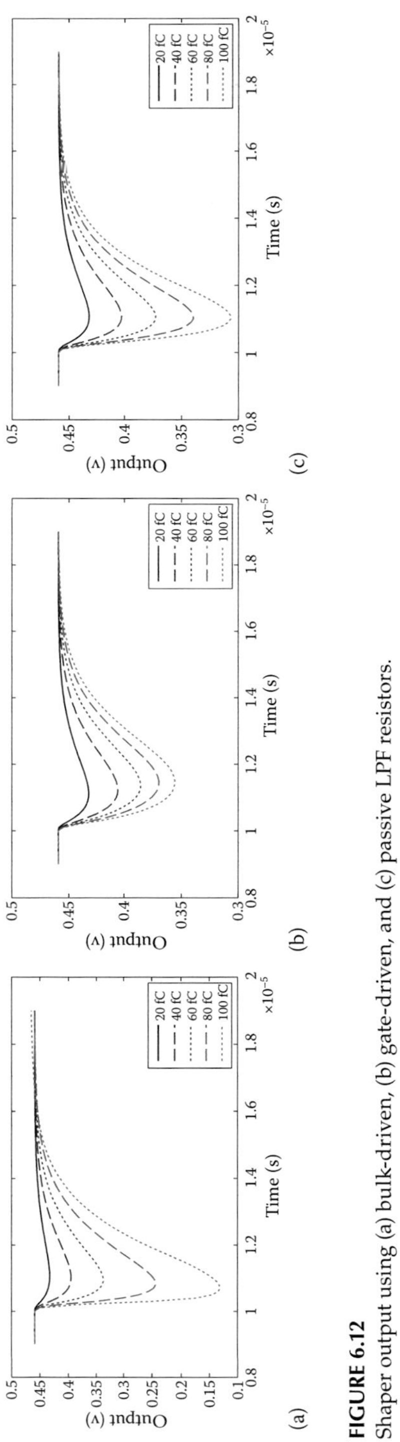

FIGURE 6.12
Shaper output using (a) bulk-driven, (b) gate-driven, and (c) passive LPF resistors.

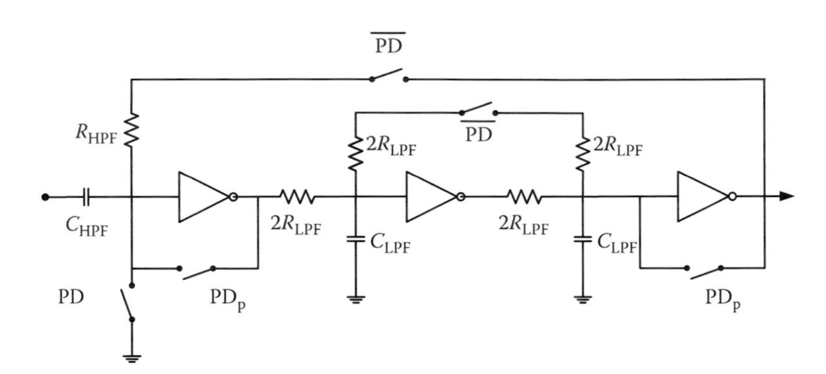

FIGURE 6.13
Complete CR-RC[2] inverter-based shaper with improved biasing.

first inverter to ground potential. In addition, the feedback loops are broken to avoid current flowing through the high-pass resistor and the resistive divider around the second inverter, as shown in Figure 6.13.

When powering up, the shaper takes too much time to settle due to the high RC constant of the HPF and extra steps must be taken if high-speed power switching is needed. A loop is formed around the first and last inverter momentarily (PD_p) in order to precharge the HPF capacitance, setting the capacitor at the inverters' bias point voltage. Using the proposed power-down (PD) scheme, the settling time to 1 mV from the final output voltage is 10.5 μs as opposed to the 114 ms needed if the HPF capacitance is not precharged, allowing the shaper to remain switched off for longer periods, saving extreme amounts of power.

6.2.5.1 Power-Down Signal Generator

The PD generator is illustrated in Figure 6.14a. An NOR operation of two delayed PD signals generates the needed PD pulse only on the negative edge of the external PD signal. It is beneficial to minimize the duration of the generated PD pulse but still turn off the inverters quickly. If the pulse is too brief, the shaper takes longer to power up. A longer pulse reduces the maximum power-down/up frequency by forcing the inverters to remain in the transition region for longer than necessary.

The generator can be optimized to achieve precise timing or reduced area. The generator of Figure 6.14a is designed to reduce the area. For that reason, Miller capacitance amplification is applied at the delay capacitors. Each delay capacitor closes a loop around an inverter. Consequently, the effective capacitance is amplified by the gain of the inverter, $C_{TOTAL} = (1 + A_{INV}) \times C_{DELAY}$. Figure 6.14b illustrates the PD generator waveforms. Ext_PD is the external PD signal, Nor_1 and Nor_2 are the inputs of the NOR gate, and PD is the generated PD pulse.

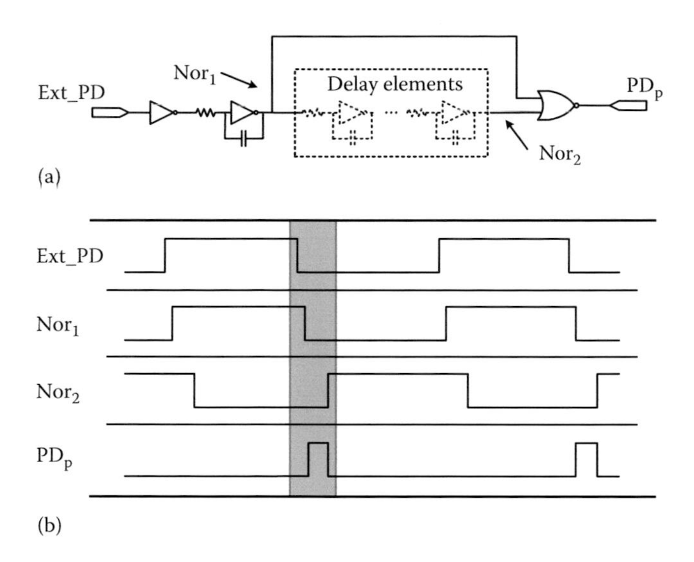

FIGURE 6.14
(a) Power-down signal generator. (b) Power-down waveforms.

6.2.6 Inverter-Based Reduced-Area Successive-Approximation ADC

The ADC used in this application is a modified successive-approximation (SA) ADC, based on a charge redistribution Digital-to-analog converter (DAC), occupying approximately half the area of typical SA ADCs of the same type and resolution.

A typical SA [18] analog/digital (A/D) converter is illustrated in Figure 6.15a. A charge redistribution DAC is the main part of the converter and uses inverters as switches for the DAC capacitors. The comparator is based on Ref. [19] and uses synthesizable inverters. In Figure 6.15b, the proposed converter is illustrated. The proposed converter is missing the MSB capacitor, downsizing the charge redistribution DAC by 2. To achieve the same step size as before, the last capacitor is replaced by a capacitance divider by 2 of the same total capacitance. In effect, now the MSB adds/subtracts $V_{ref}/4$ instead of $V_{ref}/2$, and a modified SA algorithm is needed. An extra comparator is used to accommodate the modified SA algorithm. The comparator decides whether the input is higher or lower than half of the reference voltage. If the input is lower than $V_{ref}/2$, then a typical SA loop begins. If $V_{in} > V_{ref}/2$, then a modified SA loop begins.

6.2.6.1 *Modified SA Algorithm*

The output value of a 4-bit charge redistribution DAC is given by.

$$-V_{in} + \left(\frac{V_{ref}}{2} b_0 + \frac{V_{ref}}{4} b_1 + \frac{V_{ref}}{8} b_2 + \frac{V_{ref}}{16} b_3 + \frac{V_{ref}}{16} b_c \right) \tag{6.1}$$

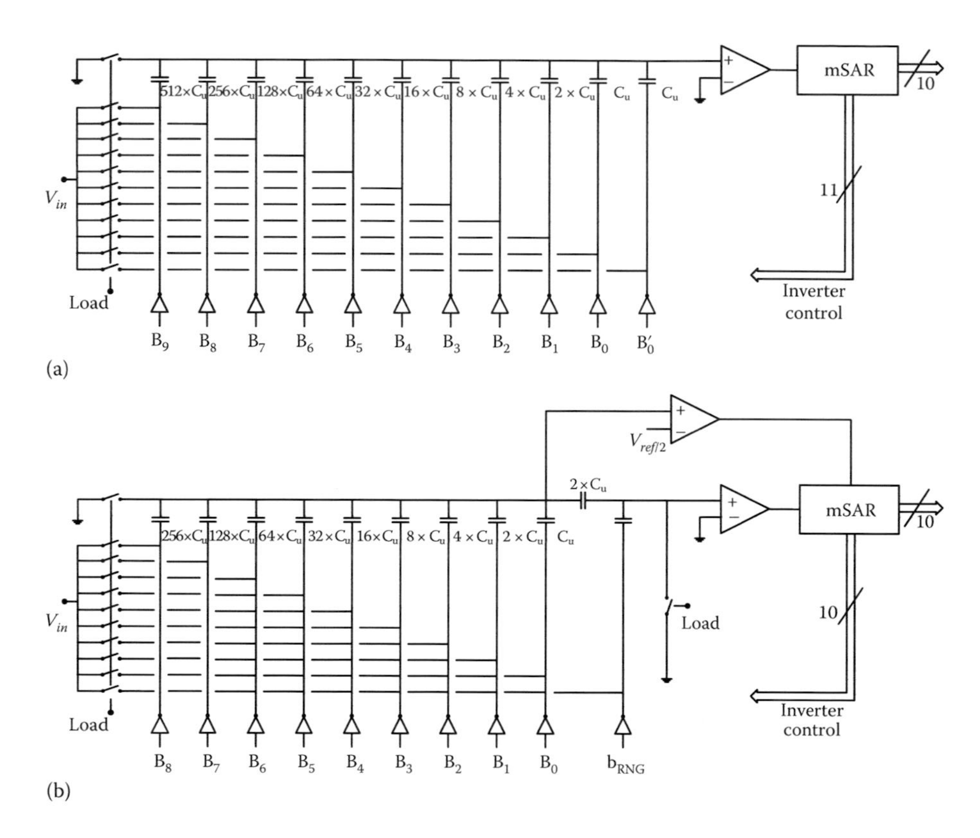

FIGURE 6.15
(a) Typical 10-bit SAR ADC. (b) Modified 10-bit SAR ADC.

It is evident that the MSB capacitor has an active role only when $V_{in} > V_{ref}/2$. In that case, the bottom plate of the capacitor is switched to V_{ref} ($b_0 = 1$), adding $V_{ref}/2$ to the output. If $V_{in} < V_{ref}/2$, the bottom plate is switched to ground ($b_0 = 0$). In that case, Equation 6.1 can be rewritten as follows, alleviating the need for the MSB capacitor:

$$-V_{in} + \left(\frac{V_{ref}}{4}b_1 + \frac{V_{ref}}{8}b_2 + \frac{V_{ref}}{16}b_3 + \frac{V_{ref}}{16}b_c\right), \quad V_{in} < V_{ref}/2 \tag{6.2}$$

By manipulating the initial value of the capacitors, we can skip the MSB capacitor switching even when $V_{in} > V_{ref}/2$.

By substituting b_0 with $1-\overline{b}_0$ in Equation 6.1, we end up with.

$$V_{ref} - V_{in} - \frac{V_{ref}}{2}\overline{b}_0 - \frac{V_{ref}}{4}\overline{b}_1 - \frac{V_{ref}}{8}\overline{b}_2 - \frac{V_{ref}}{16}\overline{b}_3 - \frac{V_{ref}}{16}\overline{b}_c \tag{6.3}$$

As stated previously, for $V_{in} > V_{ref}/2$, $b_0 = 1 = \overline{b}_0 = 0$ and Equation 6.3 can be rewritten as

$$V_{ref} - V_{in} - \left(\frac{V_{ref}}{4}\overline{b}_1 - \frac{V_{ref}}{8}\overline{b}_2 - \frac{V_{ref}}{16}\overline{b}_3 - \frac{V_{ref}}{16}\overline{b}_c \right), \; V_{in} > V_{ref}/2 \tag{6.4}$$

The b_0 term is missing from Equation 6.4, and thus, the MSB capacitor is no longer needed. The output of the charge redistribution DAC is given by.

$$V_0 = \begin{cases} -V_{in} + \dfrac{V_{ref}}{4}b_1 + \dfrac{V_{ref}}{8}b_2 + \dfrac{V_{ref}}{16}b_3, \; V_{in} < V_{ref}/2 \\ V_{ref} - V_{in} - \dfrac{V_{ref}}{4}\overline{b}_1 - \dfrac{V_{ref}}{8}\overline{b}_2 - \dfrac{V_{ref}}{16}\overline{b}_3, \; V_{in} > V_{ref}/2 \end{cases} \tag{6.5}$$

The case described by Equation 6.4 can be implemented by switching all the bottom plates of the capacitors to 1 V (instead of ground) after the initial sampling of the input voltage. Furthermore, for each SA step, the bottom plate of the appropriate capacitor is switched to ground, subtracting the appropriate value from the output.

It must be noted that while the b_0 variable is omitted, the b_1 variable, and the associated capacitor, must still change the voltage by $V_{ref}/4$. If we simply remove the b_0 MSB capacitor without introducing further changes, then the b_1 capacitor would add/subtract $V_{ref}/2$. By replacing the last capacitor with two in series capacitors of twice the unitary value and reading the output at their common terminal, the charge step is effectively divided by 2, while at the same time, the total capacitance of the DAC remains the same. This means that after removal of the MSB capacitor, switching of the b_1 capacitor yields a change of $V_{ref}/4$ at the new output node. The proposed ADC is more sensitive to the parasitic capacitance at the DAC output and thus is limited to medium resolution. Figure 6.16 illustrates the final 10-bit SAR ADC circuit. It is possible to remove the extra comparator if the speed of conversion is not critical. In this application, the extra comparator is omitted, and the SAR comparator is reused. A basic inverter is used as a comparator [19] so that a large part of the ADC can be described in HDL.

Two examples of the modified SAR operation are given in Figures 6.17 and 6.18. For these examples, a 4-bit ADC is assumed with an Least significant bit (LSB) of 62.5 mV.

Figure 6.17 illustrates the operation when $V_{in} = 410$ mV $< V_{ref}/2$. In that case, after sampling the input voltage, the bottom plates of the capacitors are set at ground potential. The output is $-V_{in}$. Both ADCs switch b_0 to decide upon the next step. For this first comparison, the modified SAR reads the voltage to the left of the capacitance divider and decides whether the input is larger or smaller than $V_{ref}/2$. In this case, it is, and the range bit, b_{RNG}, remains 0.

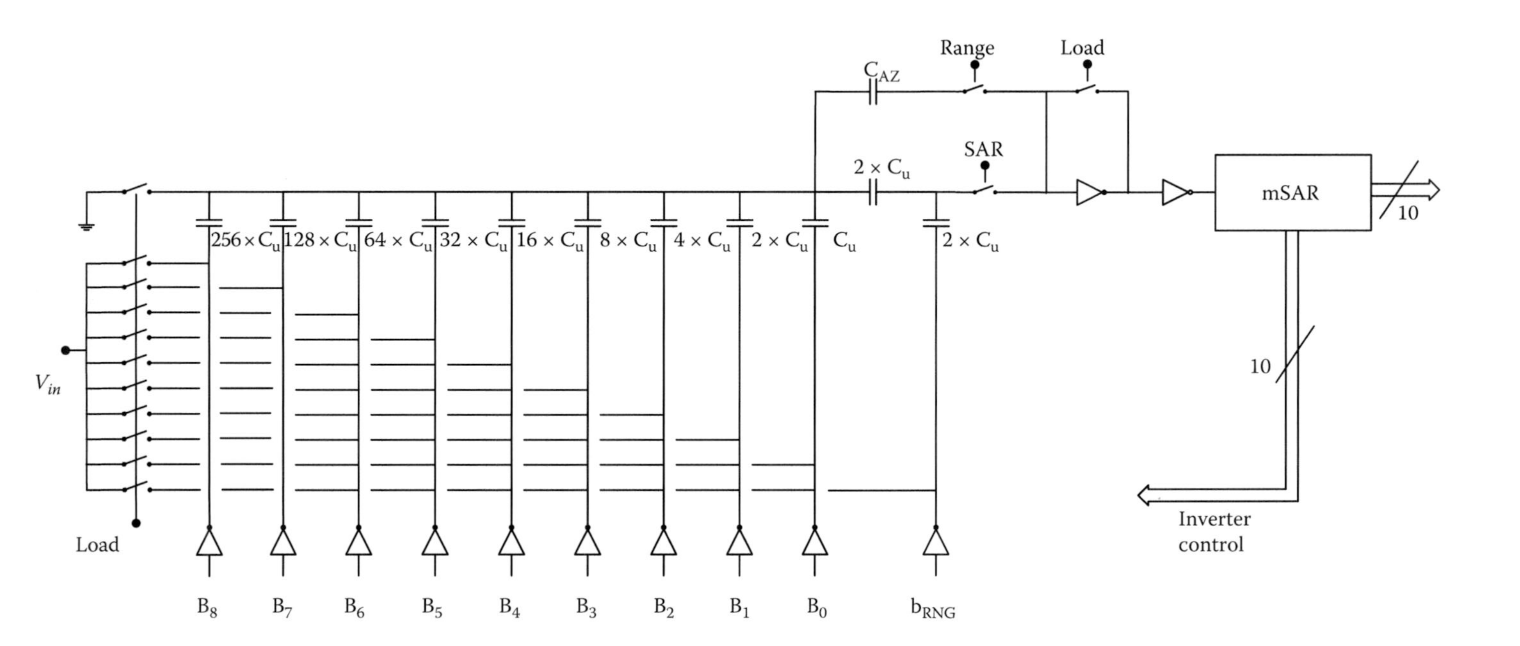

FIGURE 6.16
Modified 10-bit SAR ADC with comparator reuse.

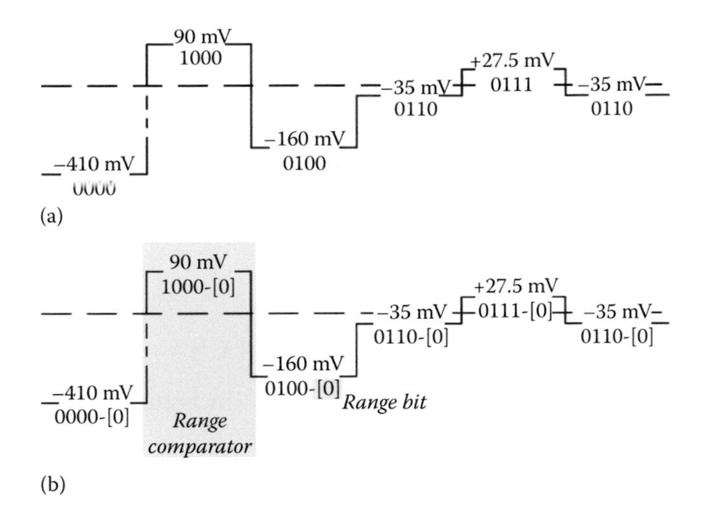

FIGURE 6.17
(a) 410 mV input example, typical 4-bit SAR. (b) Modified 4-bit SAR.

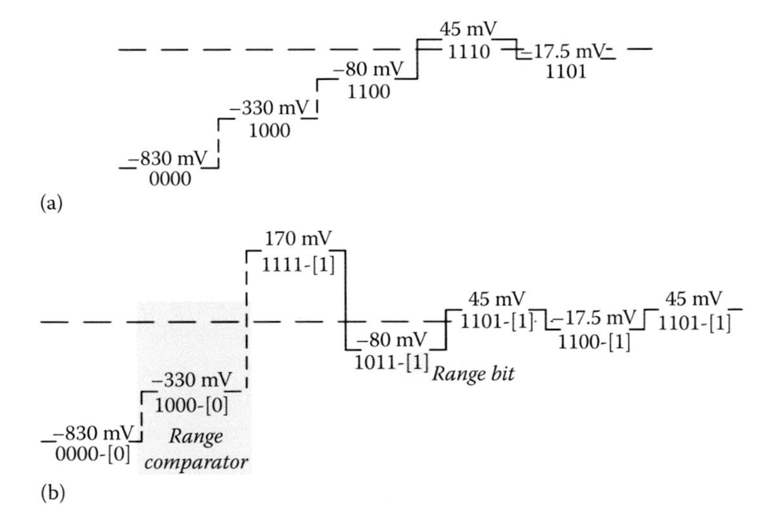

FIGURE 6.18
(a) 830 mV input example, typical SAR. (b) Modified SAR.

The typical SAR then switches b_1. The modified SAR switches b_0 again but reads the output at the capacitance divider. For the next steps, only the value after the divider is read.

Figure 6.18 illustrates the operation when $V_{in} = 830$ mV > $V_{ref}/2$. Again both ADCs switch b_0. Since the output is lower than $V_{ref}/2$, the range bit is set to V_{ref} (1 V), and the modified SAR starts the "inverted" loop. The bottom plates of the capacitors are set to V_{ref}, and the DAC output is $1 - V_{in}$. Then b_0 is switched again, and the output is read at the capacitance divider. The steps to 170 mV (1111-[1]) and the final step to 45 mV (1101-[1]) are not necessary; they are included for clarity.

In both cases, the final value of the ADC is the same. The proposed SAR provides the same resolution while occupying approximately half the area.

6.3 Simulation Results of Digital-Based CSA–Shaper System

The complete inverter-based channel is shown in Figure 6.19. It utilizes an inverter-based CSA followed by the shaper, discussed in Section 6.2.2. The shaped pulse is sampled at the capacitance bank of the modified SAR and held at the appropriate time after the arrival of a pulse by the use of a typical fast shaper/peak detector circuit [17].

The complete system was designed in TSMC 65 nm CMOS and simulated with CADENCE© SPECTRE©. The design specifications concern a radiation detector of 2 pF, 10 pA leakage current, and 125,000 e^- collected charge per event, and the time needed for the collection of 90% of the total charge is about 300 ns. In terms of the CSA–shaper design specifications, the shaper order is 2, the peaking time is 1 μs, and the temperature is 25°C. Unless otherwise noted, the shaper inverters have NMOS transistors of 120 nm width and 60 nm length. The PMOS transistors have a width of 240 nm and length of 60 nm. The shaper HPF and LPF have a time constant of 160 μs and 800 ns, respectively. The BW is 200 kHz. The modified 10-bit SAR ADC operates with a clock of 10 MHz, completing a conversion in 1.3 μs.

Figure 6.20 illustrates the shaper's gain in decibels for various input charges. The shaper provides a –1 dB gain up to a 140 fC input and –3 dB gain for about 230 fC input.

Figure 6.21 illustrates the equivalent noise charge (ENC) over different input pulse heights for the circuits shown in Figures 6.4 and 6.6. As stated previously, by omitting the biasing inverter, the ENC improves. Specifically, the ENC decreases by about 50%.

To investigate the effect of the shaper inverter transistor width (standard cells with various driving strength), the widths of both NMOS and PMOS transistors were swept. By increasing the width of the transistors, the shaper current increases linearly, (Figure 6.22). As expected, the noise decreases, and in return, the total ENC decreases with increased transistor width (Figure 6.23).

As ENC follows a $1/x$ behavior, the ENC and DC current product stays relatively constant for different shaper inverter transistor widths (Figure 6.24), and thus, both ENC and power consumption can be traded almost linearly.

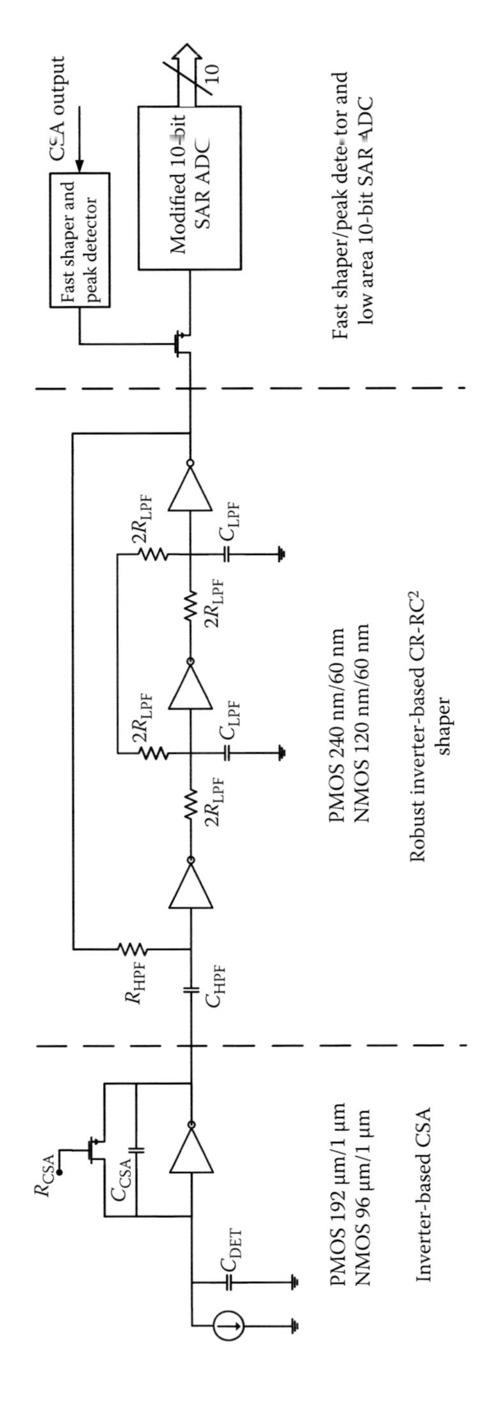

FIGURE 6.19
Complete inverter-based channel.

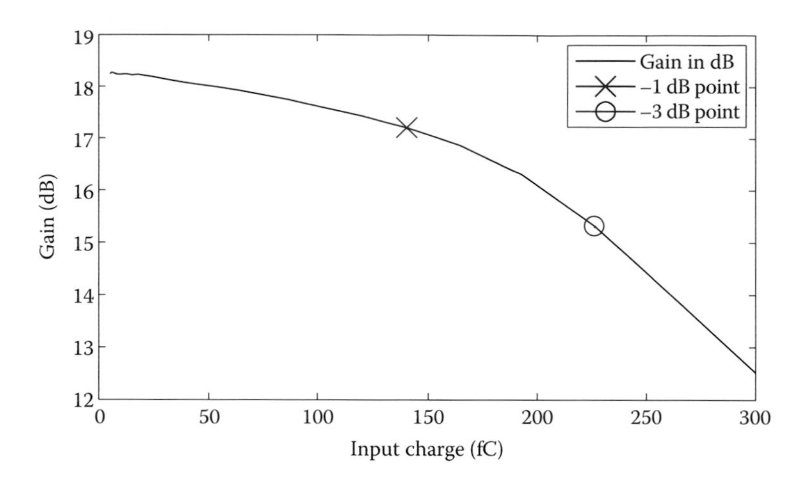

FIGURE 6.20
Detector linearity.

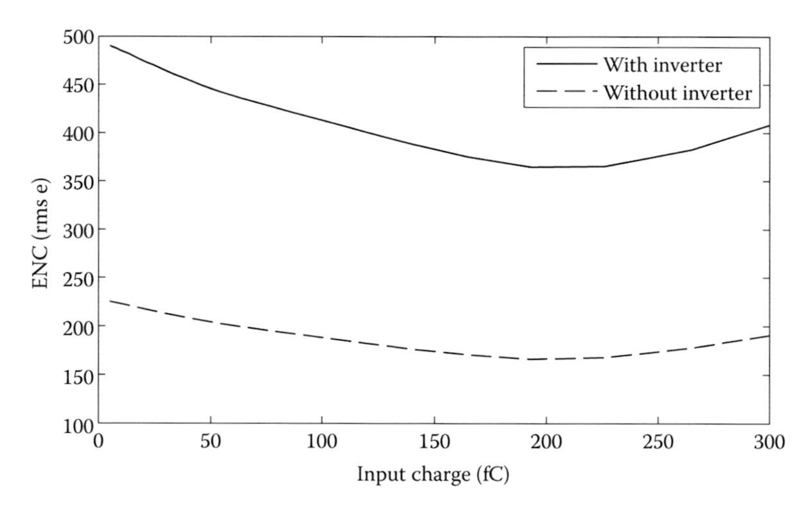

FIGURE 6.21
Detector ENC versus input pulse.

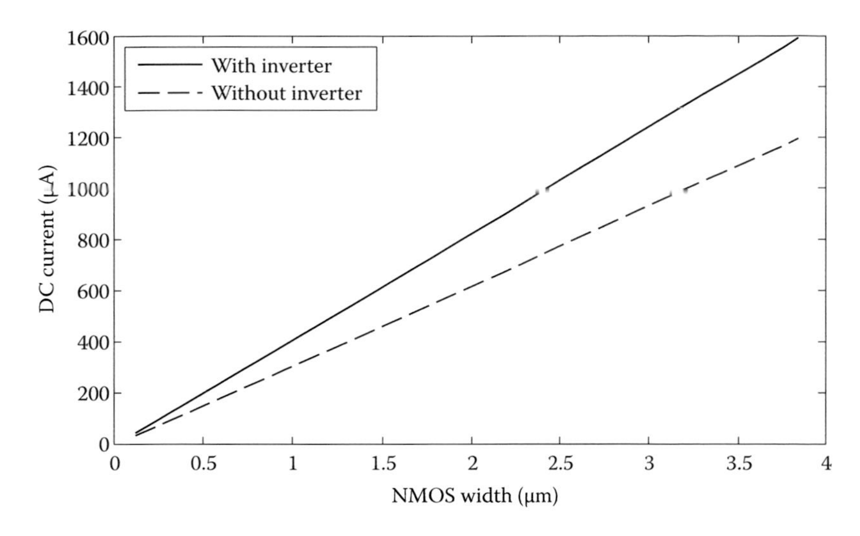

FIGURE 6.22
Shaper DC current versus shaper inverter width.

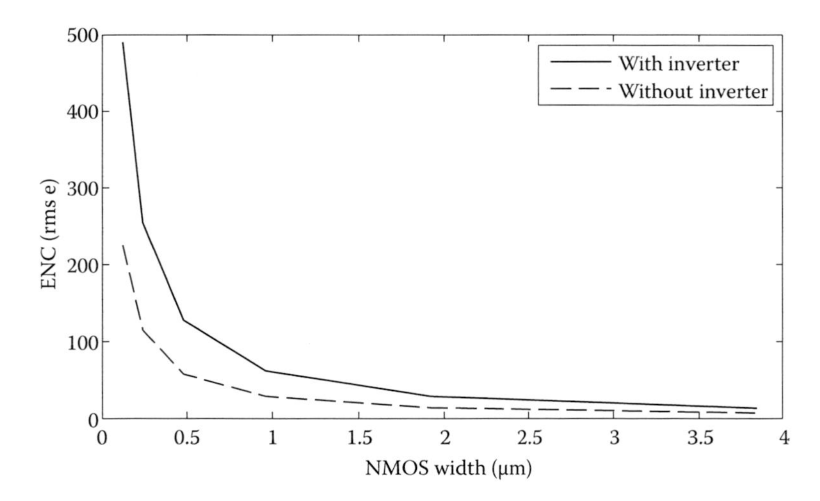

FIGURE 6.23
Detector ENC versus shaper inverters width.

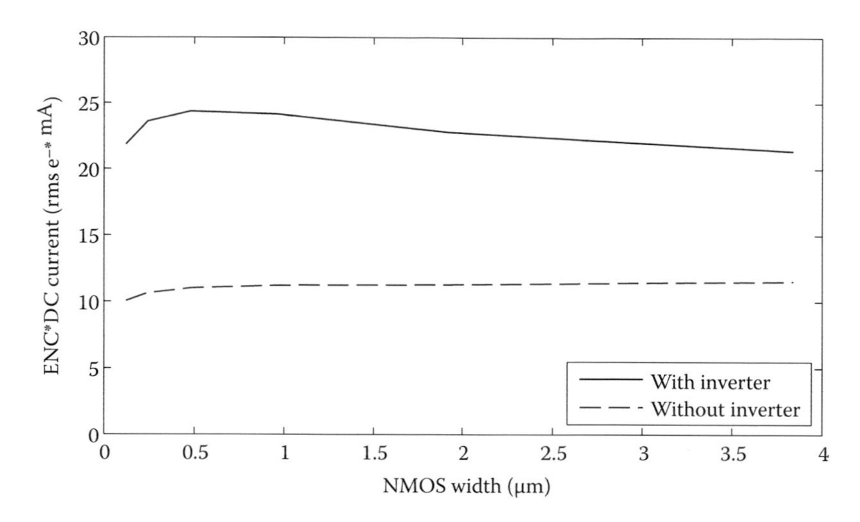

FIGURE 6.24
Detector ENC and shaper DC current product versus shaper inverter width.

The current consumption of the shaper is 33 μA. The average current consumption of the ADC is about 80 μA and, if operated at 10 MHz, can complete a conversion in 1.3 μs. The CSA current consumption is 675 μA. Assuming the same event train as in Ref. [17] with an idle time of 198 ms, the average current consumption of the whole system consumes 7.1 μW when operating at 1 V.

6.4 Conclusions

A synthesizable radiation readout system for radiation detector ASICs is presented. A new design approach to implement high-performance radiation detection processing channels is provided, implementing analog signal processing using digital cells and therefore optimizing die area and overall design cycle time. Custom design and automated design methodologies can be applied since the S-G shaper and an ADC is based on inverter cells. The CSA does not use minimum-length transistors and as such requires custom design.

Advanced design analysis is performed over optimization of mismatches and process variations, over the achievable dynamic range and total output noise, and in relation to resistive passive element replacement using BD and GD FET-based structures. Furthermore, utilizing the switching operating mode of digital logic design, a power-down architecture is implemented together with a respective signal generator structure in order to optimize the power consummation of the readout system. This way, the shaper remains switched off, saving extreme amounts of power while it is enabled according to the detector event's time constraints and the related peaking time specifications.

The proposed inverter-based self-biased shaper achieves low power consumption and significantly low silicon area. In addition, the modified SAR ADC that completes the full readout analog processor occupies approximately half the area of typical SAR ADCs.

References

1. P. Lechner, S. Eckbauer, R. Hartmann, S. Krisch, D. Hauff, R. Richter et al., Silicon drift detectors for high resolution room temperature x-ray spectroscopy, *Nucl. Instruments Methods Phys. Res. Sect. A Accel. Spectrometers, Detect. Assoc. Equip.* 377, pp. 346–351. doi:10.1016/0168-9002(96)00210-0, 1996.
2. L. Ratti, M. Manghisoni, V. Re, V. Speziali, Integrated front-end electronics in a detector compatible process: Source-follower and charge-sensitive preamplifier configurations, in: R.B. James (Ed.), *Int. Symp. Opt. Sci. Technol.*, International Society for Optics and Photonics, pp. 141–151. doi:10.1117/12.450752, 2001.
3. V. Radeka, P. Rehak, S. Rescia, E. Gatti, A. Longoni, M. Sampietro et al., Design of a charge sensitive preamplifier on high resistivity silicon, *IEEE Trans. Nucl. Sci.* 35, pp. 155–159. doi:10.1109/23.12696, 1988.
4. J.C. Lund, F. Olschner, P. Bennett, L. Rehn, Epitaxial n-channel JFETs integrated on high resistivity silicon for x-ray detectors, *IEEE Trans. Nucl. Sci.* 42, pp. 820–823. doi:10.1109/23.467786, 1995.
5. P. Rehak, S. Rescia, V. Radeka, E. Gatti, A. Longoni, M. Sampietro et al., Charge-sensitive preamplifier for integration on silicon radiation detectors: First experimental results, *Electron. Lett.* 25, 1057. doi:10.1049/el:19890706, 1989.
6. Z.Y. Chang, W. Sansen, Effect of noise on the resolution of CMOS analog readout systems for microstrip and pixel detectors, *Nucl. Instruments Methods Phys. Res. Sect. A Accel. Spectrometers, Detect. Assoc. Equip.* 305, pp. 553–560. doi: 10.1016/0168-9002(91)90156-K, 1991.
7. W.M.C. Sansen, Z.Y. Chang, Limits of low noise performance of detector readout front ends in CMOS technology, *IEEE Trans. Circuits Syst.* 37, pp. 1375–1382. doi:10.1109/31.62412, 1990.
8. J.C. Santiard, W. Beusch, S. Buytaert, C.C. Enz, E. Heijne, P. Jarron et al., Gasplex: A low noise analog signal processor for readout of gaseous detectors, in *Adv. Detect. Proceedings*, 6th Pisa Meet. La Biodola, Italy, May 22–28, 1994, http://inspirehep.net/record/382290?ln=en (accessed March 21, 2016), 1994.
9. T. Noulis, C. Deradonis, S. Siskos, G. Sarrabayrouse, Programmable OTA based CMOS shaping amplifier for x-rays spectroscopy, in *2006 Ph.D. Res. Microelectron. Electron.*, IEEE, pp. 173–176. doi:10.1109/RME.2006.1689924, 2006.
10. T. Noulis, C. Deradonis, S. Siskos, G. Sarrabayrouse, Novel fully integrated OTA based front-end analog processor for x-rays silicon strip detectors, in *MELECON 2006–2006 IEEE Mediterr. Electrotech. Conf., IEEE*, pp. 47–50. doi: 10.1109/MELCON.2006.1653032, 2006.

11. M. Pedrali-Noy, G. Gruber, B. Krieger, E. Mandelli, G. Meddeler, W. Moses et al., PETRIC-a positron emission tomography readout integrated circuit, *IEEE Trans. Nucl. Sci.* 48, pp. 479–484. doi:10.1109/23.940103, 2001.
12. B. Krieger, K. Ewell, B.A. Ludewigt, M.R. Maier, D. Markovic, O. Milgrome et al., An 8 × 8 pixel IC for x-ray spectroscopy, *IEEE Trans. Nucl. Sci.* 48, pp. 493–498. doi:10.1109/23.940105, 2001.
13. P.D. Walker, M.M. Green, A tunable pulse-shaping filter for use in a nuclear spectrometer system, *IEEE J. Solid-State Circuits.* 31, pp. 850–855. doi:10.1109/4.509873, 1996.
14. B. Krieger, I. Kipnis, B.A. Ludewigt, XPS: A multi-channel preamplifier-shaper IC for x-ray spectroscopy, *IEEE Trans. Nucl. Sci.* 45, pp. 732–734. doi: 10.1109/23.682625, 1998.
15. M. Konrad, Detector pulse shaping for high resolution spectroscopy, *IEEE Trans. Nucl. Sci.* 15, pp. 268–282. doi:10.1109/TNS.1968.4324864, 1968.
16. X. Shi, R. Dinapoli, D. Greiffenberg, B. Henrich, A. Mozzanica, B. Schmitt et al., A low noise high dynamic range analog front-end ASIC for the AGIPD XFEL detector, in *2012 19th IEEE Int. Conf. Electron. Circuits, Syst. (ICECS 2012)*, IEEE, pp. 933–936. doi:10.1109/ICECS.2012.6463508, 2012.
17. M. Bouchel, F. Dulucq, J. Fleury, C. de La Taille, G. Martin-Chassard, L. Raux, SPIROC (SiPM integrated read-out chip): Dedicated very front-end electronics for an ILC prototype hadronic calorimeter with SiPM read-out, in *2007 IEEE Nucl. Sci. Symp. Conf. Rec., IEEE*, pp. 1857–1860. doi:10.1109/NSSMIC.2007.4436519, 2007.
18. M. van Elzakker, E. van Tuijl, P. Geraedts, D. Schinkel, E. Klumperink, B. Nauta, A 1.9¿W 4.4fJ/conversion-step 10b 1MS/s charge-redistribution ADC, in *2008 IEEE Int. Solid-State Circuits Conf. - Dig. Tech. Pap., IEEE*, pp. 244–610. doi:10.1109 /ISSCC.2008.4523148, 2008.
19. A.G.F. Dingwall, V. Zazzu, An 8-MHz CMOS subranging 8-bit A/D converter, *IEEE J. Solid-State Circuits.* 20, pp. 1138–1143. doi:10.1109/JSSC.1985.1052451, 1985.

7

Semiconductor Detector Readout ASICs for Baggage Scanning Applications

Kris Iniewski and Chris Siu

CONTENTS

7.1 Introduction

Baggage scanning equipment needs to detect bulk, sheet, liquid, and slurry explosives, using x-ray to scan the objects. Traditional radiation detection systems use scintillator technology; ceramic or organic scintillators detect those photons that have passed through the object and produce a visible light signal. The received light signal is, in turn, converted by a photodiode to produce analog electrical signals that can be used to produce an image.

Most detection systems take linear projection through the luggage traveling on a conveyor belt in a so-called line scan mode. Detectors used are efficiently collimated linear arrays. For dual-energy capabilities, two solutions have been used. In the first technique, the linear scan is performed twice, with and without an x-ray filter in front of the beam. This way, an elementary technique of low- and high-energy photon separation can be obtained. It is also possible to use two accelerating voltages with high-speed switching.

In the second technique, the linear scan is performed only once, but sandwich detectors have been optimized for dual-energy scanning. They consist of two layers of scintillator-photodiode type, separated by a metal filter. The first layer absorbs the low-energy photons, and the second layer absorbs the high-energy photons.

In both cases, due to the poor energy separation of those acquisition systems, and due to significant noise resulting from the acquisition speed, the obtained accuracy only allows materials to be classified into broad categories such as inorganic and organic [1].

Modern baggage scanning equipment uses more efficient, direct method of radiation detection as shown in Figure 7.1. The two-step process involving

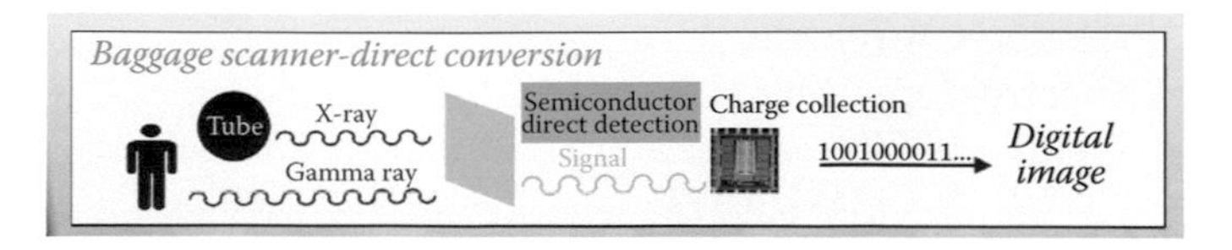

FIGURE 7.1
Semiconductor direct detection baggage scanning technology.

scintillator and photodiode is replaced by a semiconductor detector that converts x-rays directly into electric charge. Various semiconductor detectors can be used made of silicon (Si), germanium (Ge), gallium arsenide (GaAs), cadmium telluride (CdTe), or cadmium zinc telluride (CdZnTe). Silicon detectors have very poor stopping power. Germanium detectors are very expensive, while GaAs detector technology is not developed enough for commercial applications. CdTe and CdZnTe (CZT) detectors fit perfectly into baggage scanning applications due to their high stopping power, reasonable cost, high stability [2], and reliability [3].

In addition, through the use of high-energy resolution semiconductor detectors, multiple energy analysis, and multiple source points, the equipment might be able to detect thin sections and subtle differences in atomic number. The equipment might also provide full volumetric reconstruction and analysis of the object being imaged. The result of the advanced CZT technology could be an exceptional ability to detect and discriminate a wide variety of threat materials, providing enhanced overall security. One of the key motivations in technological developments is to eliminate the liquid carry-on ban currently in place. Once that ban is eliminated (already legislated in Europe), the airport operators will be forced to use better technology for liquid security detection.

An example of such technology [4] is shown in Figure 7.2 [5]. Suitcases enter the device at the right of the figure where spatial landmarks for registration purposes are measured by a pre-scanner. In the main housing center-left of the figure, a primary cone beam executes a meander scan, either of a region of interest or a suitcase in its entirety. The underlying principle of detection is x-ray diffraction imaging or XRD [6]. XRD refers to the volumetric analysis of extended, inhomogeneous objects by spatially resolved XRD and is discussed

FIGURE 7.2
Direct tomographic, energy-dispersive XRD 3500 system from Morpho-Detection (http://www.morpho.com).

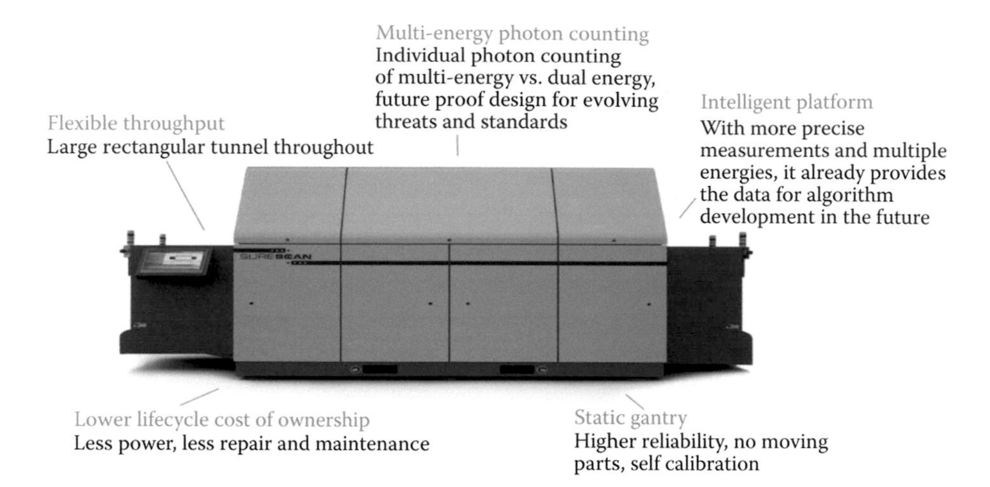

FIGURE 7.3
Direct detection photon-counting system from SureScan. (Courtesy of SureScan, http://www.surescaneds.com.)

in more detail later on in Section 7.2.7. The detector technology used in such a system requires very good energy resolution and, for this reason, only solid-state detectors such as high purity germanium (HPGe) or CZT qualify.

Another example of the semiconductor direct readout technology scanner is shown in Figure 7.3. SureScan x1000 is the first TSA certified multi-energy static gantry explosive detection system (EDS) for checked baggage screening, representing the next generation in EDS detection technology and design. With its innovative use of computed tomography (CT) and implementation of multi-energy detection for atomic number analysis, the x1000 delivers low false alarm rates and a high level of accuracy of current and emerging threat detection. x100 scanner utilizes direct x-ray detection using the photon-counting principle.

7.2 Techniques for X-Ray Detection

Let us lay some foundation of radiation detection before specific detection techniques are described. Energetic photons such as hard x-rays and gamma rays interact with matter mainly by four basic processes:

- Elastic scattering (also known as Thomson scattering) in which photons change path without changing energy. This process is useful in forwarded scattering techniques as discussed later. Elastic scattering has been widely used for crystallography and the techniques developed for powder diffraction are now being deployed into the marketplace for explosive detection.

- Photoelectric absorption where the photon disappears after transferring all its energy to an electron. CZT is a particularly suitable absorber in the energy range of interests (40 keV to 1 MeV). Many handheld instruments have been used for security detection applications.
- Compton scattering in which some fraction of the photon energy is transferred to a free electron in the material. The path of the photon is changed. This process can be used to detect the direction from which the photon arrived at the detector. Scattering direction is a function of the incoming photon's energy and the crystallographic structure and orientation of the scattering object. The commercial scanners use collimation to limit the acceptance angle of the scattered photons.
- Pair production in which the photon energy is spontaneously converted into an electron–positron pair. Pair production is dominant at the very high-energy range (well above 1 MeV) and therefore will not be discussed in this chapter as it is not applicable to baggage scanning.

From the above comparison, it is clear that elastic scattering, photoelectric absorption, and Compton scattering are physical processes that can be leveraged in radiation detection with CZT. The last two processes lead to a partial or complete transfer of the photon energy to electron energy. The charge generation processes are not dependent on temperature. Photons of high-enough energy can penetrate solid objects, but are scattered or absorbed by dense objects (or a sufficient thickness of less-dense material). This is the basis for radiography. For cargo scanning, x-rays or gamma rays are beamed through a container, and a detector on the other side records the number of photons received in each pixel.

There are several detection techniques available in the marketplace. Their principal of operation, cost, performance characteristics, and ease of use vary broadly. We are presenting here a short summary of these techniques that have been used or are in active development.

The key issue in baggage scanning application is the ability to identify explosive materials. This difficulty in explosive detection lies in the fact that the materials contained in a luggage are unknown in number and nature, and they are very chemically close to common materials. The effective atomic number (Z_{eff}) of most explosive materials ranges from 7 to 7.7 and their density from 1.4 to 1.9 [7]. Some of the techniques discussed here are capable of overcoming that limitation.

7.2.1 X-Ray Radiography

Conventional x-ray radiography (CXR) is a quick way to provide an initial screening at very low performance levels. These machines produce a 2D projection image of the integrated density through an object. Both photoelectric (PE) and Compton absorption vary with density, and that enables some material identification. However, the provided material identification and contrast are rather poor and could use additional secondary screening methods. Despite these disadvantages, CXR is used extensively due to its low cost as it does provide visual capability for conventional threats (knife, gun, etc.).

7.2.2 Dual-Energy X-Ray Radiography

Adding dual-energy capability to CXR provides higher accuracy and some rudimentary capability of energy separation. In practice, the technique is implemented by scanning the object twice: once as a normal scan and a second time with a high-energy filter added. Poor energy separation might be capable of distinguishing between organic and inorganic matter but will likely not be able to detect any serious security threats. In recent deployments, high-speed switching of accelerating voltage is the preferred method as it is cheaper than dual-layer scintillator technology.

7.2.3 Color X-Ray Radiography

Adding more energy bins leads to the so-called color x-ray technology, where multiple energy bins (typically three to five) might actually provide some useful energy separation and material identification capabilities [8]. CZT detectors are ideally suited for this application due to their ability to operate at room temperature and having sufficient energy resolution at the count rates that are required. Adding more energy bins enables clearer visualization of the baggage content. Traditional security threats (knife, gun, etc.) can be easily detected, especially when 3D scanning is adopted. However, material identification of explosives might require more sophisticated diffraction techniques as discussed below.

7.2.4 Computed Tomography

CXR, double-energy, and color x-ray techniques provide two-dimensional (2D) view of the object under investigation. Three-dimensional (3D) views can be obtained using a CT approach. CT is a well-known medical imaging modality used in virtually every hospital in the world. CT equipment can be tailored to be used in security applications by providing a 3D image of the object under the search. However, pure visualization will likely not be sufficient to detect serious security threats including various types of explosive materials [7].

7.2.5 CT with Energy Discrimination

To be truly useful in threat detection, CT-like technology needs to be coupled with energy discrimination using color x-ray concepts. Combined technology can be used to detect explosives using information of object density combined with atomic number analysis for enhanced detection. By discriminating the atomic composition of baggage contents, the enhanced CT technology can deliver high accuracy in threat detection, with very low false alarm rates. 3D CT technology with color x-ray is expensive, though, and airport operators may not be willing to pay for it. However, the key question is not the capital cost of CT scanners but rather the cost of ownership. When using automated threat detection, the cost of handling false alarms can be substantially more than the machine's capital cost.

7.2.6 Back-Scattering

Due to the limited capability of x-ray detection in identifying radiation isotopes, a different set of technology has been developed relying on analyzing scattering effects of x-rays. One class of techniques relies on back-scattering information [9]. It is more accurate, although slower than CXR, but still might be insufficient for explosive determination. The military has been using this technology for mine detection as it allows for discrimination of low- vs. high-density materials. It provides only side-only access.

7.2.7 Forward-Scattering and Diffraction

The techniques based on forward scattering or diffraction can provide ultimate performance in threat detection. XRD is a powerful analytical tool that has been used for the nondestructive analysis of a wide variety of materials for nearly 100 years. XRD is now widely applied in a variety of industries including metallurgy, photovoltaics, forensics, pharmaceuticals, semiconductors, and catalysis and can be used to analyze virtually any solid with crystalline structure. It has been recently applied to security detection. The fundamental strength of the technique is its ability to characterize the periodic atomic structure present in crystalline or polycrystalline materials. It also has some capability to distinguish liquids.

In XRD analysis, a sample is illuminated by a collimated x-ray beam of known wavelength. If the material is crystalline, it possesses a 3D ordering or "structure" with repeat units of atomic arrangement (unit cells). X-rays are elastically scattered (i.e., diffracted) by the repeating crystal plane lattice of materials, while x-rays are randomly scattered by amorphous materials. XRD occurs at specific angles with respect to the lattice spacings defined by Bragg's law. Any change or difference in lattice spacing results in a corresponding shift in the diffraction lines. It is this principle that such properties as identification (based on phase) and residual stresses are obtained.

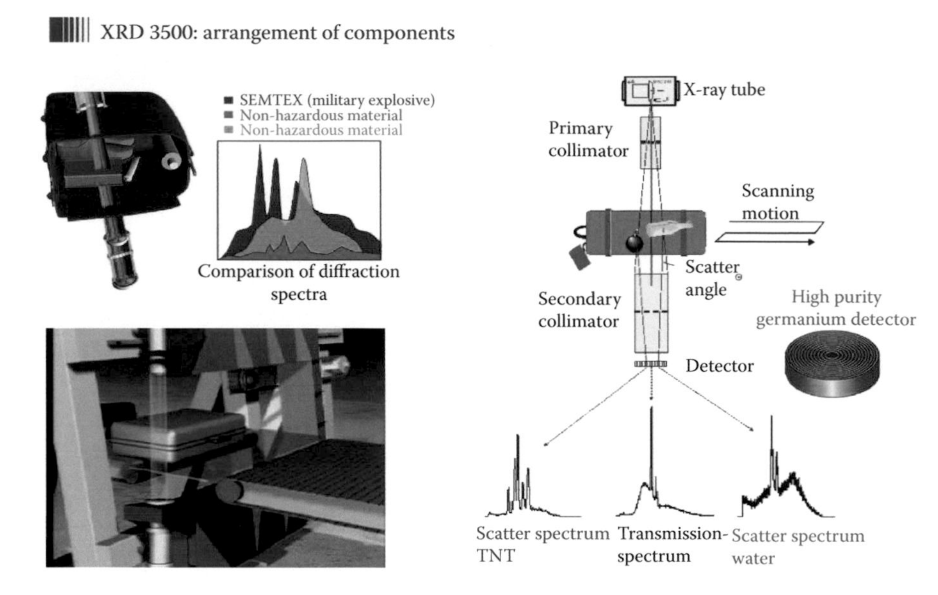

FIGURE 7.4
XRD 3500 architecture and principle of operation (http://www.morpho.com).

Application of XRD in security detection provides very powerful opportunities to detect chemical/structural property of the material under investigation. For effective baggage screening, this technique is frequently coupled with CT to visualize the object in question. Traditionally, this is done in a serial fashion making the scan time long and equipment expensive. With CZT, it is, however, possible to perform CT-like imaging and XRD-like detection simultaneously. If you are interested in discussing this opportunity, please contact the authors of this document.

An example of commercial application of the XRD technology is shown in Figure 7.4. Note the difference in obtained scattered x-ray spectra for Semtex (explosive) and some nonhazardous materials. HPGe detector used in XRD 3500 can be substituted by CZT, which offers similar energy resolution (few percent at 30–120 keV energy range). However, none of the scintillator materials can be used for this application as their energy resolution is too poor.

Application of XRD in security detection provides very powerful opportunity to detect chemical/structural property of the material under investigation [10]. For effective baggage screening, this technique is frequently coupled with CT to visualize the object in question.

7.2.8 Hyperspectral X-Ray Radiography

While the XRD technique is very powerful in material identification, it places hard demands on the detector technology used. While they can be met with Ge, CdTe, and CZT materials to provide the required spectral response, a simpler technique is highly desirable to reduce equipment costs. Although high-purity Ge detectors provide the best energy resolution, the stopping power of Ge is rather low and these detectors require cooling, typically at liquid nitrogen temperatures (77 K).

It has been shown very recently that traditional high-energy x-ray systems can capture scattered x-rays to deliver 3D images with structural or chemical information in each voxel [12]. This type of imaging can be used to separate and identify chemical species in bulk objects with no special sample preparation. Defining hyperspectral technology precisely is difficult as it is a relatively new concept that, at a minimum, should contain 100 energy bins. In addition, the hyperspectral technology takes advantage of measurement orthogonality.

The capability of hyperspectral imaging has been demonstrated by examining an electronic device where we can clearly distinguish the atomic composition of the circuit board components in both fluorescence and transmission geometries. Researchers not only were able to obtain attenuation contrast but also were able to image chemical variations in the object, potentially opening up a very wide range of applications from security to medical diagnostics [13]. The possibility of this technique being introduced to commercial applications does exist but has not been shown yet.

7.3 Introduction to Readout ASICs

7.3.1 ASIC Technology

Semiconductor pixelated detectors need to have a high level of segmented multichannel readout. Several decades ago, the only way to achieve this was via massive fan-out schemes to route signals to discrete low-density electronics. At present, CMOS technology is used to build very dense low power electronics with many channels, which can be bonded directly or indirectly (through a common carrier PCB) to the detector.

There are different requirements for the CMOS technology used for the analog front-end signal processing, as opposed to that for the digital signal processing. For the analog part of the electronics, there is a requirement for a robust technology that has low electronic noise and high dynamic range that typically requires large power supply voltages. Digital signal processing, in turn, requires very high speed and high density that is more compatible with the more modern low-voltage supply, deep submicron processes.

There seems to be a technology optimum at around 0.35 to 0.18 μm minimum feature size for the analogue requirements. The large feature size limits the complexity of circuitry that can be integrated in a pixel, but even at 0.35 μm it is possible to place a million transistors on a reasonable size silicon die. In comparison, the digital signal processing can benefit from the rapid development of deep submicron processes. Some selected research developments now take place using 90 or 65 nm processing nodes. These technologies are well suited to high-speed ADC architectures and to very fast data manipulation for data sparcification and compression. The deep submicron technologies have their own limitations in terms of gate oxide thickness, noise, and cost.

7.3.2 ASIC Attachment

Semiconductor pixel detectors require a connection from the pads on the detector material to the bond pads on the ASICs. In some cases, the pixel pitch on the readout ASIC is the same as the detector pixel pitch. It is also possible to fan out the connections on the detector with multilevel metal routing on the detector or with the use of the interposer board. This fanning out routing has to be done very carefully as there is a large danger for signal crosstalk. With integrating readout and synchronous input signals, where the signal is totally removed from the detector, this might not be a problem, but transients can still upset thresholds in these systems.

The pitch of x-ray imaging systems currently ranges from about 100 μm to 1 mm. For small pixel pitches, bump bonding is used to connect the detector pixels to the ASICs. There are many different technologies to do this depending on the requirements of the detectors and environmental constraints.

The industry-standard area bump-bonding method is to deposit solder onto under-bump metallization on the pads of the detector and ASIC, and then to align the two and heat them to reflow the solder. Various solders are used, including lead–tin, bismuth–tin, indium alloys, and silver alloys, depending on the temperature to reflow and the operating temperature required. Typically these materials require 240°C to 140°C to reflow. Indium is used either in a lower-temperature reflow process or straight compression bonding and gives good results but cannot be used if high operating temperatures will ever be experienced.

7.3.3 ASICs for Spectroscopy

CZT detectors typically operate in a single photon detection mode where an electric charge generated by one photon needs to be collected by the readout electronics. As the amount of generated charge is small (about 5 fC for a 122 keV photon), very sensitive analog circuitry is required to amplify that charge. In spectroscopic applications, the amount of charge, which directly corresponds to the photon energy, needs to be precisely determined. In photon-counting applications, only binary (or multibinary) decisions are required, but the count rate might be very high creating its related challenges. The purpose of this section is to explain some of the design considerations that are important when building semiconductor readout electronics system.

7.3.3.1 Analog Front End

Analog signal processing can be divided into the following steps:

- *Amplification.* The input charge signal is amplified and converted to a voltage signal using a charge sensitive amplifier (CSA). A main characteristic of the amplification stage is equivalent noise charge (ENC) which is required to be as low as possible in order to not degrade intrinsic detector energy resolution. Another important consideration for the CSA operation is a dark current compensation mechanism. A solution that accommodates continuous compensation for dark currents up to several nAs levels while maintaining low ENC is desired.
- *Signal shaping.* The time response of the system is tailored to optimize the measurement of signal magnitude or time and the rate of signal detection. The output of the signal chain is a pulse where the area is proportional to the original signal charge, i.e., the energy deposited in the detector. The pulse shaper transforms a narrow detector current pulse to broader pulse (to reduce electronic noise), and with a gradually rounded maximum at the peaking time to facilitate measurement of the amplitude. A solution that provides effective signal shaping while maximizing the channel count rate needs to be applied.
- *Pulse detection.* The input pulse, broadened by the shaping process, needs to be detected against a setup threshold value. The threshold level is a critical parameter that determines whether the event is recognized as a true event or false reading caused by noise. As a result, the threshold value is typically adjustable both globally and at the pixel level. The peak detection value determines energy level information. A solution that prevents temperature drift of the peak detector (PD) needs to be used.
- *Channel multiplexing.* In the case of ASIC spectroscopy, all parallel channels of the channel readout ASIC need to have their signals multiplexed at the output before being sent out to an external analog to digital converter (ADC). The key requirement to channel multiplexing and signal shaping is a maximum channel count rate determined by the given application.

7.3.3.2 Charge Sensitive Amplifier

CdTe and CZT detectors typically operate in a single photon detection mode where an electric charge generated by one photon needs to be collected by the readout electronics. As the amount of generated charge is small (about 1 fC for a 20 keV photon), very sensitive analog circuitry is required to amplify that charge. To precisely read the amount of charge that directly corresponds to the photon energy, further analog signal processing is required. The purpose of this section is to explain some of the design considerations that are important when building a CZT readout electronics system.

The current signal induced in the sensing electrode can be integrated in the pixel capacitance and read out with a high-input impedance stage, which amplifies the resulting voltage at the pixel node, or it can be read out directly with a low-input impedance stage, which amplifies the charge Q and keeps the pixel node at a virtual ground, such as a charge amplifier. The latter usually is the preferred choice since, among its other advantages, it stabilizes the sensing electrode by keeping its voltage constant during the measurement and/or the readout.

In both cases, the low-noise amplification is required to reduce the noise contribution from the processing electronics (such as the shaper, PD, and ADC) to negligible amounts; good design practice dictates maximizing this amplification while avoiding overload of subsequent stages. Also, in both cases, low-noise amplification would provide either a charge-to-voltage conversion (e.g., source follower, charge amplifier) or a direct charge-to-charge (or current-to-current) amplification (e.g., charge amplifier with compensation, current amplifier). Depending upon this choice, the shaper would be designed to accept a voltage or a current, respectively, as its input signal.

In a properly designed low-noise amplifier, the noise is dominated by processes in the input transistor. Assuming that CMOS technology is employed in the design, the input transistor is referred to as the "input MOSFET," although the design techniques can easily be extended to other types of transistors, such as the JFET, the bipolar transistor, or the heterojunction transistor. The design phase, which consists of sizing the input MOSFET for maximum resolution, is called "input MOSFET optimization" and has been studied extensively in the literature.

7.3.3.3 Equivalent Noise Charge

ENC expresses an amount of noise that appears at the chip input in the absence of a useful input signal and is a key chip parameter that affects the energy resolution of the system. Following the standard approach, the total ENC can be divided into three independent components: the white thermal noise associated with the input transistor of the CSA (ENC_{th}), the flicker noise associated with the input transistor of the CSA ($ENC_{1/f}$), and the noise associated with the detector dark leakage current (ENC_{dark}). Noise arising in other components connected to the ASIC input node, such as the bias resistor, is generally made negligible in a properly designed system. For a first-order shaper, the ENC components can be approximately expressed as

$$\mathrm{ENC}_{\mathrm{th}}^2 = (8/3)kT/\left(T_{\mathrm{peak}}^* g_{\mathrm{m}}\right) * C_{\mathrm{tot}}^2$$

$$\mathrm{ENC}_{1/\mathrm{f}}^2 = K_\mathrm{f}/2 * WL * C_{\mathrm{tot}}^2/C_{\mathrm{inp}}^2$$

$$\mathrm{ENC}_{\mathrm{dark}}^2 = 2q * I_{\mathrm{dark}} * T_{\mathrm{peak}}$$

$$\mathrm{ENC} = \left(\mathrm{ENC}_{\mathrm{th}}^2 + \mathrm{ENC}_{1/\mathrm{f}}^2 + \mathrm{ENC}_{\mathrm{dark}}^2\right)^{1/2}$$

where g_m is the transconductance of the CSA input transistor, C_{tot} is the total capacitance at the input of the CSA, T_{peak} is the shaper peaking time, K_f is the CSA input transistor flicker noise constant, W and L are input transistor width and length, respectively, and I_{dark} is the detector leakage current given by (3). Note that C_{tot} is the sum of the detector capacitance C_{det}, the gate–source and gate–drain capacitances of the input transistor C_{inp}, and any other feedback or parasitic capacitance at the CSA input originating from the chip package, ESD diodes, and PCB traces. Based on experience with ASIC design and radiation detection module manufacturing, we have assumed C_{inp} to be fractions of pF while the remaining C_{tot} components to be about 1–2 pF. Clearly, particular values are strongly dependent on the chosen technology for ASIC design and packaging as well as on the chosen connectivity scheme between the CZT detector and the chip. It can be easily shown that the optimum peaking time T_{opt} is given by the condition where $\mathrm{ENC_{th}}$ is equal to $\mathrm{ENC_{dark}}$ leading to the following expression:

$$T_{\mathrm{opt}}^2 = 4kT\, C_{\mathrm{tot}}^2/(3g_\mathrm{m} q I_{\mathrm{dark}})$$

ENC is typically measured in the lab by measuring the output noise and referring it back to the input knowing the overall gain of the system. It is also possible to measure channel performance using the scope. By acquiring the channel shaper output signal on the oscilloscope at 1 MHz sampling frequency (for example, 5 ms observed time on 5000 points), a fast Fourier transform (FFT) can be applied to the acquired data and the resulting spectrum calculated for each frequency. The noise expressed in mV was calculated as the standard deviation with respect to the shaper output average value, and expressed as the equivalent noise in terms of electrons (by considering the known nominal gain of the data channel, typically about in the order of hundreds of mV per fC of the input charge). The result of these calculations is an ENC value expressed in number of electrons, typically in the range of hundreds of electrons, depending on what electronics are used, how high the count rate is, and the loading capacitance at the detector input.

7.3.3.4 Signal Shaping

The low-noise amplifier is typically followed by a filter, frequently referred to as the shaper, responding to an event with a pulse of defined shape and finite duration ("width") that depends on the time constants and the number of poles in the transfer function. The shaper's purpose is twofold: first, it limits the bandwidth to maximize the signal-to-noise ratio (SNR); and second, it restricts the pulse width in view of processing the next event. Extensive calculations have been made to optimize the shape, which depends on the spectral densities of the noise and system constraints (e.g., available power and count rate).

Optimal shapers are difficult to realize, but they can be approximated, with results within a few percent from the optimal, either with analog or digital processors, the latter requiring analog-to-digital conversion of the charge amplifier signal (anti-aliasing filter may be needed). In the analog domain, the shaper can be realized using time-variant solutions that limit the pulse width by a switch-controlled return to baseline, or via time-invariant solutions that restrict the pulse width using a suitable configuration of poles. The latter solution is discussed here as it minimizes digital activity in the front-end channels.

In a front-end channel, the time-invariant shaper responds to an event with an analog pulse, the peak amplitude of which is proportional to the event charge, Q. The pulse width, or its time to return to baseline after the peak, depends on the bandwidth (i.e., the time constants) and the configuration of poles. The most popular unipolar time-invariant shapers are realized either using several coincident real poles or with a specific combination of real and complex-conjugate poles. The number of poles, n, defines the order of the shaper. Designers sometimes prefer to adopt bipolar shapers, attained by applying a differentiation to the unipolar shapers (the order of the shaper now is n-1). Bipolar shapers can be advantageous for high-rate applications, but at expenses of a worse SNR.

In a typical readout system, the shaping time varies from fraction of a microsecond up to several microseconds. The shaping time is defined as the time equivalent of the standard deviation of the Gaussian output pulse. In the laboratory, it is the full width of the pulse at half of its maximum value (FWHM) that is typically being measured. The FWHM value is greater than the shaping time by a factor of 2.35.

The DC component of the shaper from which the signal pulse departs is referred to as the output baseline. Since most extractors process the pulses' absolute amplitude, which reflects the superposition of the baseline and the signal, it is important to properly reference and stabilize the output baseline. Nonstabilized baselines may fluctuate for several reasons, like changes in temperature, pixel leakage current, power supply, low-frequency noise, and the instantaneous rate of the events. Nonreferenced baselines also can severely limit the dynamic and/or the linearity of the front-end electronics, as in high-gain shapers where the output baseline could settle close to one of the two rails, depending on the offsets in the first stages. In multiple front-end channels sharing the same discrimination levels, the dispersion in the output baselines can limit the efficiency of some channels.

7.3.3.5 Peak Detection

PD is one of the critical blocks in the radiation signal detection system as accurate photon energy is determined by the detected peak amplitude. Standard PDs may be sampled or asynchronous solutions. Sampled PDs are more precise but suffer from high-circuit complexity and high-power dissipation. Asynchronous PDs have simpler structures but suffer from lower output precision.

7.3.4 ASICs for Photon Counting

One of the major advantages of photon-counting detectors is electronics noise rejection. Well-designed photon-counting detectors allow for ASIC electronics thresholds high enough to reject noise pulses while still counting useful signals. Therefore, quantum limited operation of the photon-counting detector can be achieved as image noise is determined by only statistical variations of x-ray photons. On the other hand, energy integrating detectors suffer from electronics noise, which is mixed with useful photon signals, and separating it from statistical noise is not possible. Electronics noise rejection is important because its magnitude for currently used digital x-ray detectors is not negligible.

After converting the CZT generated charge to voltage by the CSA amplifier and subsequent shaping by the shaping amplifier, the signal, given enough gain in the system, is ready for digitization. Typically, the signal is compared against user selected threshold voltage (discriminator box in Figure 7.5) to

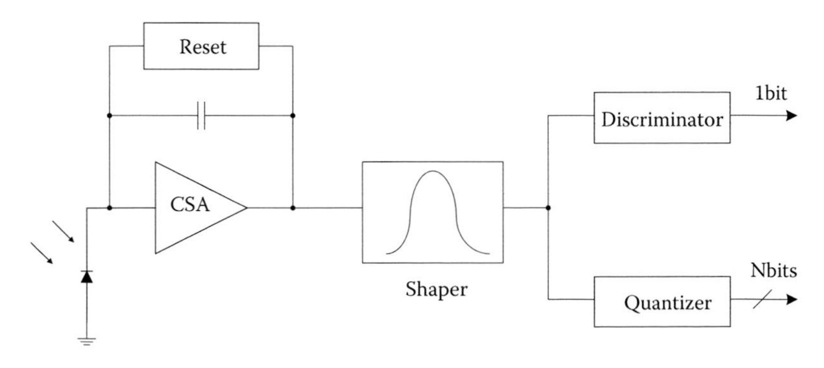

FIGURE 7.5
Detector readout signal chain.

produce 1-bit trigger signal indicating detection of the pulse. In parallel, the value of the shaped signal is sent to an ADC (or time over threshold [ToT] processor) with n-bit accuracy. The conversion resolution n is typically between 8 and 16 bits, depending on the system accuracy, noise levels, and degree of signal precision achieved.

One important consideration in the practical system is CSA reset. As the feedback capacitor C_f is charged by the input signal, there must be some means of discharging this capacitor in order for the CSA to be ready for the next signal. This circuitry is schematically shown as the reset block in Figure 7.5.

There are two possible implementations for the reset block: digital and analog. The digital one involves using a switch that will discharge the feedback capacitor quickly. Unfortunately, this process typically creates too much disturbance for the sensitive CSA. The analog solution involves using a resistor (or MOSFET operating in the triode region) and provides continuous discharging during the entire process. The discharge cannot be too slow (in which case the capacitor will not be fully discharged before the next event) or too fast (as that will affect signal formation). While the CZT readout scheme shown in Figure 7.5 is typical, let us point out that it is possible to directly sample the anode (and cathode) signals without producing any trigger signal to obtain timing and amplitude information.

On a final note, let us point out that while the principle of CSA signal amplification, pulse shaping, and ADC conversions outlined above are fairly simple, practical implementations can be very challenging due to the very small input signals involved (below 1 mV). One has to worry about system noise, power supply decoupling, ESD protection, EMI radiation, and op-amp stability issues.

A typical photon-counting ASIC implementation contains hundreds of channels frequently implemented with multiple energy bins. One of the early 128-channel ASICs is shown in Figure 7.6. A clear advantage of the photon-counting detectors over the integrating detectors is the ability to perform at higher SNR at low photon counts.

7.3.5 Photon Counting vs. Spectroscopy

There are two ways of signal processing for baggage scanning with energy sensitive semiconductor detectors: photon counting and spectroscopy. While a precise difference between the two is hard to establish as all analog signals eventually become digital at some point in the readout system, we would like to suggest the following practical definition. Photon counting relies on energy binning, with the term "binning" implying the use of comparators inside the ASIC chip. Spectroscopy, in turn, preserves the analog nature of the signal representing photon energy and uses ADC after the ASIC signal processing has been accomplished.

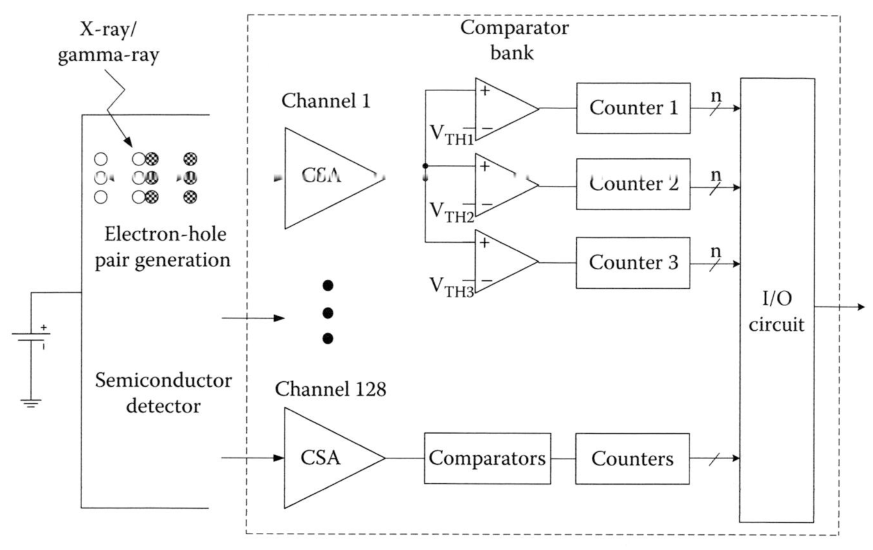

FIGURE 7.6
Block diagram of 128-channel photon-counting ASIC.

As a result of this architectural change, photon-counting systems can achieve a very high count rate while sacrificing energy resolution (ER), while spectroscopic systems can have very good noise properties but face limitations with a maximum count rate. To achieve specific design objectives, either count rate of the spectroscopic system is maximized or energy resolution of the photon-counting systems is maximized.

7.4 Examples of Readout ASICs

There are literally hundreds of readout ASICs that have been published in the literature. This section summarizes the design and performance characteristics of three photon counting and three spectroscopic devices.

7.4.1 Photon-Counting ASICs

Fast readout electronic circuits have been developed to reach count rates of several million counts per second [15–19]. These systems provide coarse energy resolution given by a limited number of discriminators and counters. This section provides some information about the most important photon-counting devices.

7.4.1.1 TIMEPIX

TIMEPIX ASIC is a CZT pixel detector developed in the framework of the MEDIPIX2 collaboration [22]. The pixel matrix consists of 256 × 256 pixels with a pitch of 55 μm, which gives a sensitive area of about 14 × 14 mm^2. TIMEPIX is designed in a 0.25 μm CMOS process and has about 500 transistors per pixel.

The chip has one threshold and can be operated in photon counting, ToT, or time of arrival (ToA) modes. The principles of the different operating modes are described in detail in the literature [22].

In the photon-counting mode, the counter is incremented once for each pulse that is over the threshold, while for the ToT mode, the counter is incremented as long as the pulse is over the threshold. In the ToA mode, the pixel starts to count when the signal crosses the threshold, and keeps counting until the shutter is closed.

7.4.1.2 MEDIPIX-3

While TIMEPIX is a general-purpose chip, the MEDIPIX-3 is aimed specifically at x-ray imaging [22]. It can be configured with up to eight thresholds per pixel and features analog charge summing over dynamically allocated 2 × 2 pixel clusters. The intrinsic pixel pitch of the ASIC is 55 μm as in TIMEPIX. Silicon die can be bump-bonded at this pitch (fine pitch mode), and the chip can be run with either four thresholds per pixel in the single pixel mode or with two thresholds per pixel in the charge summing mode. Optionally, the chip can be bump-bonded with a 110 μm pitch, combining counters and thresholds from four pixels. Then operation is possible in the single pixel mode with eight thresholds per pixel or in the charge summing mode having four thresholds and summing charge of a 220 × 220 μm^2 area.

Being a very versatile and configurable chip, there is also a possibility to utilize the two counters per pixel and run in the continuous read/write mode where one counter counts while the other one is being read out. This eliminates the readout dead time but comes at a cost of losing one threshold, since both counters need to be used for the same threshold. Finally, the charge summing mode is a very important feature to combat contrast degradation by charge sharing in semiconductor detectors with small pixels.

7.4.1.3 ChromAIX

A proprietary multi-energy resolving ASIC called ChromAIX has been designed to support studying spectral CT applications. In order to enable K-edge imaging, at least three spectrally distinct measurements are necessary; for a photon-counting detector, the simplest choice is to have at least the same number of different energy windows. With more energy windows, the spectrum of incident x-ray photons is sampled more accurately, thus improving the separation capabilities.

The ChromAIX ASIC accommodates a sufficient number of discriminators to enable K-edge imaging applications. Postprocessing allows separating the photo effect, Compton effect, and one or possibly two contrast agents with their corresponding quantification. The ChromAIX ASIC is a pixelated integrated circuit that has been devised for direct flip-chip connection to a direct converting crystal like CZT. The design target in terms of observed count rate performance is 10 Mcps/pixel, which corresponds to approximately 27.2 MHz/pixel periodic pulses, assuming a paralyzable dead-time model. Although the pixel area in CT is typically about 1 mm^2, both the ASIC and direct converter feature significantly smaller pixel, or sub-pixel. In this way, significantly higher rates can be achieved at an equivalent CT pixel size, while further improving the spectral response of the detector via exploiting the so-called small-pixel effect. The subpixel should not be made too small, since charge sharing affects, and then starts to deteriorate, the spectral performance. Very small pixels would need countermeasures as implemented in MEDIPIX-3, the effectiveness of which at higher rates remains doubtful due to charge-sharing effects.

The ChromAIX ASIC consists of a CSA and a pulse shaper stage, as any other photon-counting device. The CSA integrates the fast transient current pulses generated by the direct converter, providing a voltage step-like function with a long exponential decay time. The shaper stage represents a band-pass filter that transforms the aforementioned step-like function into voltage pulses of a defined height. The height of such pulses is directly proportional to the charge of the incoming x-ray photon. A number of discriminator stages are then used to compare a predefined value (i.e., energy threshold) with the height of the produced pulse. When the amplitude of the pulse exceeds the threshold of any given discriminator, the associated counter will increment its value by one count.

In order to achieve 10 Mcps observed Poisson rates, which would typically correspond to incoming rates exceeding 27 Mcps, a very high bandwidth is required. The two-stage approach using a CSA and a shaper allows achieving such high rates while relaxing the specification of its components. The design specification in terms of noise is 400 e-, which corresponds to approximately 4.7 keV FWHM. Simulations of the analogue front end have been carried out to evaluate the noise performance of the channel. According to these simulations, the complete analogue front-end electronic noise (CSA, shaper, and discriminator input stage) amounts to approximately 2.51 mV_{RMS}, which in terms of energy resolution corresponds to approximately 4.0 keV FWHM for a given input equivalent capacitance.

7.4.2 Spectroscopic ASICs

7.4.2.1 IDEF-X

IDeF-X HD is the last generation of low-noise radiation-hard front-end ASICs designed by CEA/Leti for spectroscopy with CZT detectors [20,21]. The chip, as shown in Figure 7.7, includes 32 analog channels to convert the impinging charge into an amplified pulse-shaped signal, and a common part for slow control and readout communication with a controller.

The first stage of the analog channel is a charge sensitive preamplifier (CSA) based on a folded cascode amplifier with an inverter input amplifier. It integrates the incoming charge on a feedback capacitor and converts it into voltage; the feedback capacitor is discharged by a continuous reset system realized with a PMOS transistor. The increase in drain current in this transistor during the reset phase is responsible for a nonstationary noise; to reduce the impact of this noise on the ENC, a so-called nonstationary noise suppressor was implemented for the first time in this chip version using a low-pass filter between the CSA output and the source of the reset transistor to delay this noise.

The second stage is a variable gain stage to select the input dynamic range from 10 fC (250 keV) to 40 fC (1 MeV). The third stage is a pole zero cancellation (PZ) implemented to avoid long-duration undershoots at the output and to perform a first integration. The next stage of the analog channel is a second-order low-pass filter (RC^2) with variable shaping time. To minimize the influence of the leakage current on the signal baseline, a so-called baseline holder (BLH) was implemented by inserting a low-pass filter in the feedback loop between the output of the RC^2 filter and the input of the PZ stage. The DC level at the output is stabilized for leakage current up to 7 nA

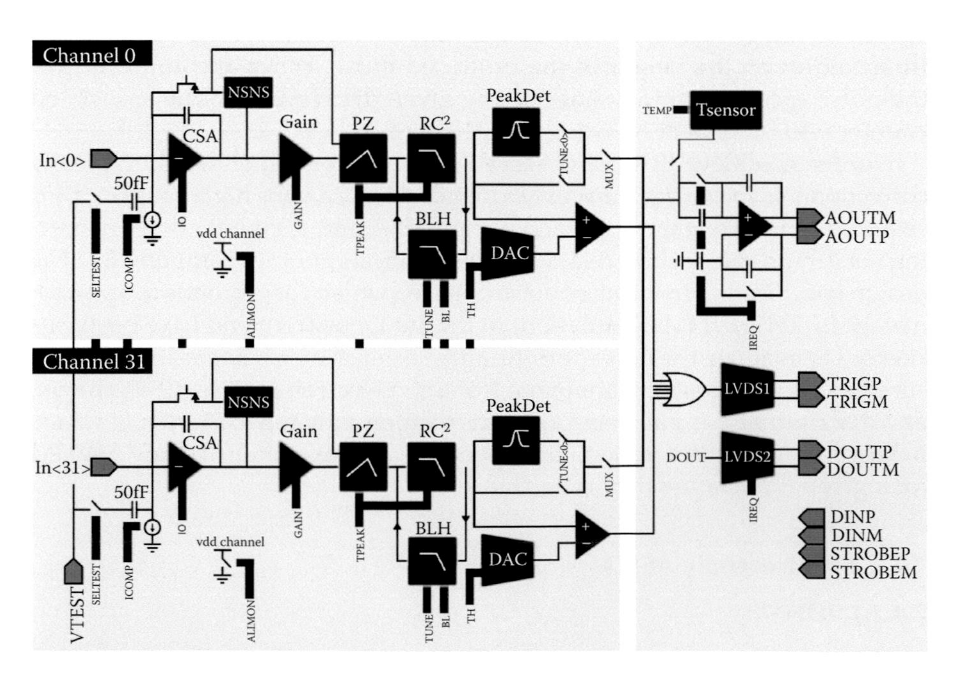

FIGURE 7.7
32-channel IDEF-X ASIC architecture.

per channel. The output of each analog channel feeds a discriminator and a stretcher. The discriminator compares the amplitude with an in-pixel reference low-level threshold to detect events. The stretcher consists of a PD and a storage capacitor to sample and hold the amplitude of the signal, which is proportional to the integrated charge and hence to the incident energy. In addition, each channel can be switched off by slow control programming to reduce the total power consumption of the ASIC when using only few channels of the whole chip.

The slow control interface was designed to minimize the number of signals and to get the possibility to connect together up to eight ASICs and address them individually. This optimization has allowed reducing the electrical interface from 49 pins in Caliste 256 to 16 pins in Caliste-HD for the same number of channels and using low-voltage differential signals (LVDS). When an event is detected by at least one channel, a global trigger signal (TRIG) is sent out of the chip. The controller starts a readout communication with three digital signals (DIN, STROBE, and DOUT) to get the address of the hit ASIC and then the hit channels. Then the amplitudes stored in the PDs of the hit channels are multiplexed and output using a differential output buffer (AOUT). The whole readout sequence lasts between 5 and 20 μs, according to the set delays and clock frequencies and the number of channels to read out.

7.4.2.2 VAS UM/TAT4

The VAS UM/TAT4 ASIC chip is used to read out both the amplitude of charge induction and the electron drift time independently for each anode pixel [24]. The ASIC has 128 channels, each with a charge-sensitive preamp and two CR-RC unipolar shapers with different shaping times. The slow shaper has 1 μs peaking time and is coupled to a peak-hold stage to record pulse amplitude. The fast shaper has a 100 ns shaping time and is coupled to simple level discriminators for timing pickoff.

Of the 128 channels, 121 are connected to the pixels, 1 is connected to the grid, and 1 is connected to the cathode. Compared to the anodes, the polarity of the signals is reversed for the cathode and the grid. The peak-hold properties, signal shaping, ASIC noise, and triggering procedures are included in the ASIC readout system model. The fast shaper can trigger off pulses as small as 30 keV for the anode and 50 keV for the cathode. Only the pixels with slow-shaped signals greater than a noise discrimination threshold of 25 keV are typically used in operation.

VAS UM/TAT4 is particularly well suited for 3D imaging and detection using thick CZT detectors (>10 mm) with high-energy photons (>1 MeV). 3D position-sensing techniques enable multiple-pixel events of pixelated CZT detectors to be used for 4π Compton imaging. Multiple-pixel events occur by either multiple gamma-ray interactions or charge sharing from a single electron cloud between adjacent pixels. To perform successful Compton imaging, one has to correct for charge sharing. There is a large research effort at the University of Michigan under the direction of Professor Zhing He to resolve these complicated signal processing issues and to reconstruct the trajectory of incoming photons for dirty bomb detection.

7.4.2.3 HEXITEC

HEXITEC was a collaborative project between the Universities of Manchester, Durham, Surrey, Birkbeck, and The Science and Technology Facilities Council (STFC). The objective of the program was to develop a new range of detectors such as CZT for high-energy x-ray imaging applications. The project has been funded by EPSRC on behalf of RCUK under the Basic Technology Program.

The HEXITEC ASIC consists of an 80 × 80 pixel array on a pitch of 0.25 mm [23]. Each pixel contains a 52 μm bond pad that can be gold stud bonded to a CZT detector. Figure 7.8 shows a block diagram of the electronics contained in each HEXITEC ASIC pixel. The charge is read from each of the CZT detector pixels using a charge amplifier, which has a selectable range and a feedback circuit that compensates for detector leakage currents up to 50 pA.

The output from each charge amplifier is filtered by a 2 μs peaking circuit comprising a CR–RC shaper followed by a second-order low-pass filter, as schematically shown in Figure 7.8. A peak hold circuit maintains the voltage at peak of the shaped signal until it can be read out. Three track-and-hold buffers are used to sample the shaper and peak hold voltages sequentially prior to the pixel being read.

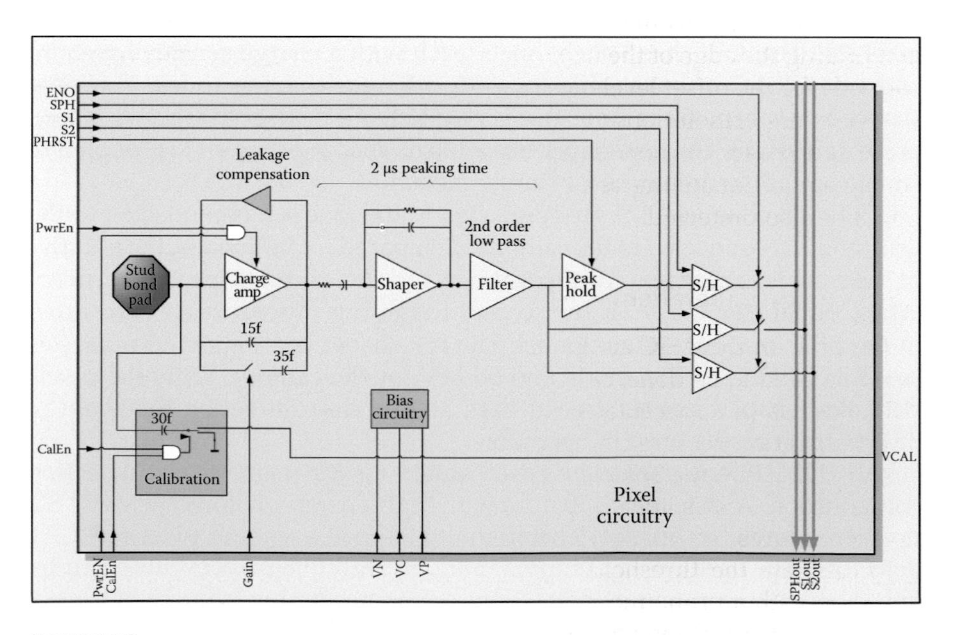

FIGURE 7.8
Block diagram of the HEXITEC architecture.

The HEXITEC is read out using a rolling shutter technique. A row select register is used to select the row that is to be read out. The data from each pixel become available on all column outputs at the same time, and at this point, the peak hold circuits in that row can be reset to accept new data. The data being held on the column output are read out through a column multiplexer. The column readout rate is up to 25 MHz, and the total frame rate depends on the number of pixels being read out. The main limitation of the HEXITEC is a maximum count rate due to 10 kHz frame readout scheme.

7.5 Operational Issues

7.5.1 Threshold Equalization

Before calibration of a CZT pixel detector, the chip has to be equalized in order to minimize the threshold dispersion between pixels. This requirement results from the fact that the threshold that the pixel sees is applied globally, but the offset level of the pixel can be slightly different due to process variations affecting the baseline of the preamplifier.

The equalization is performed with a threshold adjustment DAC in each pixel. The resolution of the adjustment DAC is usually in the range of four bits, depending on particular ASIC implementation. The standard way to calculate the adjustment setting for each pixel is by scanning the threshold and finding the edge of the noise, then aligning the noise edges. This adjusts correctly for the offset level of the pixel, but gain variations can still deteriorate the energy resolution at a given energy. To correct for the gain mismatch, either test pulses or monochromatic x-ray radiation has to be used for the equalization. Equalizing at the energy of interest instead of the zero level might be also preferred.

7.5.2 Energy Calibration

Depending on the ASIC architecture, there are two types of energy calibration that need to be done: calibration of the threshold and calibration of the ToT response (if applicable). For photon-counting chips as MEDIPIX-3, the only calibration required is the one of the threshold, while in ToT ASICs, such as TIMEPIX, the ToT response has to be calibrated as well. Virtually all spectroscopic ASICs, including HEXITEC, need to undergo energy calibration procedures.

To calibrate the threshold, we need monochromatic photons or at least radiation with a pronounced peak. These can be obtained from radioactive sources, by x-ray fluorescence, or from synchrotron radiation like Am241 and/or Co57 point sources. To find the corresponding energy for a certain threshold, the threshold is scanned over the range of the peak obtaining an integrated spectrum. The data are then either directly fitted with an error or sigmoid function or first differentiated and then fitted with a Gaussian function. From this fit, the peak position and energy resolution can be extracted. Repeating the procedure for multiple peaks, the result can then be fitted with a linear function, and the relation between threshold setting in DAC steps or mV and deposited energy in the detector is found.

7.5.3 Charge Sharing Corrections

When the pixel size is starting to approach the size of the charge cloud, the input signal is subjected to charge sharing. Charge sharing creates a characteristic low-energy tail and leads to a reduced contrast and distorted spectral information. To counteract this problem, there are two possibilities: either to use larger pixels (reduced spatial resolution) or to implement charge summing on a photon by photon basis.

For lower rates and with detectors that stores the energy information in each pixel either using ToT as TIMEPIX or a peak-and-hold circuit and an ADC as HEXITEC, the charge summing can be done offline. However, this requires that you do not have a second hit in the same pixel before you read the first one out. Using this approach, you also lose charge that is below the threshold.

Another approach is to sum the charge in the detector as implemented in MEDIPIX-3, where the analog charge is summed in a 2×2 cluster before being compared to the threshold. The advantage of this approach is that it can handle much higher interaction rates and that even charge below the threshold is summed as long as one pixel is triggered. However, since this correction has to be implemented in the ASIC architecture, this complicates the chip design and is less flexible.

7.5.4 Pileup Effects

Given that the processing of each photon takes time, there will be problems with pileup effects at high count rates. The pileup happens when a second photon arrives in the same pixel before the first one is processed. Depending on the system architecture, the second photon could either be lost or added to the signal of the first photon. The result will be a deviation from linear behavior for the count rate. This deviation can be corrected for up to a certain limit in photon-counting devices, but more problematic are the spectral distortions due to pileup that cannot be corrected for. For this reason, the operation of the HEXITEC, or any other spectroscopic ASIC, is limited to the maximum count rate that is not causing any pileup effects.

Different detectors will have different responses, and it is important that the detector is characterized and suitable for the flux in a specific application.

Since the flux is measured per area, smaller pixels offer an advantage leading to a smaller number of photons per second per pixel.

7.6 Conclusions

Key ASIC challenges are present in spectroscopic and photon-counting systems due to high flux and pileup that affect both the count rate linearity and the spectral response. One way to counteract that problem is to use smaller pixels, but smaller pixels will lead to more charge sharing. In this respect, the various chips offer an interesting combination of relatively small pixels and still very good energy resolution.

References

1. R. Macdonald, "Design and implementation of a dual-energy x-ray imaging system for organic material detection in an airport security application," *Proc. of SPIE*, vol. 4301, 2001.
2. Y. A. Boucher, F. Zhang, W. Kaye, and Z. He, "Study of long-term CdZnTe stability using the Polaris System," *IEEE NSS-MICS*, 2012.
3. H. Chen, F. Harris, S. A. Awadalla, P. H. Lu, G. Bindley, H. Lenos, and B. Cardoso, "Reliability of pixellated CZT detector modules used for medical imaging and Homeland Security," *SPIE Invited Paper*, 2012.
4. G. Harding, H. Strecker, S. Olesinski, and K. Frutschy, "Radiation source considerations relevant to next-generation x-ray diffraction imaging for security screening applications, penetrating radiation systems and applications," edited by F. Patrick Doty, H. Bradford Barber, Hans Roehrig, Richard C. Schirato, *Proc. of SPIE*, vol. 7450, 2009.
5. S. Skatter, H. Strecker, H. Fleckenstein, and G. Zienert, "Morpho detection, CT-XRD for improved baggage screening capabilities," *IEEE NSS MICS Conference*, 2009.
6. G. Harding, H. Strecker, and P. Edic, "Morpho detection and GE global research, high-throughput, material-specific, next-generation x-ray diffraction imaging (XDi)," *IEEE NSS-MICS Conference*, 2010.
7. V. Rebuffel, J. Rinkel, J. Tabary, and L. Verger, "New perspectives of x-ray techniques for explosive detection based on CdTe/CdZnTe spectrometric detectors," *International Symposium on Digital Industrial Radiology and Computed Tomography*, Berlin, 2012.
8. Y. Tomita, Y. Shirayanagi, S. Matsui, M. Misawa, H. Takahashi, T. Aoki, and Y. Hatanaka, "X-ray color Scanner with multiple energy differentiate capability," *IEEE*, 0-7803-8701-5/04, 2004.

9. A. A. Faust, R. E. Rothschild, P. Leblanc, and J. E. McFee, "Development of a coded aperture x-ray backscatter imager for explosive device detection," *IEEE Transactions on Nuclear Science*, vol. 56, no. 1, February 2009.
10. G. Harding and B. Schreiber, "Coherent x-ray scatter imaging and its applications in biomedical science and industry," Philips Research, *Radiation Physics and Chemistry*, 56 pp. 229–245, 1999.
11. http://www.morpho.com/IMG/pdf/X-ray_Diffraction_Technology_Presentation.pdf
12. S. D. M. Jacques, C. K. Egan, M. D. Wilson, M. C. Veale, P. Seller, and R. J. Cernik, "A laboratory system for element specific hyperspectral x-ray imaging," *Analyst*, Issue 3, pp. 755–759, 2013.
13. E. Cook, R. Fong, J. Horrocks, D. Wilkinson, and R. Speller, "Energy dispersive x-ray diffraction as a means to identify illicit materials: A preliminary optimisation study," *Appl. Radiat. Isotopes* 65, 959, 2007.
14. L. Verger, E. Gros d'Aillon, O. Monnet, G. Montemont, and B. Pelliciari, "New trends in x-ray imaging with CdZnTe/CdTe at CEA-Leti," *Nucl. Instrum. Methods A*, vol. 571, pp. 33–43, February 2007.
15. J. S. Iwanczyk, E. Nygard, O. Meirav, J. Arenson, W. C. Barber, N. E. Hartsiugh, N. Malakhov, and J. C. Wessel, "Photon counting energy dispersive detector arrays for x-ray imaging," in *Proc. IEEE Nucl. Sci. Symp. Rec., 2007*, pp. 2741–2748, 2007.
16. C. Szeles, S. A. Soldner, S. Vydrin, J. Graves, and D. S. Bale, "CdZnTe semiconductor detectors for spectrometric x-ray imaging," *IEEE Trans. Nucl. Sci.*, vol. 55, no. 1, pp. 572–582, Feb. 2008.
17. S. Mikkelsen, D. Meier, G. Maehlum, P. Oya, B. Sundal, and J. Talebi, "An ASIC for multi-energy x-ray counting," in *Proc. IEEE Nucl. Sci. Symp. Rec., 2008*, pp. 294–299, 2008.
18. O. T. Tümer, V. B. Cajipe, M. Clajus, S. Hayakawa, and A. Volkovskii, "Multichannel front-end readout IC for position sensitive solid-state detectors," in *Proc. IEEE Nucl. Sci. Symp. Rec., 2006*, pp. 384–388, 2006.
19. X. Wang, D. Meier, B. M. Sundal, B. M. Oya, P. Maehlum, G. E. Wagenaar, D. J. Bradley, E. Patt, B. M. W. Tsui, and E. C. Frey, "A digital line-camera for energy resolved x-ray photon counting," in *Proc. IEEE Nucl. Sci. Symp. Rec., 2009*, pp. 3453–3457, 2009.
20. A. Brambilla, C. Boudou, P. Ouvrier-Buffet, F. Mougel, G. Gonon, J. Rinkel, and L. Verger, "Spectrometric performances of CdTe and CdZnTe semiconductor detector arrays at high x-ray flux," in *Proc. IEEE Nucl. Sci. Symp. Rec., 2009*, pp. 1753–1757, 2009.
21. J. Rinkel, G. Beldjoudi, V. Rebuffel, C. Boudou, P. Ouvrier-Buffet, G. Gonon, L. Verger, and A. Brambilla, "Experimental evaluation of material identification methods with CdTe x-ray spectrometric detector," *IEEE Trans. Nucl. Sci.*, vol. 58, no. 2, pp. 2371–2377, October 2011.
22. L. Tlustos, "Spectroscopic x-ray imaging with photon counting pixel detectors," *Nucl. Instrum. Methods A*, vol. 623, no. 2, pp. 823–828, Nov. 2010.
23. L. Jones, P. Seller, and M. Wilson, Alec Hardie, "HEXITEC ASIC—A pixellated readout chip for CZT detectors," *Nucl. Instr. Methods A*, 2009.
24. F. Zhang and Z. He, "New readout electronics for 3-D position sensitive CdZnTe/HgI2 detector arrays," *IEEE Trans. Nucl. Sci.*, vol. 53, no. 5, pp. 3021–3027, 2006.

8

Applications of Hybrid Pixel Detectors for High Resolution Table-Top X-Ray Imaging

Jan Dudák

CONTENTS

8.1 Introduction

In the last decade, x-ray imaging has undergone enormous technological progress. The imaging technique previously known mostly as a medical diagnostic tool has expanded to a wide range of industrial and research fields. Thanks to the rapid development of compact x-ray sources and detection technologies, table-top, or at least laboratory-scale, x-ray inspection systems providing spatial resolution at the level of tens of microns or better have become widely available and affordable. The market nowadays provides constantly increasing numbers of micro-CT systems designed for different purposes.

Although photon counting detector (PCD) technology was originally developed for particle tracking and for use in large research facilities for high-energy physics, it has also proved its potential if used as an imaging camera. Despite the fact that PCDs can offer unique properties important for imaging applications, they are still not widely used in this field and remain mostly in laboratory systems.

This chapter overviews applications of PCDs in the field of high-resolution imaging using ionizing radiation in lab-scale systems. The most common application is using the PCD as a detector for high-resolution x-ray radiography and tomography. Both mentioned imaging techniques are the main subject of this chapter. Nevertheless, PCD technology has found its use even in other imaging modalities, which will be mentioned as well. The chapter is divided into several parts. First of all, the technology of PCDs is briefly described and available PCDs are introduced. Furthermore, the reader will find basic information about the principles of x-ray radiography and tomography and a summary of state-of-the-art micro-CT system construction and technology. Important parameters of detectors, radiation sources, and image quality evaluation criteria will be discussed. Finally, applications of PCD technology in different fields of science will be addressed. Conclusions summarize the content of the chapter and bring the author's opinion of PCD technology perspectives in the near future.

The author of this chapter has a master's degree from biomedical engineering. His institute, the Institute of Experimental and Applied Physics, Czech Technical University in Prague (IEAP)-is one of the contributing facilities to the development of Medipix-type detectors. As a member of the Department of Applied Physics and Technology, he leads a research team focused on applications of PCDs for imaging in the field of natural sciences. Since the author's expertise is in the biomedical field, it is also emphasized in this chapter.

8.2 Photon Counting Semiconductor Detectors

Photon counting (or single particle counting) detectors are a CMOS-based technology that can be used for detection of ionizing radiation. A characteristic feature of all PCDs is the integration of electronics in each pixel. The device works with an adjustable energy threshold, since the electronic part of each individual pixel contains a pre-amplifier, comparator, and counter. The charge deposited and collected in the semiconductor sensor is amplified, compared to a custom-adjusted threshold, and if the threshold level is exceeded, the event is digitally counted. Occurrence of dark current in the detected signal is avoided thanks to the adjustable threshold. Therefore, PCDs operate in a noise-free event-by-event counting mode and are capable of achieving virtually unlimited dynamic range. The importance of individual PCD properties, making them a powerful x-ray imaging detection technology, is discussed in the third section of this chapter. In the following subsections, a description of PCD technology is provided and available PCDs are introduced.

8.2.1 Anatomy of PCDs

PCDs are built using two different processing technologies. Either monolithic or hybrid pixel detectors can be produced. In the case of monolithic devices, the whole detector–sensor and readout electronics are integrated into the same semiconductor substrate (see Figure 8.1, left). The monolithic approach is advantageous, e.g., if an extremely thin detector is required. On the other hand, the sensor/readout integration brings limitations as well. Obviously, the sensor material must be the same as the readout part. Therefore, the sensor material of monolithic PCDs is restricted to silicon only. A representative of monolithic PCDs is the macropixel detector based on DEPFET technology (depleted P-channel field effect transistor) developed in the MPI Semiconductor Laboratory in Germany (Zhang et al. 2008).

Hybrid pixel detectors are assembled from an independent sensor layer and readout chip. Both parts are interconnected using so-called bump-bonding technology (use of solder microballs to conductively connect matching

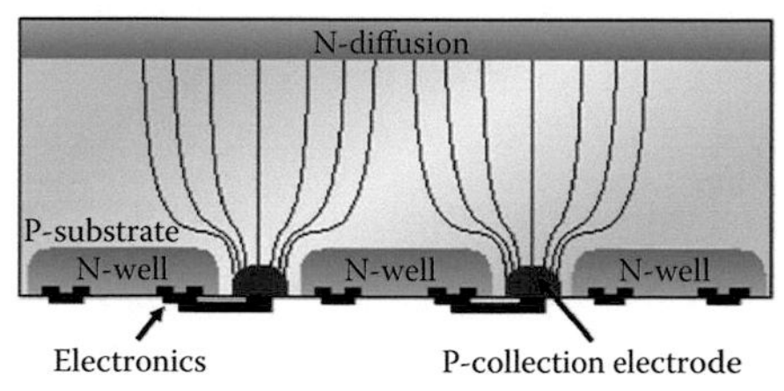

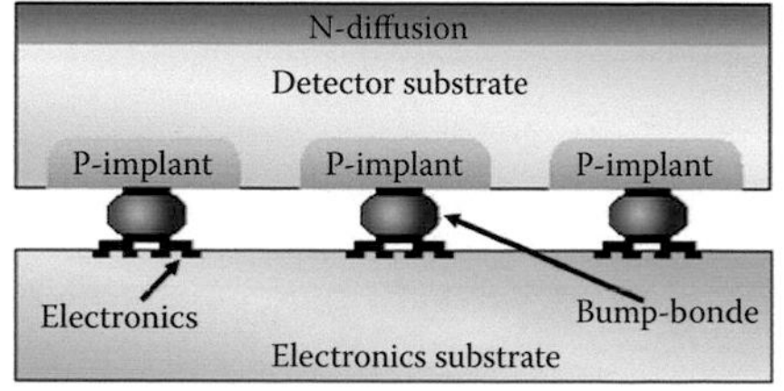

FIGURE 8.1
Comparison of construction of a monolithic PCD (left) and a hybrid pixel detector (right). Monolithic construction allows for the construction of extremely thin assemblies; on the other hand, the sensor material is restricted to silicon only. In the case of hybrid pixel construction, a variety of different sensor materials can be used. (From M. Platkevic, Signal Processing and Data Read-Out from Position Sensitive Pixel Detectors. PhD Thesis, Czech Technical University in Prague, 2014.)

electrode patterns; see Figure 8.1, right). Hybrid pixel construction allows for the use of a variety of sensor materials. Although silicon is still the most widely used, several alternative sensor materials are under development and testing (GaAs, CdTe, CdZnTe, and others). The aim of these efforts is to improve quantum efficiency of the sensor of the semiconductor detector to higher photon energies. The rest of this chapter is dedicated purely to hybrid pixel PCDs.

8.2.2 Sensor Materials Utilized for PCDs

The most frequently used sensor material is silicon. Generally speaking, silicon is the building material of all electronic components worldwide. Obviously, growing high-purity silicon crystals as well as fabricating and processing wafers is at a very high level. As a sensor material, silicon is, therefore, easily accessible and provided at the highest quality. On the other hand, with sensor thickness of typically 100–1000 μm, the detection efficiency of silicon sensors drops rapidly for energies higher than approximately 20 keV. Considering x-ray imaging, PCDs with silicon sensors are not suitable for the imaging of large objects or objects composed of highly attenuating materials.

Options to overcome the mentioned limitation of silicon as the sensor material are being intensively investigated. Currently, several alternative semiconductor or semi-insulating sensor materials are available. All of them contain high-Z elements (GaAs, CdTe, CdZnTe) and make PCDs suitable for detecting photons up to 100 keV. The calculated detection efficiency of a 1000 μm CdTe sensor for 100 keV photons is approximately 56%. This improvement of sensor efficiency has opened a wide range of new applications—mostly in the field of material science and engineering. Particular applications are described in the dedicated section of this chapter. Unfortunately, regular use of the mentioned high-Z sensor materials is not very common because, besides their great detection efficiency, there are drawbacks compared to silicon, as well. Sensor inhomogeneity and temporal and temperature instability of sensor are the major issues to be solved in the case of high-Z semiconductive sensor materials. Characterization and improvements of these materials for use as PCD sensors are subjects of a number of recent scientific publications.

8.2.3 Hybrid Pixel PCDs Available for High-Resolution Imaging

As hybrid pixel technology became more widely used, several different PCD families have been developed. Some of them are now commercially available for imaging applications. The next sections provide basic information about currently available PCDs suitable for x-ray microradiography and micro-CT.

8.2.3.1 Medipix

The Medipix Collaboration founded at CERN has already developed three generations of PCDs (Medipix1, Medipix2/Timepix, and Medipix3/Timepix3).

The Medipix1 chip was an array of 64 × 64 pixels with 170 μm pitch. It was introduced in 1997 (Bisogni et al. 1998). Currently, the most widely used Medipix-type detectors for high-resolution imaging are based on Medipix2 and Timepix technology (Llopart et al 2007). The basic chip assembly provides an array of 256 × 256 pixels with a 55 μm pixel pitch. The sensitive area of a single chip is approximately 1.4 × 1.4 cm^2. Since Medipix2 development, the pixel size and chip layout have not changed. Except for the electronics described at the beginning of this chapter, the Timepix chip pixels also have an integrated timer. The timer allows the detector to directly measure the energy deposited at each pixel. The per-pixel fully spectroscopic response makes the Timepix detector suitable for energy-sensitive x-ray imaging. Energy-sensitive radiography can be acquired with the Medipix3 detector as well. However, in the case of Medipix3, the response is not truly spectroscopic as a system of several thresholds is only used.

Medipix3 can be operated either with 55 μm pixels with two thresholds in each pixel or so-called superpixel mode with 110 μm pixels and eight thresholds per pixel. Furthermore, the charge sharing effect between adjacent pixels is compensated by dedicated processes at the pixel level (Gimez et al. 2011). The latest generation, Timepix3, brings back the fully spectroscopic response and, furthermore, it detects time of arrival of each individual particle with a nanosecond precision together with the hit position and energy. Despite the high pixel granularity, the applicability of Medipix-type detectors was limited due to the small area of a single detector chip. Several attempts have been made to increase the field of view (FOV) of Medipix detectors. The first step in the way was the Timepix Quad detector configuration, utilizing a common sensor and an array of 2 × 2 Timepix readout chips increasing the sensitive area to 2.8 × 2.8 sq. cm (512 × 512 pixels). A similar solution is the Hexa configuration built of the array of 3 × 2 Timepix chips (Zuber et al. 2014) or LAMBDA with the array of $N \times 2$ chips (Pennicard et al. 2011). The maximal size of all mentioned configurations is limited by the size of the common sensor layer since it can only be fabricated in limited dimensions. A significant step forward had come with the development of three-side buttable edgeless sensors that enabled the modular assembling of larger detector arrays from individual detector assemblies—WidePIX technology. Two different WidePIX assemblies as well as a Quad detector are shown in the first row of Figure 8.2. The technology was developed at IEAP and now is offered by Advacam Company (Advacam, Prague. http://advacam.com/.). The largest WidePIX detector ever built was assembled from 100 individual Timepix chips precisely aligned to a squared array. The size of the detector is larger than 14 × 14 sq. cm and contains a matrix of 2560 × 2560 pixels (Jakubek et al. 2014). As the construction of WidePIX detectors is modular, numerous different sizes and shapes can be built. Currently, WidePIX detectors are the only PCD technology enabling the assembly of PCD modules into a 2D array without significant insensitive areas between individual tiles.

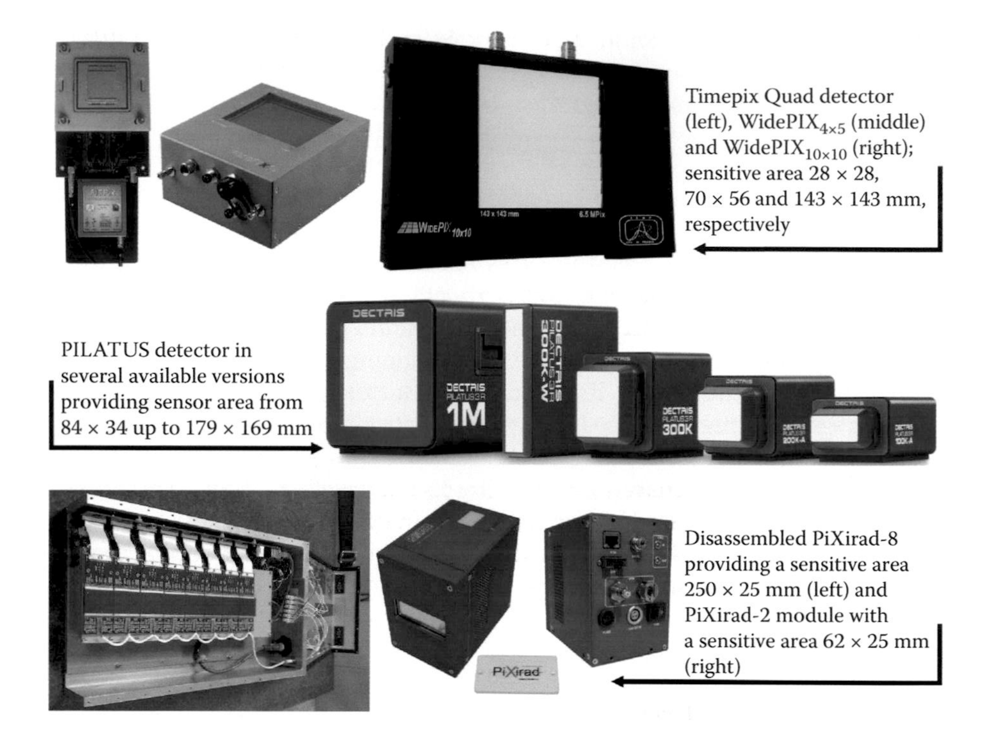

FIGURE 8.2
Examples of different types of hybrid pixel PCDs available for high-resolution imaging. From top to bottom: Timepix (Courtesy of IEAP and Advacam.), Pilatus (Courtesy of DECTRIS Ltd.), and PiXirad (Courtesy of PiXirad IC s.r.l.).

8.2.3.2 PILATUS, EIGER

Other widely known hybrid PCDs are PILATUS and EIGER, both developed by the PSI-SLS group, Switzerland (Dectris, Baden-Daettwil. https://www.dectris.com/). Both detectors are offered by the DECTRIS Company. Their development has been aimed for use at synchrotron sources, but they are suitable for laboratory-scale x-ray imaging as well. A series of detectors has been designed directly for laboratory and industrial use. The laboratory series of PILATUS detectors is shown in the second row of Figure 8.2. While PILATUS, the older generation, uses 172 μm pixels, its successor, EIGER, has a pixel pitch of 75 μm. A basic readout chip composes of an array of 256 × 256 pixels. Chips are typically bump-bonded to a common semiconductor sensor in a 2 × 4 layout producing a detector with sensitive area of about 40 × 80 mm (512 × 1024 pixels). Since such modules provide continuous sensitive area, they are suitable for x-ray radiography. Detector modules are further assembled into larger arrays to increase the sensitive area. In the case of PILATUS and EIGER, assembling detector modules into larger arrays brings a drawback in insensitive gaps between modules. Therefore, the dominant field of use of these large area detectors is in diffraction measurements, i.e., in crystallography at synchrotron facilities.

8.2.3.3 PiXirad

An interesting sensor solution is provided by PiXirad PCDs manufactured by Pixirad Imaging Counters s.r.l. ("PiXirad Imaging Counters Website" 2017, http://www.pixirad.com/). Unlike the other PCD technologies, PiXirad uses hexagonal pixels with 60 μm pixel pitch (Delogu et al. 2016). Thanks to this unusual pixel arrangement, an almost isotropic spatial resolution in all directions is achieved, while in the case of a square pixel grid, the resolution in the diagonal direction is obviously decreased. A basic PiXirad assembly is composed of an array of 512 × 476 pixels. Each pixel has two independently adjustable thresholds enabling dual energy imaging. As PiXirad sensors are two-side buttable, the sensitive area can be increased by assembling several basic units into a row. The largest available PiXirad detector built this way is PiXirad-8 providing sensitive area of 250 × 25 mm^2 (4096 × 476 pixels, shown in the bottom row of Figure 8.2). The dead gap between neighboring modules is equivalent to two pixel columns. The successor of PiXirad detectors is also already available. The latest generation of PCD developed by Pixirad Imaging Counters s.r.l. is PIXIE III (Bellazzini et al. 2015). Compared to previous versions, the PIXIE III readout chip uses a square pixel grid with 62 μm pixel pitch and again two energy thresholds in each pixel. The basic assembly provides an array of 512 × 402 pixels. Like Medipix3, PIXIE III addresses the issue of charge sharing and uses processes to avoid these effects and maintain the highest spatial and energetic resolution.

8.2.3.4 XPAD

The European Synchrotron Radiation Facility (ESRF) successfully developed another PCD dedicated to imaging in material science called XPAD (Basolo et al. 2005; Delpierre et al. 2007a; XPAD Pixel Detector). The latest generation is XPAD3 (Pangaud et al. 2007; Berar et al. 2009). The XPAD3 chip consists of an array of 120 × 80 pixels with 130 μm pixel pitch. Up to eight chips can be aligned together to produce a detector module. Further, up to eight modules can then be assembled together; however, gaps between modules are inevitable. The largest available detector based on XPAD3 technology provides sensitive area 150 × 150 mm. The older version, XPAD2, provides a chip with an array of 24 × 25 pixels with 330 μm pixels. Although XPAD technology was, similarly as PILATUS and EIGER, originally developed for use at synchrotron facilities, XPAD technology has found its applications even in laboratory-scale x-ray imaging systems.

8.3 High-Resolution X-Ray Imaging Principles and Device Construction

PCDs are used mostly for x-ray microradiography and micro-CT, but several different applications are under development (i.e., SPECT imaging and use in multimodal imaging systems or x-ray fluorescence imaging). X-ray transmission radiography is an imaging technique based on irradiation of an investigated object by an x-ray photon beam and detection of the transmitted beam intensity behind the object by a proper imaging unit. In this way, a radiographic projection of the object containing information of its inner structures is recorded.

The following section describes basic important parameters affecting radiographic data quality and construction of radiographic systems. Further, the state-of-the-art detection technologies conventionally used in the field of high-resolution x-ray imaging are briefly mentioned, and key PCD properties important for imaging are discussed.

8.3.1 Computed Tomography

Microtomography (or micro-CT) is an advanced radiographic approach. It provides 3D information about inner structures of an investigated object. During data acquisition, a large set of 2D projections (typically from hundreds to thousands) is captured under different angles. The obtained 2D projections are then used as an input for dedicated mathematical algorithms that perform the tomographic reconstruction—the process of creating a model of the scanned object by back-projecting the acquired 2D projection into a 3D space. The obtained voxel-based 3D model of the scanned object provides a number of possibilities for further analysis of the data compared to a simple 2D radiographic projection.

8.3.2 Construction of Micro-CT Systems

The construction of micro-CT systems usually follows the same pattern—the radiation source and detector unit are mounted steadily, and a multiaxial sample positioning stage is situated between them. Typically, the sample rotates around a vertical axis during the scan. The described construction has very adaptable geometry and provides maximal stability and precision as the heavy-weight components are immobile during a measurement. Nevertheless, such design is not suitable for all applications. For example, in the field of biomedical or preclinical research, rotation of the sample is inconvenient. Micro-CT systems dedicated to small animal imaging, therefore, use construction derived from human-scale medical CT systems. The sample—e.g., a living animal—is placed steadily on a horizontal bed, and the source and the detector are mounted on a gantry system rotating around the sample during data acquisition. Micro-CT systems constructed with a rotating gantry generally provide lower resolution, but they are still irreplaceable since they allow small animals to be scanned in a natural position and maximal stability of the sample is maintained this way.

8.3.3 Detector Technologies Conventionally Used for High-Resolution X-Ray Imaging

Currently, the most common x-ray imaging digital sensors are based on a combination of a scintillation crystal, converting incoming x-ray photons to visible light, with several different types of readout electronics.

Systems dedicated for the imaging of volumes smaller than a few cubic centimeters usually utilize CCD-based x-ray detectors. Such x-ray cameras provide extremely high pixel granularity. The pixel size of such devices is often smaller than 10 μm. Together with micro- or nanofocus x-ray sources, CCD x-ray detectors are suitable for imaging with an extremely high resolution. To avoid undesirable diffusion of the light within the scintillation sensor leading to degradation of spatial resolution, the scintillation sensor must be kept very thin—typically tens of micrometers. For this reason, these detectors are not, similarly to PCD with silicon sensors, very suitable for imaging using hard radiation.

In the case of systems designed for scanning large objects, where the use of highly penetrating radiation is necessary, or if scan time is a crucial parameter, a flat-panel detector is typically used. As flat-panel detectors are based on CMOS technology, they provide very fast data readout. Further, they provide a large FOV, pixel size usually in tens or hundreds of microns, and high detection efficiency for hard radiation. It is widely used in industry or even in medical radiographic systems. Shared limitations of all conventional scintillation-based x-ray imaging cameras are the occurrence of the dark current decreasing the signal-to-noise ratio of the data and the diffusion of the light within the sensor causing image blur. Both mentioned drawbacks are avoided in the case of a PCD.

8.3.4 Spatial Resolution and Contrast of Radiographic Data

The state-of-the-art laboratory x-ray imaging systems today are capable of visualizing a sample even with submicron spatial resolution. People frequently ask how it is possible to achieve resolution of a few microns or less if detector pixel size is typically tens of microns. The answer to this question is hidden behind the way a radiographic projection is formed. The spatial resolution of an x-ray imaging system not only depends on detector pixel size but also is strongly influenced by properties of the x-ray source and depends on imaging geometry. Typically, laboratory high-resolution x-ray imaging systems use magnifying projection geometry—a radiation source with a point-like focal spot produces a divergent photon beam (also called a cone beam) that enables the magnification of an object projection by a factor of M = FDD/FOD, where FDD is the focal-spot-to-detector-distance and FOD is the focal-spot-to-object-distance (see Figure 8.3, left).

This way, as the object projection on the detector is enlarged, the sampling by detector pixels becomes effectively finer. Under such conditions, the term effective pixel size (*EPS*) is commonly used. *EPS* is a ratio of the detector's actual pixel size and magnification factor M. *EPS* actually characterizes the sampling density of the imaged object. Furthermore, to maintain projection sharpness, it is necessary to minimize the focal spot size of the x-ray source as with higher M the penumbra effect, image blur caused by radiation source spot dimensions, becomes significant. The penumbra effect is demonstrated in the right part of Figure 8.3 at an example of two x-ray tubes with different spot dimensions. Therefore, the quality of an x-ray tube is just as important as the detector unit for high-resolution approaches. X-ray sources designed for high-resolution imaging are usually called microfocused (with focal spot smaller than approximately 15 μm) or nanofocused (focal spot smaller than 1 μm).

Beside spatial resolution, crucial parameters of radiographic projection quality are contrast and occurrence of noise in the data. In the case of radiography, both contrast and noise are simultaneously described by the contrast-to-noise ratio (CNR). The CNR basically describes the detectability of certain regions of interest within the data based on its contrast and actual noise level. While the contrast of the radiographic image mainly depends on elemental composition of the sample and on the used x-ray spectrum, there are several different sources of noise. The most significant is the dark current—a signal integrated by the detector regardless of whether it is irradiated or not. Therefore, avoiding the dark current significantly improves the data quality.

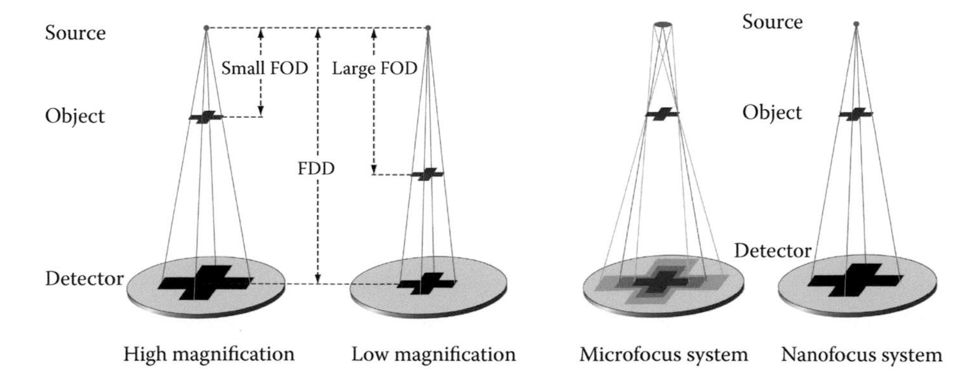

FIGURE 8.3
Scheme showing basic principles of radiographic imaging using divergent (cone) beam. The magnifying effect (left) allows achieving high resolution of the obtained data; however, the maximal resolution is limited by penumbra effect caused by source spot size (right). (Courtesy of GE Inspection Technologies, "Solder Joint Inspection and Analysis," 2011.)

8.3.5 PCD Features Important for Imaging

Although PCD technology today is not widely used for high-resolution x-ray imaging, it has several unique features that make PCDs powerful tools for imaging applications. The previously mentioned adjustable energy threshold, which is a characteristic property of all PCDs, allows dark-current-free data acquisition. As was stated in the previous section, avoiding image noise sources significantly improves the data quality. Only quantum noise obeying Poisson distribution is present in the case of PCD technology. Under such circumstances, the CNR of acquired data is limited only by the number of detected photons.

Despite the fact that contrast is given mainly by the composition of the sample and beam spectrum, it can be partially influenced by the mechanism of detected signal processing. The signal produced in a scintillation crystal is dependent on deposited energy. Highly energetic photons produce a stronger signal; therefore, their contribution to image formation is stronger compared to low-energy photons. In the case of a PCD, each photon—regardless of its energy—is handled as an equal event. Considering that low energies provide higher contrast, PCD technology should, thanks to equal weighing of all photon energies, achieve higher contrast of radiographic projection with a given x-ray spectrum. Thanks to features described in the two previous paragraphs, PCD technology is especially suitable for the imaging of samples with low intrinsic contrast.

As the current generation of PCDs provides two or more adjustable energy thresholds in many cases, PCDs are suitable for dual-energy x-ray imaging or even imaging with several energy bins within a single acquisition. Energy-sensitive imaging brings new opportunities to the field of x-ray imaging. As attenuation coefficients are characteristic for various materials and also energy-dependent, energy-sensitive x-ray radiography and CT can provide valuable information on the material composition of the imaged object. In the case of the Timepix detector, fully spectral response in each pixel is achievable. For imaging purposes, such capability can be used, i.e., for imaging using x-ray fluorescence, where photons of characteristic energies are emitted from the sample.

8.4 Biomedical Applications

In the biomedical field, PCDs are applied at preclinical imaging, where small animals (mice, rats, rabbits, etc.) are used as a model for the human body. Successful visualization of inner structures of small animals in fine detail is an important diagnostic tool for cancer treatment research, genomics, developmental biology, and many others. The most common application is using PCD as an imager in a micro-CT scanner. However, there are also different approaches like a PCD-based SPECT scanner or using PCDs in multimodal imaging systems (PET-CT, SPECT-MRI).

Key features in this case are noiseless counting (especially if visualization of soft tissue structures is required) and the energy sensitive response enabling acquisition of so-called color radiography/tomography.

Considerable efforts in the application of PCD technology for high-resolution imaging in biomedicine are carried out at IEAP, where the author of this chapter works and carries out research in this field. IEAP, as a member of Medipix collaboration, has been contributing to the field of high-resolution x-ray imaging since the Medipix technology was introduced. Before the development of WidePIX technology, Medipix detectors provided a very limited FOV; therefore, the first publications focused on biological samples were dedicated to the imaging of insects or other small biological objects. The emphasis was put on the imaging of low absorption contrast objects, in vivo imaging of insects (Dammer et al. 2009), phase-contrast enhanced x-ray imaging (Jakubek et al. 2007), or energy sensitive radiography with Timepix (Dammer et al. 2011). An approach using the Timepix detector for x-ray phase-contrast imaging with a single grating was very promising—a novel experimental way of phase-contrast radiography acquisition that significantly reduces the radiation dose and necessary acquisition time compared to double-grating methods (Krejci et al. 2011).

Research in the biomedical field at IEAP currently runs in collaboration with the Third Faculty of Medicine, Charles University (FM UK) and the Faculty of Biomedical Engineering CTU in Prague (FBME). The micro-CT laboratory of IEAP is equipped with three different micro-CT systems of its own construction using Timepix detector technology. Two of them are designed for scanning with high magnification factors and provide a spatial resolution of approximately 5 and 1 μm, respectively. To provide sufficient FOV for wide range of applications, both systems are equipped with WidePIX detectors with silicon sensors (currently $\text{WidePIX}_{10\times5}$ providing a fully sensitive area of approximately 140 × 70 mm and $\text{WidePIX}_{5\times4}$ with sensitive area of 70 × 56 mm). The third micro-CT system is dedicated to small animal imaging purposes. It is based on MARS-CT system (detailed description in the section dedicated to in vivo imaging) construction with additional modification performed at IEAP (Dudak et al. 2015). The modified system situated at joint laboratory of IEAP, FM UK, and FBME is equipped with Timepix Quad detector and minifocus x-ray tube with 35 μm spot size. Compared to the original design, it provides better spatial resolution thanks to the smaller focal spot, and it is capable of fully spectroscopic measurements.

In 2012, IEAP was awarded by honorable mention the 2012 International Science & Engineering Visualization Challenge for a contribution "X-ray micro-radiography and microscopy of seeds" (Sykora et al. 2013). Recent research is focused on utilizing noiseless photon counting to achieve sufficient contrast of soft biologic tissue even without the application of dedicated high-Z contrast agents (Dudak et al. 2016). Results of such research can be seen in Figures 8.4 and 8.5. While Figure 8.4 shows CT reconstruction of a PlastiMouse™ phantom scanned with an EPS of about 20 μm, Figure 8.5 demonstrates results of a CT-scan of an ex vivo ethanol-preserved mouse

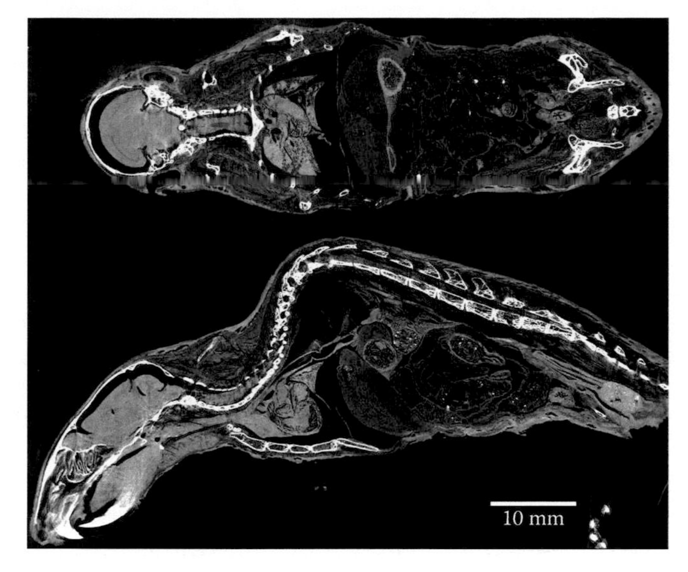

FIGURE 8.4
Frontal (top) and sagittal (lower) slices of a CT reconstruction of a PlastiMouse™ mouse phantom. The phantom was scanned at IEAP with approximately 20 μm EPS using a micro-CT setup equipped with a WidePIX$_{10\times5}$ PCD and Hamamatsu microfocus x-ray tube. Along with the skeletal system, even soft tissue structures are visualized in high contrast and detail.

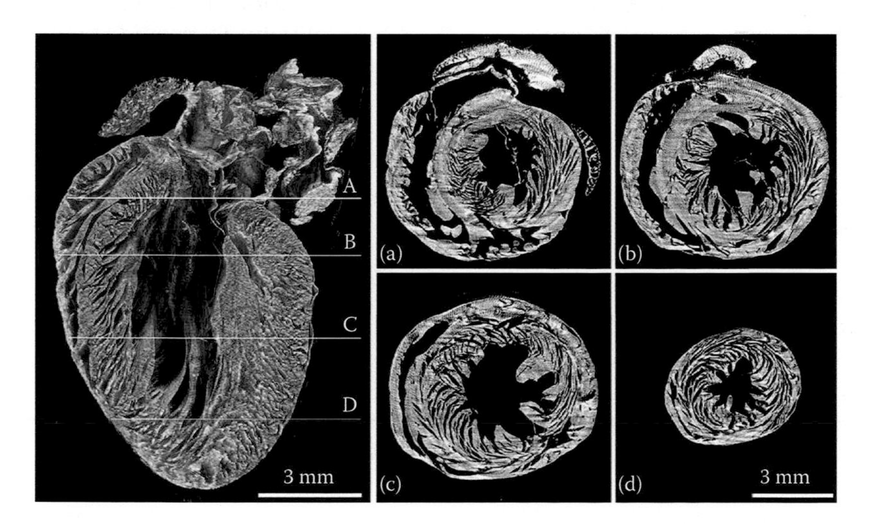

FIGURE 8.5
CT reconstruction of an ex vivo ethanol-preserved mouse heart scanned using a micro-CT setup equipped with WidePIX$_{10\times5}$ PCD and Hamamatsu microfocus x-ray tube; volume rendering presented in false colors system (left) and four selected transversal slices (a–d). The phantom was scanned at IEAP with about 7.2 μm EPS. (From Dudak, J. et al., *Scientific Reports* 6:30385, 2016.)

heart scanned with EPS 7.2 μm. In the case of ex vivo scans, higher resolution and overall higher data quality can be achieved compared to in vivo measurements, as scan time and the delivered dose are no longer key parameters. In the future—with continuous technology development—ex vivo x-ray imaging of soft biologic tissue may become one of the complementary tools to conventional tissue histology.

8.4.1 In Vivo X-Ray Imaging

The first commercially available micro-CT system based on PCD dedicated for biomedical imaging is MARS-CT (Medipix All Resolution System) developed by the University of Canterbury, New Zealand (Zainon et al. 2010; Walsh et al. 2011). Since MARS-CT used the Medipix3 chip in Quad configuration, it has also been the first available spectral x-ray scanner using PCD technology. In the so-called spectroscopic mode, the detector provides an array of 256 × 256 pixels with 110 μm pixel pitch. The gantry construction allows for the adjustment of the magnification factor within the range of approximately 1.2–2. The FOV of a Quad sensor is, therefore, about 14–23 mm. Nevertheless, the detector position is also adjustable, and the FOV can be effectively widened up to 100 mm. Obviously, the scanning of several projections with different detector positions is necessary in each angular position of the gantry. The MARS-CT system was successfully tested using both phantom objects and real biomedical samples. Today, MARS-CT exploits the capability of the Medipix3 detector to acquire data in several energy bins, and it is being used for preclinical research (Aamir et al. 2014; Rajendran et al. 2017).

Another CT system dedicated to small animal imaging utilizing PCD technology was developed in France as a cooperative project of four research facilities. The system is called PIXSCAN and it is based on the use of the XPAD3 detector (Delpierre et al. 2007b). The first prototype was built with an XPAD2 detector module composed of eight one-by-eight detector modules (total area of 68 × 65 mm). Gaps between individual chips, unfortunately, could not be avoided as each sensor is surrounded by a guard ring. Since the sample holder is positioned in the center between the source and the detector, the EPS of obtained projections is 165 μm. The construction of the prototype is unusual, as the sample is mounted to a rotation stage. Conventional CT systems dedicated to small animal imaging use a rotating gantry to maintain natural and stable position of the sample. In the case of PIXSCAN, the sample—a mouse—is hung along a vertical carbon axis mounted to the rotation stage. Despite the unusual construction, the PIXSCAN system has been used for noninvasive in vivo visualization of the development of lung cancer in mice (Debarbieux et al. 2010).

8.4.2 Multimodal Imaging Systems

Multimodal imaging systems have been grabbing increased attention in biomedical and preclinical research as they simultaneously provide both anatomical and functional information. Several attempts utilizing PCDs in multimodal imaging systems have been made. XPAD technology was used for the design and development of a hybrid PET/CT system (Khodaverdi et al. 2007; Nicol et al. 2009; Hamonet et al. 2015). A CT unit based on XPAD3 was combined with modified ClearPET–positron emission tomography system for scanning small animals developed by the Crystal Clear Collaboration (Ziemons et al. 2005). The system uses a partial ring of PET detectors shielded against the scattered x-radiation and an x-ray CT unit sharing the same FOV. The installed XPAD3 detector consists of an array of 560 × 960 pixels with 130 μm pixel pitch. In contrast to the PIXSCAN, the hybrid PET/CT system was designed with a rotating gantry construction, which is much more convenient for in vivo measurements.

As was stated in the first paragraph of the application section, PCD technology is also used for different imaging applications than x-ray radiography. A prototype of a magnetic resonance imaging (MRI) compatible single photon emission computed tomography (SPECT) unit utilizing PCD was developed and built at IEAP (see left part of Figure 8.6; Zajicek et al. 2013). In contrast to various common multimodal combinations (i.e., PET/CT, SPECT/CT), the design of a multimodal imaging system containing an MRI unit is a much more complicated task due to the presence of an intensive magnetic field (approximately 5–18 Tesla in the small animal scanners). Conventional SPECT detectors with scintillation crystals and photomultipliers are not capable of working in such magnetic fields. Moreover, the collimators used

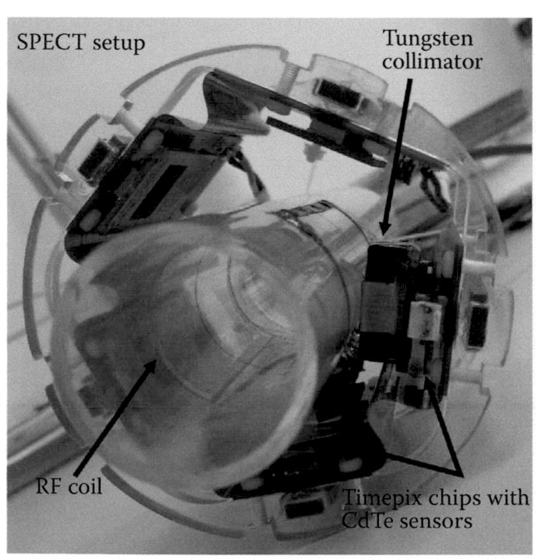

FIGURE 8.6
Prototype of an MRI-compatible SPECT module based on Timepix detection technology developed at IEAP (left). The SPECT unit inserted in the Bruker BioSpec 47/20 MR animal scanner gantry (right). (Courtesy of Jiří Zajíček, IEAP.)

for SPECT measurements are normally fabricated from conductive materials and that would cause disturbances of MRI signals.

The SPECT unit for the MRI-SPECT system developed at IEAP is built of plastic components holding a set of Timepix detectors with 1 mm thick CdTe sensors equipped with MRI-compatible collimators fabricated from tungsten and an integrated radiofrequency (RF) coil. The whole SPECT unit was designed to be compatible with the Bruker BioSpec 47/20 MR animal scanner. During the testing it was found that the high-power RF pulses of the MRI system influence the electronics of the Timepix readout system. Also DC power sources and frequencies used in Timepix readout electronics interfere with resonance frequencies sensed by the MRI system disturbing the reception of the MRI signal. Therefore, it has not yet been possible to perform simultaneous MRI and SPECT measurements. The functionality of the designed prototype has been successfully tested by consecutive measurements using phantom objects and is now in further development with optimization aimed mostly to avoid mutual interferences of both MRI and SPECT units in the multimodal imaging system.

8.5 Material Sciences

Fast data readout and especially energy discrimination capability of PCD open wide application fields in material engineering and industry. In these fields, the use of novel sensor materials (CdTe, CdZnTe, GaAs) for PCDs is extremely favorable, as the use of higher photon energies is frequently necessary, e.g., 1 mm CdTe sensor provides high quantum efficiency for photons up to 100 keV. Such properties open new possibilities as high-energetic and penetrating radiation can be used for scanning. In the field of material engineering, x-ray micro-CT has become a popular tool used for inspection of composite materials. Composites are used, i.e., in the aeronautic industry as extremely lightweight and strong components can be built by combination of several materials into a composite. Micro-CT techniques can help with the evaluation of production quality as any inner imperfections can be easily visualized. Further, energy sensitive imaging is very helpful in this field, as it enables material decomposition of investigated objects (Pichotka et al. 2015).

Nevertheless, even in the field of material sciences, the capability to obtain enormously high CNR is still of high importance. Interesting results on the edge between biomedicine and material engineering achieved using high-resolution x-ray imaging with PCDs have come from the Centre of Excellence Telč (CET), Czech Republic, where the largest ever built Timepix detector (WidePIX$_{10\times10}$ consisting of 100 individual Timepix chips) is operated. Results of research focused on x-ray microtomography of biocompatible bone scaffolds fabricated from nanoparticulate bioactive glass reinforced gelan-gum (GG-BAG) have been carried out (Kumpová et al. 2016). Since GG-BAG is a material with very low x-ray absorption, PCD technology seems to be an ideal tool for achieving sufficient contrast during the CT scan. Results of the research show that PCD technology gains three times better detail detectability than flat-panel detectors and, therefore, provides much more precise information about the inner structure of biocompatible bone scaffolds.

8.6 Cultural Heritage

In the field of cultural heritage, x-ray radiography is an extremely valuable tool for the inspection of historical artifacts or artworks. Since these objects are usually very rare, only noninvasive investigation methods are allowed. X-ray radiographic methods are, therefore, ideal techniques to obtain valuable information about their internal structures. Knowledge about internal structures of pieces of art helps restorers to reveal, e.g., repaints, previous restoration interventions, inner damages, etc. Frequently, radiographic measurements of pieces of art are performed on medical or industrial radiographic and CT systems since there have not been any imaging systems available dedicated directly to imaging of cultural heritage artifacts. In many cases, the results are not satisfactory, as these systems are optimized for different types of objects.

The use of a large area PCD offers all required properties—large FOV, high contrast, and high resolution. Proof of concept measurements had been carried out at IEAP and CET using their high-resolution x-ray imaging setups (Zemlicka et al. 2014). Later, two x-ray imaging systems dedicated to high-resolution x-ray radiography of painted artwork were designed and built at IEAP in cooperation with the Academic Materials Research Laboratory of Painted Artworks (ALMA) of the Academy of Fine Arts in Prague (AFA). Both setups provide a FOV of up to 100 × 100 cm. One of them is meant to be stable and is situated in a shielded cabinet at ALMA. The second has a modular portable construction and is dedicated for measurements in situ because, in some cases, paintings cannot be transferred to the laboratory. Figure 8.7 shows the scanner operated at ALMA (top right) and a high-resolution radiography of a technological copy of seventeenth century Flemish painting with partially overpainted original motive (Zemlicka 2016). The radiographic projection presented in the bottom row of Figure 8.7 restores the information from hidden layers of overpainted parts of the artwork. The obtained image consists of an array of 10942 × 8358 pixels (approximately 91.45 megapixels) with spatial resolution of about 50 μm.

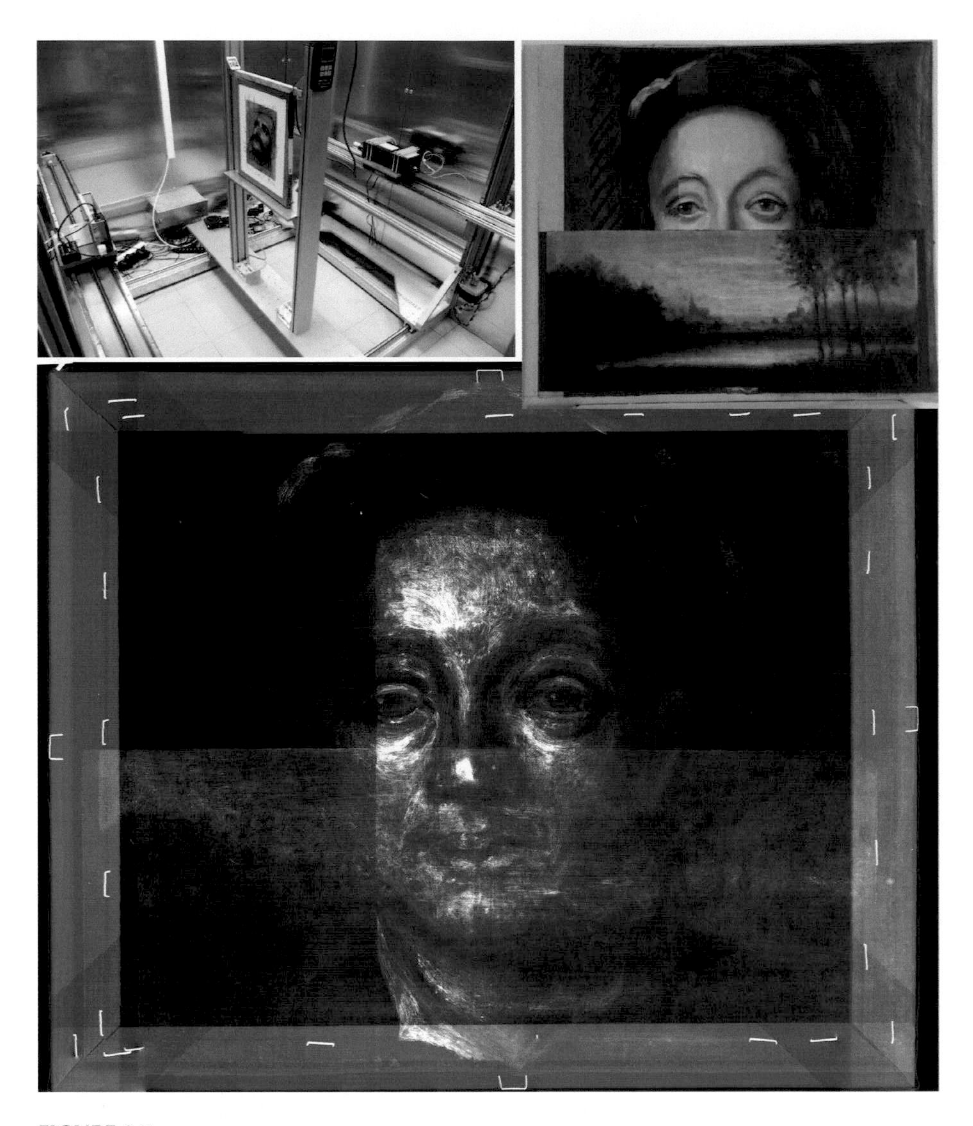

FIGURE 8.7
Scanning of a painting in the shielded cabinet dedicated to x-ray imaging of painted art operated at ALMA, AFA Prague (top left). (Photo courtesy of J. Hradilova.) Technological copy of seventeenth century painting with partial overpainting designed and created in ALMA, AFA Prague (top right). High-resolution x-ray radiography of the technological copy (bottom). (Courtesy of J. Hradilova, ALMA, AFA Prague; and J. Zemlicka, X-ray radiography/tomography combined with x-ray fluorescence spectroscopy for position sensitive elemental analysis of studied objects," PhD Thesis, Czech Technical University in Prague, 2016.)

8.7 Nanoscale Imaging

Besides a number of widely tested micro-CT systems, PCD technology has also been used in the development of x-ray imaging devices with spatial resolution deeply below the 1 μm level. German company Fraunhofer IIS and Technology has been working on the development of a laboratory-scale x-ray microscope and two different nano-CT x-ray imaging systems. All developed devices provide a spatial resolution at a level of 150 nm. As the sources with an extremely small focal spot provide low photon flux, all developed systems have been equipped with PCDs as PCD technology is, thanks to the noiseless operation, a perfect solution for data acquisition with long exposure times.

The first attempt was an x-ray microscope (Nachtrab et al. 2011). The nanofocused source was built using an electron gun with electron optics from a microprobe analyzer and a thin transmission tungsten target. According to a simulation using a 0.1 μm thick target, focal spot at a level of 50 nm had been expected; however, the experimentally achieved value was 154 nm. The x-ray microscope was equipped with a Medipix2 MXR in Quad configuration. The system was designed to be able to scan samples with magnification factors up to 1000 and to provide an EPS of 55 nm.

The next two approaches have been aimed at bringing a CT-system with submicron resolution and a large FOV. As a large FOV is required, magnifying geometry is again used instead of x-ray optics. The NanoXCT is equipped with a commercially available nanofocus x-ray tube with a 100 nm focal spot (Nachtrab et al. 2015). It was designed with a modular detector composed of four Timepix Hexa assemblies. The full detector, therefore, provides a sensitive area approximately 170 × 28 mm (3072 × 512 pixels). The FOV is adjustable within the range from 0.15 to 16 mm (from 50 nm to 5.5 μm effective pixel size).

Later, another nano-CT setup had been introduced (Stock et al. 2016). Since it was designed for 3D characterization of metal structures, a Pixirad PCD with a CdTe sensor was used. The sensor area is 61.8 × 25 mm (1024 × 476 pixels). Similar to the x-ray microscope, it is based on use of an electron gun as the source of electrons. As a target, tungsten film or a tungsten needle with fine-etched tip is used. Both target and sample are placed in a vacuum chamber while the detector is outside the vacuum behind a thin beryllium window (see the scheme in the top left corner of Figure 8.8). The top right and lower row of Figure 8.8 present the results obtained using the nano-CT system ("Nano CT of an Alloy"). Results of tomographic reconstruction revealing the crystal structure of an AlCu21 (mass%) outperform conventional microfocus systems by means of spatial resolution. Thanks to an enormously high resolution, three different metal phases are clearly distinguishable within the sample. The measurement was taken with a 150 nm voxel size and 450 nm spatial resolution (FWHM of the PSF). The dataset consisted of 600 projections with 15 min acquisition time per single projection. As a tungsten film

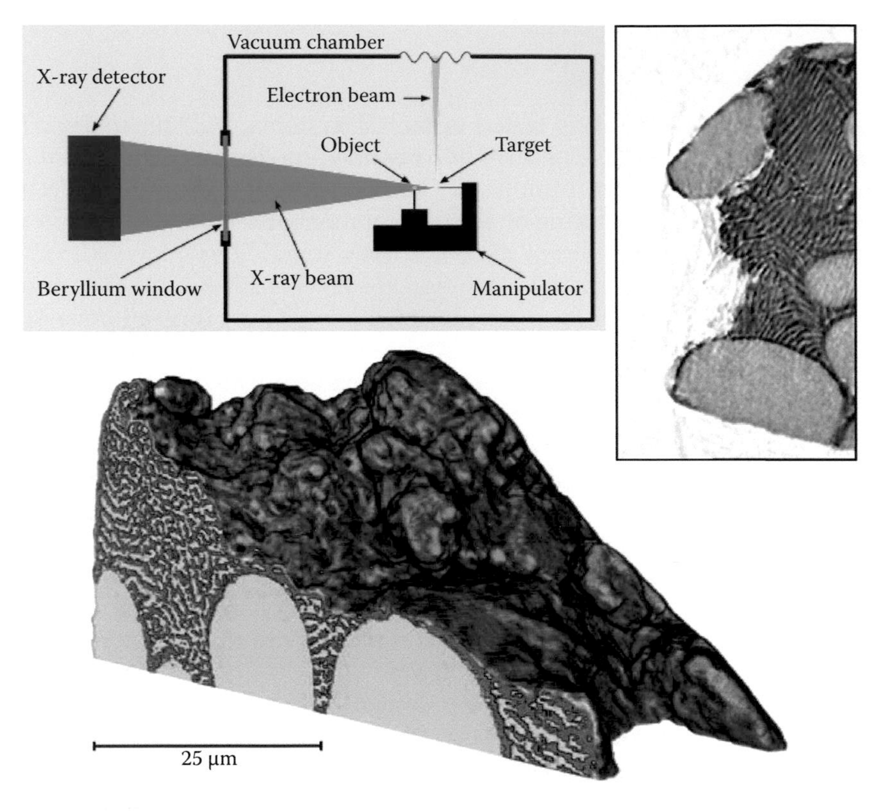

FIGURE 8.8
Scheme of the nano-CT scanner (top left) and reconstruction of a sample of AlCu21 (mass%) alloy—selected slice from the reconstructed dataset (top right) and volume rendering of segmented data (bottom row). Colors of the segmented data represent the α-aluminum phase (gray), aluminum in eutectic (blue), and Al_2Cu intermetallic phase (orange). (Courtesy of Fraunhofer ISS and Technology, Nano CT of an Alloy.)

was used as the target in this case, the spatial resolution could be further improved by using the tungsten needle instead.

As has already been mentioned earlier, the inevitable drawback of a source with extremely small focal spot is very low beam intensity. In the case of presented imaging systems, the photon flux is just tens of counts in each pixel per second. That leads to very long acquisition times compared to common microresolution systems. To achieve sufficient CNR, acquisition time from minutes up to tens of minutes is necessary. PCD technology is especially usable for such applications as the detected signal is not negatively affected by dark current and other noise sources. Regardless of the extremely long scan times, all presented devices achieve extreme spatial resolution compared to widely available x-ray micro-CT systems. Currently, such devices are perfectly suitable for material science objects, since the samples in this field are typically stable in time and can be prepared in proper size and shape suitable for a nano-CT scan.

8.8 Conclusions

Although PCD technology was originally developed for particle tracking and for use in large-scale experiments in the field of high-energy physics, it has proven to be a powerful tool for high-resolution x-ray imaging as well. The latest generation of PCDs is fully competitive with conventional x-ray imaging detectors by means of spatial resolution or sensitive area. Furthermore, PCDs feature additional unique properties making them especially suitable for some applications—namely, x-ray imaging of low contrast objects, scanning with low beam intensity, or energy sensitive radiography and tomography. A number of built and experimentally tested x-ray imaging systems show that PCDs fulfill requirements for in vivo biomedical imaging, where the crucial parameters are scan time and delivered dose, as well as for imaging with resolution deeply below 1 μm. Since PCD technology is very versatile, it has also been successfully used for other imaging modalities like single photon emission tomography.

Further research and development of PCD technology is still being carried out. An intensively investigated question is the substitution of silicon sensors, still being the first and most common choice, by sensor materials consisting of high-Z elements to improve quantum efficiency for higher photon energies. On the other hand, the production of CdTe or GaAs sensors is nowadays complicated compared to silicon, and even their stability and response uniformity is still worse than that of a silicon monocrystal. The compensation of undesirable behavior of these new sensor materials is being intensively investigated.

Currently PCDs achieve great results in the field of biomedical and preclinical research where a virtually unlimited dynamic range and single photon counting helps to improve the CNR within radiographic projections. Energy sensitivity enables recognizing different materials within a complex object and performing material decomposition. It is possible to clearly identify different types of soft tissue, calcified structures, recognize specific contrast agents in the biological sample, etc. Energy sensitive x-ray imaging is extremely useful even for material scientists since a number of various composite materials are being developed for use in the industry, and 3D analyses of their inner structures are highly demanded.

Although PCDs used to be purely experimental technology a few years ago, recent development has resulted in an establishment of several companies offering various PCD-based detectors as commercial products and, moreover, a complete Medipix3-based micro-CT system has become available as well. The unique properties of PCD technology have been proven as advantageous and are in demand in application field. As continual research and development constantly improves parameters of PCDs, further expansion in the field of imaging can be expected in the future.

References

Aamir, R., A. Chernoglazov, C. J. Bateman, A. P. H. Butler, P. H. Butler, N. G. Anderson, S. T. Bell et al. 2014. "Mars spectral molecular imaging of lamb tissue: Data collection and image analysis." *Journal of Instrumentation* 9(2):P02005. doi:10.1088/1748-0221/9/02/P02005.

Basolo, S., J.-F. Berar, N. Boudet, P. Breugnon, B. Caillot, J.-C. Clemens, P. Delpierre et al. 2005. "Xpad: Pixel detector for material sciences." *IEEE Transactions on Nuclear Science* 52(5):1994–1998. doi:10.1109/TNS.2005.856818.

Bellazzini, R., A. Brez, G. Spandre, M. Minuti, M. Pinchera, P. Delogu, P. L. de Ruvo, and A. Vincenzi. 2015. "Pixie III: A very large area photon-counting CMOS pixel ASIC for sharp X-ray spectral imaging." *Journal of Instrumentation* 10(1):C01032. doi:10.1088/1748-0221/10/01/C01032.

Berar, J.-F., N. Boudet, P. Breugnon, B. Caillot, B. Chantepie, J.-C. Clemens, P. Delpierre et al. 2009. "Xpad3 hybrid pixel detector applications." *Nuclear Instruments and Methods in Physics Research Section A: Accelerators, Spectrometers, Detectors and Associated Equipment* 607(1):233–235. doi:10.1016/j.nima.2009.03.208.

Bisogni, M. G., M. Campbell, M. Conti, P. Delogu, M. E. Fantacci, E. H. M. Heijne, P. Maestro et al. 1998. "Performance of a 4096-pixel photon counting chip." *Proceedings of SPIE* 3445:298–304. doi:10.1117/12.330288.

Dammer, J., P. M. Frallicciardi, J. Jakubek, M. Jakubek, S. Pospisil, E. Prenerova, D. Vavrik et al. 2009. "Real-time in-vivo M-imaging with Medipix2." *Nuclear Instruments and Methods in Physics Research Section A: Accelerators, Spectrometers, Detectors and Associated Equipment* 607(1):205–207. doi:10.1016/j.nima.2009.03.154.

Dammer, J., F. Weyda, J. Benes, V. Sopko, J. Jakubek, and V. Vondracek. 2011. "Microradiography of Biological Samples with Timepix." *Journal of Instrumentation* 6(11):C11005. doi:10.1088/1748-0221/6/11/C11005.

Debarbieux, F., A. Bonissent, P. Breugnon, F. C. Brunner, P. A. Delpierre, C. Hemmer, J.-C. Clemens et al. 2010. "Repeated imaging of lung cancer development using Pixscan, a low dose micro-CT scanner based on Xpad hybrid pixel detectors." *IEEE Transactions on Nuclear Science* 57(1):242–245. doi:10.1109/TNS.2009.2037319.

Delogu, P., P. Oliva, R. Bellazzini, A. Brez, P. L. de Ruvo, M. Minuti, M. Pinchera et al. 2016. "Characterization of Pixirad-1 photon counting detector for X-ray imaging." *Journal of Instrumentation* 11(1):P01015. doi:10.1088/1748-0221/11/01/P01015.

Delpierre, P., S. Basolo, J.-F. Berar, M. Bordesoule, N. Boudet, P. Breugnon, B. Caillot et al. 2007a. "Xpad: A photons counting pixel detector for material sciences and small-animal imaging." *Nuclear Instruments and Methods in Physics Research Section A: Accelerators, Spectrometers, Detectors and Associated Equipment* 572(1):250–253. doi:10.1016/j.nima.2006.10.315.

Delpierre, P., F. Debarbieux, S. Basolo, J. F. Berar, A. Bonissent, N. Boudet, P. Breugnon et al. 2007b. "Pixscan: Pixel detector CT-scanner for small animal imaging." *Nuclear Instruments and Methods in Physics Research Section A: Accelerators, Spectrometers, Detectors and Associated Equipment* 571(1–2):425–428. doi:10.1016/j.nima.2006.10.126.

Dudak, J., J. Zemlicka, J. Karch, M. Patzelt, J. Mrzilkova, P. Zach, Z. Hermanova et al. 2016. "High-contrast X-ray micro-radiography and micro-CT of ex-vivo soft tissue murine organs utilizing ethanol fixation and large area photon-counting detector." *Scientific Reports* 6:30385. doi:10.1038/srep30385.

Dudak, J., J. Zemlicka, F. Krejci, S. Polansky, J. Jakubek, J. Mrzilkova, M. Patzelt et al. 2015. "X-ray micro-CT scanner for small animal imaging based on Timepix detector technology." *Nuclear Instruments and Methods in Physics Research Section A: Accelerators, Spectrometers, Detectors and Associated Equipment* 773:81–86. doi:10.1016/j.nima.2014.10.076.

Gimenez, E. N., R. Ballabriga, M. Campbell, I. Horswell, X. Llopart, J. Marchal, K. J. S. Sawhney et al. 2011. "Study of charge-sharing in Medipix3 using a micro-focused synchrotron beam." *Journal of Instrumentation* 6(1):C01031. doi:10.1088/1748-0221/6/01/C01031.

Hamonet, M., M. Dupont, T. Fabiani, F. Cassol, Y. Boursier, A. Bonissent, F. Debarbieux et al. 2015. "The ClearPET/XPAD prototype: Development of a simultaneous PET/CT scanner for mice." In *2015 IEEE Nuclear Science Symposium and Medical Imaging Conference (NSS/MIC)*, 1–3. San Diego: IEEE. doi:10.1109/NSSMIC.2015.7582074.

Jakubek, J., C. Granja, J. Dammer, R. Hanus, T. Holy, S. Pospisil, R. Tykva et al. 2007. "Phase contrast enhanced high resolution X-ray imaging and tomography of soft tissue." *Nuclear Instruments and Methods in Physics Research Section A: Accelerators, Spectrometers, Detectors and Associated Equipment* 571(1–2):69–72. doi:10.1016/j.nima.2006.10.031.

Jakubek, J., M. Jakubek, M. Platkevic, P. Soukup, D. Turecek, V. Sykora, and D. Vavrik. 2014. "Large area pixel detector Widepix with full area sensitivity composed of 100 Timepix assemblies with edgeless sensors." *Journal of Instrumentation* 9(4):C04018. doi:10.1088/1748-0221/9/04/C04018.

Khodaverdi, M., S. Nicol, J. Loess, F. Cassol Brunner, S. Karkar, and C. Morel. 2007. "Design study for the ClearPET/XPAD small animal PET/CT scanner." In *2007 IEEE Nuclear Science Symposium Conference Record*, 4300–4302. Honolulu: IEEE. doi:10.1109/NSSMIC.2007.4437067.

Krejci, F., J. Jakubek, and M. Kroupa. 2011. "Single grating method for low dose 1-D and 2-D phase contrast X-ray imaging." *Journal of Instrumentation* 6(1):C01073. doi:10.1088/1748-0221/6/01/C01073.

Kumpová, I., D. Vavřík, T. Fíla, P. Koudelka, I. Jandejsek, J. Jakůbek, D. Kytýř et al. 2016. "High resolution micro-CT of low attenuating organic materials using large area photon-counting detector." *Journal of Instrumentation* 11(2):C02003. doi:10.1088/1748-0221/11/02/C02003.

Llopart, X., R. Ballabriga, M. Campbell, L. Tlustos, and W. Wong. 2007. "Timepix, a 65K programmable pixel readout chip for arrival time, energy and/or photon counting measurements." *Nuclear Instruments and Methods in Physics Research Section A: Accelerators, Spectrometers, Detectors and Associated Equipment* 581(1–2):485–494. doi:10.1016/j.nima.2007.08.079.

Nachtrab, F., T. Hofmann, C. Speier, J. Lučić, M. Firsching, N. Uhlmann, P. Takman et al. 2015. "Development of a Timepix based detector for the NanoXCT Project." *Journal of Instrumentation* 10(11):C11009. doi:10.1088/1748-0221/10/11/C11009.

Nachtrab, F., T. Ebensperger, B. Schummer, F. Sukowski, and R. Hanke. 2011. "Laboratory X-ray microscopy with a nano-focus X-ray source." *Journal of Instrumentation* 6(11):C11017. doi:10.1088/1748-0221/6/11/C11017.

Nano CT of an Alloy. Online. In *Fraunhofer Iis Website*. Erlangen. https://www.iis.fraunhofer.de/content/dam/iis/de/doc/ezrt/20160630_Produktblatt_NanoCT_2S_en_web.pdf.

Nicol, S., S. Karkar, C. Hemmer, A. Dawiec, D. Benoit, P. Breugnon, B. Dinkespiler et al. 2009. "Design and construction of the ClearPET/XPAD small animal PET/CT scanner." In *2009 IEEE Nuclear Science Symposium Conference Record (NSS/MIC)*, 3311–3314. Orlando: IEEE. doi:10.1109/NSSMIC.2009.5401740.

Pangaud, P., S. Basolo, N. Boudet, J.-F. Berar, B. Chantepie, P. Delpierre, B. Dinkespiler, et al. 2007. "XPAD3: A new photon counting chip for X-ray CT-scanner." *Nuclear Instruments and Methods in Physics Research Section A: Accelerators, Spectrometers, Detectors and Associated Equipment* 571(1–2):321–324. doi:10.1016/j.nima.2006.10.092.

Pennicard, D., S. Lange, S. Smoljanin, J. Becker, H. Hirsemann, M. Epple, and H. Graafsma. 2011. "Development of LAMBDA: Large area Medipix-based detector array." *Journal of Instrumentation* 6(11):C11009. doi:10.1088/1748-0221/6/11/C11009.

Pichotka, M., J. Jakubek, and D. Vavrik. 2015. "Spectroscopic micro-tomography of metallic-organic composites by means of photon-counting detectors." *Journal of Instrumentation* 10(12):C12033. doi:10.1088/1748-0221/10/12/C12033.

PiXirad Imaging Counters Website. Online. Pisa. http://www.pixirad.com/.

Platkevic, M. 2014. "Signal processing and data read-out from position sensitive pixel detectors." Ph.D. Thesis, Czech Technical University in Prague.

Rajendran, K., C. Löbker, B. S. Schon, C. J. Bateman, R. A. Younis, N. J. A. de Ruiter, A. I. Chernoglazov et al. 2017. "Quantitative imaging of excised osteoarthritic cartilage using spectral CT." *European Radiology* 27(1):384–392. doi:10.1007/s00330-016-4374-7.

Solder Joint Inspection and Analysis. 2011. Online. In *Ge Measurements*. https://www.gemeasurement.com/sites/gemc.dev/files/pcba_brochure_english_2.pdf.

Stock, S. R., B. Müller, G. Wang, P. Stahlhut, K. Dremel, J. Dittmann, J. M. Engel et al. 2016. "First results on laboratory nano-CT with a needle reflection target and an adapted toolchain." *Proceedings of SPIE* 9967:996701. doi:10.1117/12.2240561.

Sykora, V., J. Zemlicka, F. Krejci, and J. Jakubek. 2013. "X-ray micro-radiography and microscopy of seeds." *Science* 339(6119):510–511. doi:10.1126/science.339.6119.510.

Walsh, M. F., A. M. T. Opie, J. P. Ronaldson, R. M. N. Doesburg, S. J. Nik, J. L. Mohr, R. Ballabriga et al. 2011. "First CT using Medipix3 and the MARS-CT-3 spectral scanner." *Journal of Instrumentation* 6(1):C01095. doi:10.1088/1748-0221/6/01/C01095.

XPAD Pixel Detector. Online. *European Synchrotron Radiation Facility Website*. Grenoble. http://www.esrf.eu/UsersAndScience/Experiments/CRG/BM02/detectors/xpad.

Zainon, R., A. P. H. Butler, N. J. Cook, J. S. Butzer, N. Schleich, N. de Ruiter, L. Tlustos et al. 2010. "Construction and operation of the MARS-CT scanner." *Internetworking Indonesia Journal* 2(1):3–10.

Zajicek, J., J. Jakubek, M. Burian, M. Vobecky, A. Fauler, M. Fiederle, and A. Zwerger. 2013. "Multimodal imaging with hybrid semiconductor detectors Timepix for an experimental MRI-SPECT system." *Journal of Instrumentation* 8(1):C01022. doi:10.1088/1748-0221/8/01/C01022.

Zemlicka, J., J. Jakubek, J. Dudak, J. Hradilova, and O. Trmalova. 2014. "X-ray radiography of paintings with high resolution I. Testing and measurements with large area pixel detector." In *Acta Artis Academica 2014: Interpretation of Fine Art'S Analyses in Diverse Contexts*, 51–61. Prague: Academy of Fine Arts Prague.

Zemlicka, J. 2016. "X-ray radiography/tomography combined with X-ray fluorescence spectroscopy for position sensitive elemental analysis of studied objects." Czech Technical University in Prague.

Zhang, C., P. Lechner, G. Lutz, J. Treis, S. Wölfel, L. Strüder, and S. Nan Zhang. 2008. "Development of X-type DEPFET macropixel detectors." *Nuclear Instruments and Methods in Physics Research Section A: Accelerators, Spectrometers, Detectors and Associated Equipment* 588(3):389–396. doi:10.1016/j.nima.2008.02.002.

Ziemons, K., E. Auffray, R. Barbier, G. Brandenburg, P. Bruyndonckx, Y. Choi, D. Christ et al. 2005. "The ClearPET project: Development of a 2nd generation high-performance small animal PET scanner." *Nuclear Instruments and Methods in Physics Research Section A: Accelerators, Spectrometers, Detectors and Associated Equipment* 537(1–2):308–311. doi:10.1016/j.nima.2004.08.032.

Zuber, M., T. Koenig, E. Hamann, A. Cecilia, M. Fiederle, and T. Baumbach. 2014. "Characterization of a 2 × 3 Timepix assembly with a 500 μm thick silicon sensor." *Journal of Instrumentation* 9(5):C05037. doi:10.1088/1748-0221/9/05/C05037.

9

Conebeam CT for Medical Imaging and Image-Guided Interventions

Tianye Niu and Yu Kuang

CONTENTS

9.1 Introduction

Conebeam CT (CBCT) is an imaging technique that employs a divergent x-ray source and usually a large-area flat-panel detector to form a cone-shaped exposure to the imaged object. Such a configuration guarantees the volume coverage of one large object and achieves the volumetric imaging within one single scan. The first CBCT scanner entered the market in 1996 as the dental scanner by NewTom Corp. Subsequently, CBCT was extended to multiple clinical applications, including implantology, orthopedics, interventional radiology, and radiation oncology.

In the past two decades, continuous efforts have been devoted to improving the image quality of CBCT and reducing the imaging dose used. As one of the major issues seen in the development of CBCT, low-frequency shading artifacts are commonly observed in CT images and give rise to inaccurate CT numbers and spatial non-uniformity. These artifacts may lead to degraded image quality, which hinders the clinical application of CBCTs from image guidance and diagnosis. Shading artifacts are the contradiction between ideal reconstruction assumptions and realistic physical complications in the CT imaging process. These physical complications include scatter contamination (Siewerdsen et al., 2006), beam-hardening effect (Hsieh et al., 2000), photon starvation (Mori et al., 2013), detector lags (Tanaka et al., 2010), etc.

Among all these complications, scatter contamination, beam hardening, and photon starvation effects are the major error sources (Grimmer and Kachelriess, 2011). As the number of scatter photons increases significantly with the enlarged illuminated volume, scatter contamination becomes more severe in the CT images acquired with large cone-angle geometry. For example, on-board CBCTs using large-area flat-panel detector (e.g., $40 \times 30\ \text{cm}^2$) is widely utilized in image-guided radiation therapy (IGRT) for patient setup. The *z*-axis extent of CBCT images is over 15 cm and image quality is greatly degraded by high scatter-to-primary ratio (SPR). Without scatter correction, the SPR was reported to reach approximately 2 on a midsize volume, and the CT number error was as high as 300 HU around the pelvis region (Niu et al., 2010). The beam-hardening effect also leads to low-frequency shading artifacts, especially in the area composed of the same or comparable material due to the nonlinear attenuation of x-ray beam penetrating the object (Hsieh et al., 2000). Low-dose CT imaging using low tube current and/or low kVp protocols will suffer from photon starvation when dense areas (e.g., bony structure) are scanned (Mori et al., 2013).

Accumulated imaging dose is another concern for the clinical use of CBCT. For example, low-dose imaging techniques can substantially facilitate the use of on-board CBCT imaging in IGRT for treatment guidance, as the accumulative imaging dose during the radiation treatment course of 4–6 weeks may reach an unacceptable level of 100–300 cGy using a conventional reconstruction method (Niu and Zhu, 2012). Thus, the algorithms integrating the imaging physics and prior information of the CT image into the reconstruction are currently under investigation in low-dose CBCT imaging.

In this chapter, CBCT equipment and imaging schemes for medical application were reviewed. This chapter was focused on the hardware structure, relevant reconstruction algorithms, and the related improvement methods in image quality.

9.2 Hardware Implementation of a Table-Top CBCT System

The structure of a CBCT scanner usually includes an x-ray tube and a generator, a flat-panel detector, and the mechanics and electronic control system as shown in Figure 9.1. For simplicity of mechanical design, a table-top CBCT for lab use applies rotational stage instead of complicated mechanical gantry

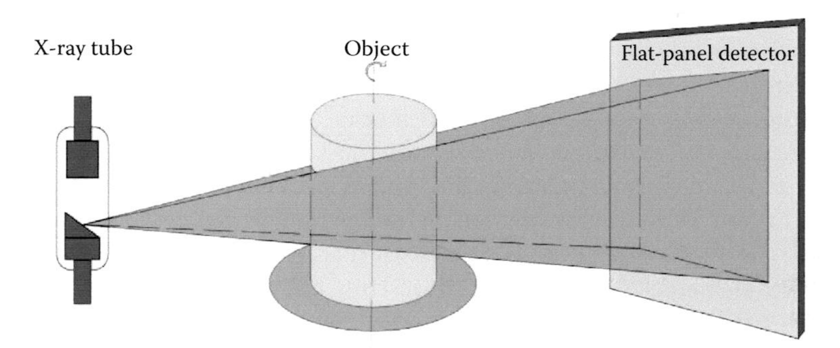

FIGURE 9.1
Illustration of the CBCT system.

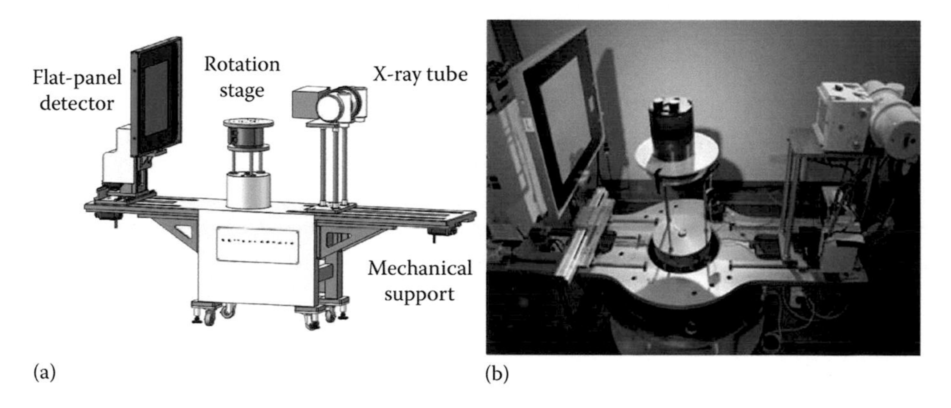

FIGURE 9.2
Table-top CBCT scanner installed in the medical imaging lab at Zhejiang University: (a) mechanical drawing; (b) photograph.

as shown in Figure 9.2. During the scan, the x-ray tube and the flat-panel detector keep stationary while the stage is rotating to acquire the incident photons from different angles.

Parameters of an x-ray tube, including tube voltage, tube current, exposure mode, and exposure time, are controlled by the console of the high-voltage generator. In the scan process, the x-ray tube generates large amounts of x-ray photons. After the attenuation of the object, the remaining photons are detected by the flat-panel detector and converted into electrical signals. Each pixel in the detector accumulates the total energy of x-ray photons penetrating the object and deposited into the pixel. Intensity distribution on the detector is, hereafter, referred to as projection. Projections from different x-ray incident angles are required to generate high-quality CT images according to the CT reconstruction theory. As such, the object is rotated by the stage, and sequential projection data are acquired in the table-top CBCT setting. Once sufficient projections are measured, the data are transferred into the computer for image reconstruction.

Due to the high volume of projection data acquired, the computer in the reconstruction is usually equipped with high-performance graphics processing unit (GPU) to improve the computation speed. To achieve more flexibility, the system allows linear movements of the x-ray tube to adjust its location with respect to the isocenter of the CBCT system. The position of the flat-panel detector is more flexible and is adjustable in three dimensions. Stepper motor can be operated in the motion control system with its easy operation and satisfactory spatial resolution accuracy. The instructions of the motion control are sent via serial port to tune the rotation speed of the stage, and the positions of the detector and x-ray tube. The major components of a typical table-top CBCT scanner are described as follows.

9.2.1 X-Ray Tube

The x-ray tube is used to generate x-ray photons for CT imaging. In the aforementioned table-top CBCT scanner, an x-ray tube with rotating anode and a 14° tungsten rhenium molybdenum graphite target was used. Two focal spots with the dimensions of 0.4/0.8 mm are available. The maximum input DC voltage is 150 kV and the maximum heat capacity is 450 kJ (equivalent to 600 kHU). The tube can be operated in two modes: pulsed or continuous exposure mode. In the pulsed mode, the x-ray tube produces x-ray photons within the pulse time interval to minimize the radiation dose. The pulse length and scan time can be appropriately adjusted dependent on the object size and weight.

The tube is powered by a high-voltage generator, which is triggered by the signals from the flat-panel detector to achieve synchronous exposure. The high voltage is applied to the anode and cathode of the x-ray tube through high-voltage cables. The current in the x-ray tube is limited by the damping resistance. The filament controller provides a variable voltage to the large and small filaments of the x-ray tube to produce a variable filament current

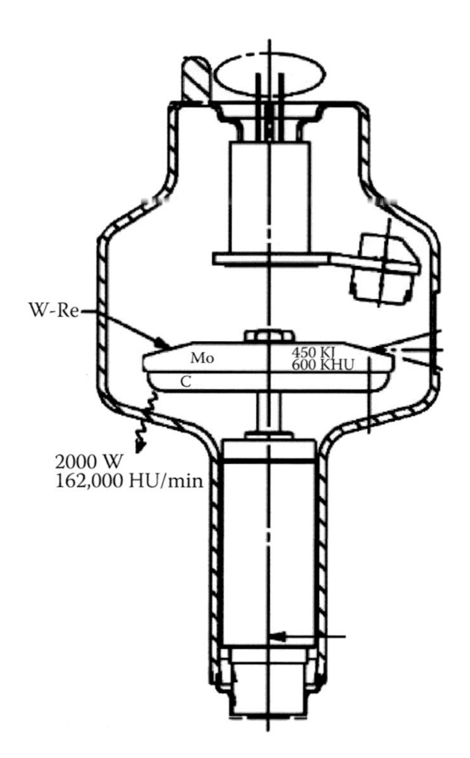

FIGURE 9.3
Structure of the x-ray tube. (Courtesy of Varian Medical System Inc., 2007.)

to adjust the tube current and dose. The controller provides a variable three-phase drive voltage for the motor of the x-ray tube to control acceleration and deceleration of the rotary anode.

In addition to the filament of the x-ray tube, a focus position control coil controls the position of the focus in accordance with the selected exposure mode during the scan. The cooling of the x-ray tube is accomplished by an "oil-gas" cooling device called heat exchanger. In this device, the oil pump circulates the insulating cooling oil between the x-ray tube and the heat exchanger. The heat is taken out of the tube and dissipates into the surrounding environment. The tube is also equipped with a sensor to measure the real-time temperature of the tube to invoke an alarm when the temperature or the cooling oil pressure is too high. A diagram of the x-ray tube is shown in Figure 9.3.

9.2.2 High-Voltage Generator

The purpose of the generator is to supply DC high voltage for the x-ray tube. The generator applied in the system can generate a maximum voltage of 150 kV. As shown in Figure 9.4, high-voltage generators consist of three major

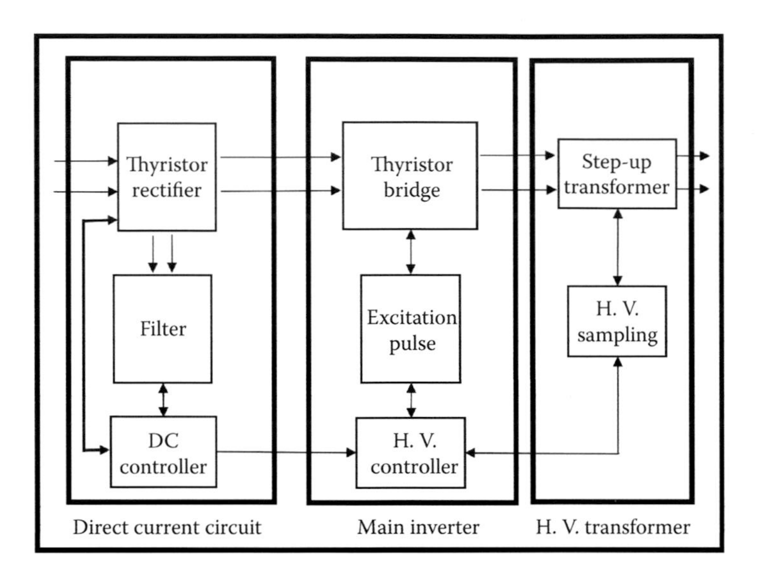

FIGURE 9.4
Block diagram of the high-voltage generator.

components: DC circuit, main inverter, and high-voltage transformer. The DC circuit converts the three-phase 380 V, 50 Hz AC voltage to DC voltage by a thyristor rectifier and filter. The DC circuit controller generates corresponding excitation pulses to control the level of the DC voltage by controlling the silicon controlled rectifier (SCR) conduction angle and conduction time. The inverter converts the DC voltage into a high-frequency alternating current through the thyristor bridge circuit and transports the alternating current to the primary of the high-voltage transformer. The SCR excitation pulse frequency is usually 10–20 kHz.

The high-voltage transformer mainly consists of the step-up transformer, high-voltage rectifier circuit, and high-voltage sampling circuit. The step-up transformer and the high-voltage rectifier circuit convert the high-frequency alternating current from the inverter into high voltage. The high-voltage sampling circuit samples the actual tube voltage and tube current, and then feeds back to the high-voltage controller. Thus, the voltage and current can be kept at a constant level to maintain the desired x-ray flux output.

9.2.3 Flat-Panel Detector

The detector used in the table-top system works with the radiation energy ranging from 40 to 160 keV and electronics housed in one package of 409.6 × 409.6 mm^2 sensitive area. The flat-panel detector captures more optical photons from the scintillator compared with the conventional technique, including the reducing lenses. The electronics are placed at the periphery of the sensor to avoid the direct exposure of the x-ray beam. This design

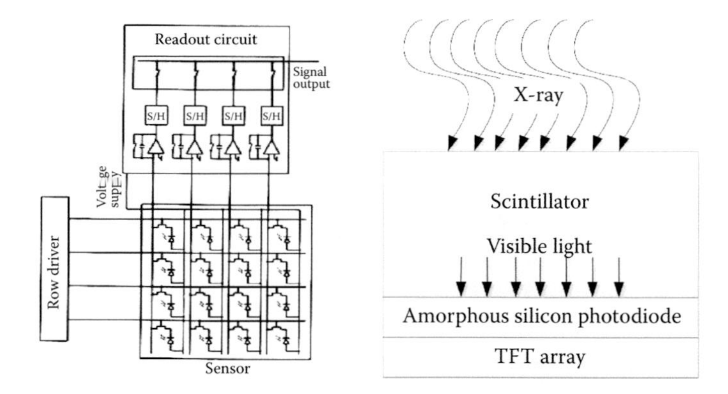

FIGURE 9.5
Structure of the flat-panel detector: (a) electronic arrangement (Courtesy of Perkin Elmer Inc., 2014.); (b) layout of the detector.

also increases the durability of this device in the MV source application. As shown in Figure 9.5, the sensors are fabricated using thin-film technology and placed on a single panel due to their small physical size. The sensors work in a similar way to the conventional photodiode array. Each pixel in the array consists of a light-sensing photodiode and a switching thin-film transistor (TFT) in the same electronic circuit (Perkin Elmer Inc., 2014).

The incident x-rays are converted by the scintillator material into visible light. The amorphous silicon photodiode is sensitive to visible light with sensitivity similar to human vision. The charge carriers are stored in the capacity of the photodiode. By turning on the gate of the TFT line in the matrix, the charges of all the columns are transferred in parallel to the signal output (Perkin Elmer Inc, 2014). All of the signals in the column are amplified in a customized readout multiplexer for further processing.

As shown in Figure 9.5, charge amplifiers for readout and row drivers for addressing the rows are placed at the boundary of the sensors. The readout circuit is connected to the A/D conversion component, which converts analog signals into digital ones. The digital signals are then processed in the control unit where sophisticated field-programmable gate array (FPGA) is applied to minimize the noise. The converted digital signals are then transferred to the computer for signal processing and image reconstruction.

9.2.4 Mechanics and Electronic Control System

For the simplicity of scan, the table-top CBCT for lab use applies compact rotation stage instead of large gantry. The stage is movable in vertical position to adjust the height of object position. For the rotation, the stage is driven by a turbine-shaft structure. The structure allows the stage to maintain a stable rotational speed. The positions of the x-ray tube and flat-panel detector are also movable. The x-ray tube is arranged on a guide rail and driven by a stepper motor to adjust the distance between the tube and the isocenter.

The position of the flat-panel detector is more flexible with three degrees of freedom. For adjustment in the horizontal plane, the detector is set on a single-axis robot, which is accurate and precise in motion control. By manually adjusting the guide screw supporting the detector, the height of the detector can be fine-tuned to align the center of the detector, the isocenter of the table-top system, and the focal spot of the x-ray tube.

A stepper motor is applied in the scanner as the force power for its easy operation and high accuracy of spatial resolution. Once a pulse is received ("step"), the motor rotates for a fixed amount of angle. Users can set the number of subdivisions of the step. The higher the subdivision, the higher the moving fraction. All stepper motors used in the table-top system are controlled by a stepper motor controller and a driver. The controller sends out the pulse signals to the driver. The driver converts the signals into current with high capability to drive the stepper motor.

Additionally, an encoder can be installed on the rotation axis of the stage to measure the rotation direction and speed. The measured codes are sent into the motor controller for feedback control. The stepper motor controllers are connected to the host computer. The instructions of the motion control are sent via serial port to tune the rotation speed of the stage and the positions of the detector and x-ray tube.

9.3 Reconstruction

The methods of CT reconstruction can be generally classified into two categories: analytical and iterative techniques. Analytical methods are mainly based on the Fourier central slice theorem. One practical analytical method in CBCT imaging is the filtered back projection (FBP) algorithm and its three-dimensional extension named the Feldkamp–Davis–Kress (FDK) algorithm, due to its reasonable balance between high computation efficiency and acceptable image quality.

Analytical methods require the completeness of measured projection data and thus are sensitive to the fluctuation due to non-ideal physical issues, including the statistical noise, incomplete projection, etc. To address the drawbacks of the analytical methods, iterative methods have been developed in the past two decades. This type of method models the reconstruction as a mathematical optimization problem. The physical constraints are thus readily incorporated into the reconstruction. Iterative techniques have several categories, including the algebraic reconstruction technique (ART), compressed sensing-based iterative reconstruction, statistical reconstruction, expectation-maximization maximum likelihood (EM-ML) algorithms, etc.

Although iterative methods can suppress the noise and resultant artifacts, their applications in clinics are still under investigation mainly due to the nontraditional image texture from nonlinear iteration and the low computation speed.

9.3.1 Algebraic Reconstruction Techniques

ARTs have been proposed for over 40 years. The main idea of ARTs is to solve a linear system of equations iteratively. To understand the algorithm in an easy way, the diagram illustrating CT imaging is shown in Figure 9.6. For simplicity, the object is assumed to have only two dimensions and divided into $N \times N$ pixels.

The image of the object is defined as a vector f with each element f_j, where j is the index of the element in the vector. The banding with shaded lines in Figure 9.6 shows the x-ray passing through the object. The projection data obtained from the flat-panel detector are discrete. Each element in the projection vector is defined as p_i, where i is its index. The mathematical relationship between f_j and p_i is defined as the forward projection model and written as

$$\sum_{j=1}^{N_d} w_{i,j} f_j = p_{n+1} \quad i = 1,2,3, \ldots M, \tag{9.1}$$

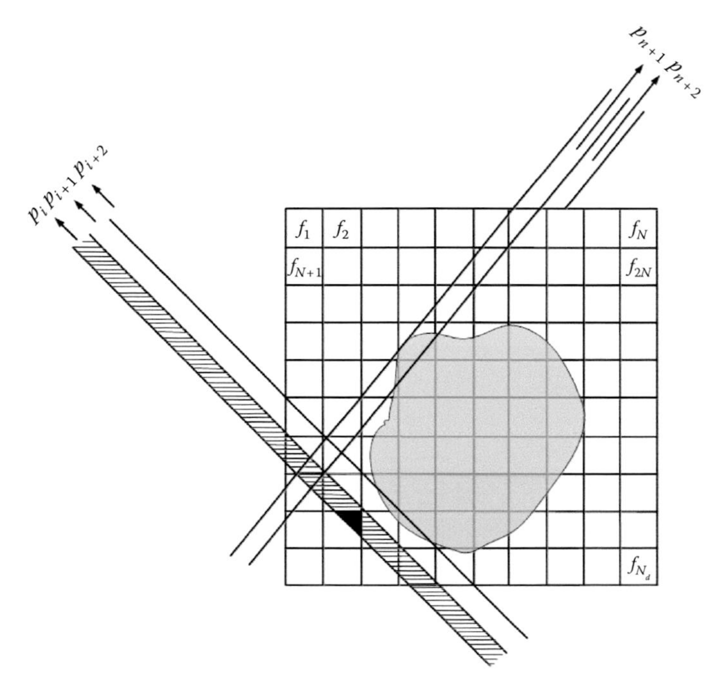

FIGURE 9.6
Diagram of CT imaging.

where M is the total number of the detector elements, and N_d is the number of object pixels (in this case, $N_d = N \times N$). $w_{i,j}$ is a weighting factor to represent the contribution of the object pixel to the forward projection. $w_{i,j}$ plays an important role in the computation efficiency and accuracy of CT reconstruction.

Many methods are proposed to calculate the weighting factor $w_{i,j}$ in the literature. Three major models were briefly introduced in this chapter. In the first model, the x-ray is considered to be a "fat" line passing through the object image (Figure 9.7). BCDFGH is the intersection region of the x-ray and the discrete pixel. T is the physical width of the pixel. The ratio between the area of BCDFGH and the area of the discrete square is applied as the weighting factor:

$$w_{i,j} = \frac{S_{BCDFGH}}{T^2} \tag{9.2}$$

In the second model, as shown in Figure 9.8, the width of the x-ray beam is disregarded. The length of intersection line segment AB is considered as the weighting factor:

$$w_{i,j} = l_{AB} \tag{9.3}$$

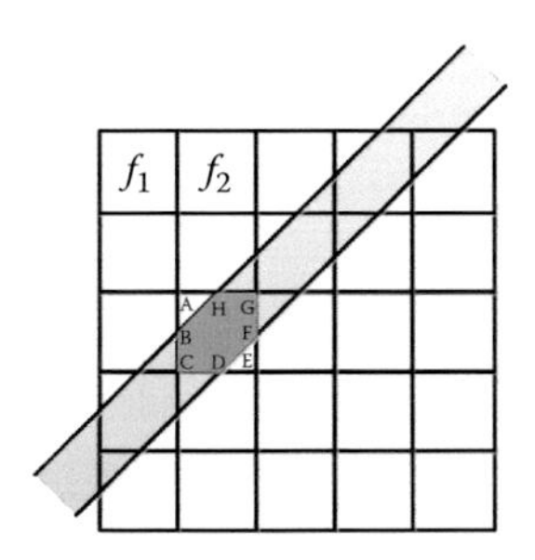

FIGURE 9.7
Illustration of the first model.

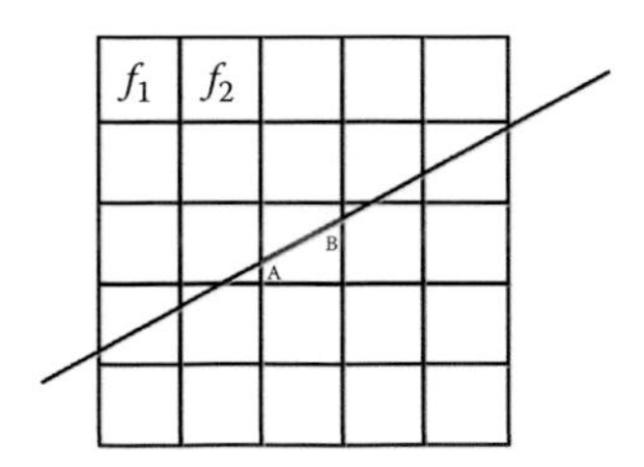

FIGURE 9.8
Illustration of the second and the third model.

The third model is a simplified version of the second one. As shown in Equation 9.4, the weighting factor is set to be 1 when the x-ray passes through the square and 0 otherwise.

$$w_{i,j} = \begin{cases} 1 \text{ if x-ray passes the pixel} \\ 0 \text{ otherwise} \end{cases} \tag{9.4}$$

For the comparison of the reconstruction time and image quality using the three models, Shepp–Logan phantom is applied in the simulation and the normalized mean square error is used to reflect the reconstruction accuracy. The normalized mean square error is defined as $q = \left[\frac{\sum_{i=1}^{N} \sum_{j=1}^{N} (x_{i,j} - y_{i,j})^2}{\sum_{i=1}^{N} \sum_{j=1}^{N} (x_{i,j} - \bar{x})^2} \right]^{\frac{1}{2}}$, where $x_{i,j}$, $y_{i,j}$ refers to the gray value of the reference image and the reconstructed image in the ith row and the jth column, respectively. Smaller q indicates better image quality.

As shown in Table 9.1, the reconstruction result using the first model has the best image quality, while it takes the longest computation time since each beam intersects with many pixels and most of the intersection region is irregular. In the third model, the reconstruction time is greatly reduced by simplifying the calculation of the weighting factor, while the image quality is the poorest due to the inaccurate approximations of the weighting factor. The second model shows a good balance between computation speed and image quality. The length of each ray passing through each pixel can be quickly implemented in the computer, and the calculation burden is much lower than that in the first model. These properties make the second model widely applied in CT reconstruction.

The linear equation system is solved when the weighting factor is obtained. When N is small, f can be solved using the direct inversion of Equation 9.1. Nevertheless, N is huge in CT imaging leading to the ill-posed system matrix. The direct inversion is thus difficult to calculate. For example, if the size of the

TABLE 9.1

Reconstruction Time and Quality of the Three Models

Method	1	2	3
Reconstruction time (s)	23.09	15.98	14.21
Normalized mean square error	0.7749	0.7904	1.581

Source: L. Chen, *Optical Instruments*, 36, 142–146, 2014.
Note: The results are evaluated using the Shepp–Logan phantom with the dimension of 128 × 128 pixels in the 180° rotation and a total of 180 projections.

CT image is 512 × 512, then $N \times N$ is 262,144. A typical scan requires 328 projections with 1,024 detector elements in each projection, which means we need to solve 335,872 equations with 262,144 unknown variables for a two-dimensional image. For the three-dimensional reconstruction, the number of data set will be 200 times larger and the direct solution to Equation 9.1 is almost impossible.

Under this condition, the Kaczmarz method is proposed to solve the ill-posed system matrix. To explain this algorithm, Equation 9.1 is rewritten in an expanded form as

$$\begin{aligned} w_{11}f_1 + w_{12}f_2 + w_{13}f_3 + \ldots w_{1N}f_{N_d} &= p_1 \\ w_{21}f_1 + w_{22}f_2 + w_{23}f_3 + \ldots w_{2N}f_{N_d} &= p_2 \\ &\ldots \\ w_Mf_1 + w_Mf_2 + w_{M3}f_3 + \ldots w_{MN}f_{2N_d} &= p_M \end{aligned} \tag{9.5}$$

In Equation 9.5, the image is described as the form of $\left(f_1, f_2, f_{3\ldots}f_{N_d}\right)$. The image is considered as a single point in an N_d-dimensional space, and each equation represents a hyperplane in this space. The intersection point of all the hyperplanes is thus the solution of the equations as shown in Figure 9.9. To simplify the question, two equations are considered as an example in

$$\begin{aligned} w_{11}f_1 + w_{12}f_2 &= p_1 \\ w_{21}f_1 + w_{22}f_2 &= p_2 \end{aligned} \tag{9.6}$$

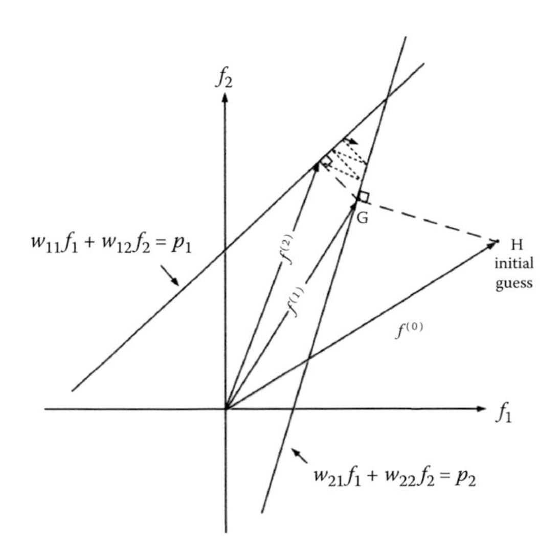

FIGURE 9.9
Diagram of Kaczmarz method.

As shown in Figure 9.9, two straight lines represent the two equations in Equation 9.6. Finding the solution starts with an initial guess, denoted by $f^{(0)}$, which in most cases is assigned by zero value. The initial guess will be projected onto the line represented by the first equation, and the corresponding point is chosen to be the next starting point. Projecting this new point to another line will generate the third point. Tanabe (1971) proved that if a unique solution exists, the iterations will always converge to that point. These steps are continued until the convergence is achieved (Tanabe, 1971). In practice, there exists a positive integer K for a given small positive number ε, such that $\left| f_j^{(K)} - f_j^{(K-1)} \right| < \varepsilon$, $(j = 1,2,3 \ldots N_d)$, when $i > k$. Here ε is given according to the required accuracy.

A mathematical expression is used to show this process:

$$f_j^{(i)} = f_j^{(i-1)} - \frac{f^{(i-1)} * w_i - p_i}{w_i * w_i} w_i \tag{9.7}$$

where $w_i = (w_{i1}, w_{i2}, \ldots w_{iN_d})$. To make the equation more concise, we rewrite Equation 9.7 as

$$f_j^{(i)} = f_j^{(i-1)} + \frac{p_i - q_i}{\sum_{k=1}^{N_d} w_{ik}^2} w_{ij} \tag{9.8}$$

where

$$q_i = f^{(i-1)} * w_i = \sum_{k=1}^{N_d} f_k^{(i-1)} * w_{ik} \tag{9.9}$$

The error between two adjacent iterations is written as

$$\Delta f_j^{(i)} = f_j^{(i)} - f_j^{(i-1)} = \frac{p_i - q_i}{\sum_{k=1}^{N_d} w_{ik}^2} w_{ij} \tag{9.10}$$

We assume that the weighting factor is 1 if the x-ray passes through the pixel, and 0 otherwise. Thus, $\sum_{k=1}^{N_d} w_{ik}^2 = N_i$. N_i is the number of pixels passed through by the ray. And the form of Equation 9.10 can be written in a simplified way as

$$\Delta f_j^{(i)} = \frac{p_i - q_i}{N_i} * w_{ij} \tag{9.11}$$

Then we have the steps of the iterative algorithm:

1. Give the initial guess to the image to be reconstructed:

$$f_j = f_j^{(0)} \quad (j = 1,2,3\ldots,N_d)$$

2. Calculate the estimated value of the *i*th forward projection:

$$q_i = \sum_{j=1}^{N_d} f_j^{(0)} * w_{ij}.z$$

3. Calculate the error:

$$\Delta f_{ij} = \frac{p_i - q_i}{N_i} * w_{ij}$$

   where N_i is the total number of pixels (voxels) passed through by the *i*th x-ray.

4. Correct for the *j*th pixel (voxel):

$$f_j = f_j^{(0)} + \Delta f_{ij}$$

5. Substitute the corrected value f_j into the next equation and repeat steps 2–5 until all incident beams are calculated.
6. Apply the corrected result of last iteration as the initial guess of the next iteration. Repeat the iteration till the stopping criteria mentioned above are achieved.

As shown in Figure 9.10, a common situation in image reconstruction is that of an inconsistent system with measurement noise. That is, the number of projections M is greater than the number of pixels N_d ($M > N_d$), and p_1, p_2, p_3 ... p_m are distorted by noise. Then the solution does not converge to a unique point and may oscillate in the neighborhood of the intersections of the hyperplanes (Kak and Slaney, 2001).

When the number of projections is smaller than that of pixels ($M > N_d$), an infinite number of solutions are possible. In this case, Tanabe (1971) has rigorously proven that the iterative approach described above converges to a solution named as f_s', such that $\left|f^{(0)} - f_s'\right|$ is minimized. The regularization-based iterative reconstruction method is proposed to solve the ill-posed problem, and compressed sensing (CS) technique is applied in the regularization.

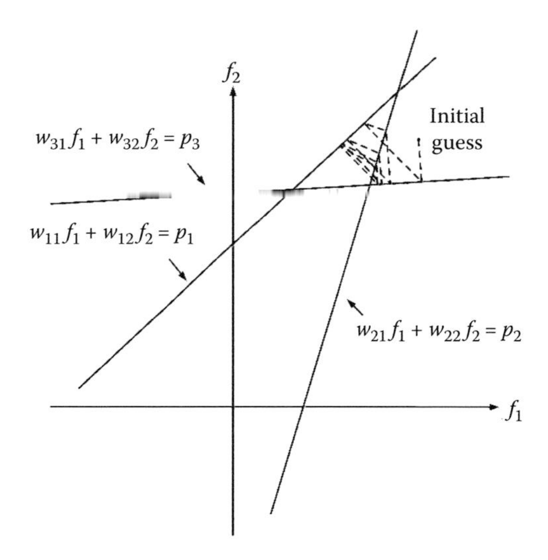

FIGURE 9.10
Illustration of Kaczmarz method when $M > N_d$.

9.3.2 Compressed Sensing-Based Iterative Reconstruction Algorithm

CS is an advanced sampling theory that is capable of keeping a good image quality while reducing the data acquisition by 90%. Therefore, it benefits those clinical applications where radiation dose is a big concern.

The CS technique takes advantage of the sparsity of signal to recover the signal in a small number of observations precisely. The properties of CS technique provide a new way to the CT reconstruction with incomplete projection data as a result of the sparsity of the CT image. CS-based iterative reconstruction method has good performance in coping with the situation where the projection data are heavily undersampled and highly noisy. Many methods used for CT imaging are based on a CS technique such as accelerated barrier optimization CS (ABOCS), alternating direction total variation minimization (ADTVM), adaptive steepest descent projection onto convex sets (ASD-POCS), and prior image constrained CS (PICCS). In this section, we mainly depict an advanced iterative reconstruction method with an optimization framework of ABOCS.

A CT image can be sparsified through the derivative operation for the reason of piecewise constant properties of the image. Total variation (TV, i.e., the L1 norm of the first-order derivative) is usually chosen as the objective to be minimized in iterative reconstruction. The image can be obtained by solving the mathematical optimization problem as follows:

$$\vec{f^*} = \arg\min \left\| \vec{f} \right\|_{TV} \text{ s.t.}, \left\| M\vec{f} - b \right\|_2^2 \leq \varepsilon, \quad \left\| \vec{f} \right\| \geq 0 \tag{9.12}$$

This is a nonlinear problem constrained by the data fidelity and image nonnegativity. $\vec{f}$ represents the vectorized image to be reconstructed. M is the forward projection matrix. And b is the line integral measurement, which can be obtained by the logarithmic operation on the projection. $M\vec{f} - \mathrm{b}_2^2$ indicates the L2 norm of data fidelity term $M\vec{f} - \mathrm{b}$. $\vec{f}_{\mathrm{TV}}$ means the TV term of the image and is defined as the L1 norm of the spatial gradient image. The parameter ε can be explained as the difference between the predicted projection and the raw projection. Many factors lead to the projection errors. If errors caused by scatter and beam-hardening effects are corrected, the projection errors left are mostly from Poisson statistics of the incident photons. The error ε can thus be estimated from the projections.

In ABOCS, Equation 9.12 can be rewritten as a new form using a logarithmic barrier method:

$$\overrightarrow{f^*} = \arg\ \min\left[\left\|\vec{f}\right\|_{\mathrm{TV}} - \log\left(\varepsilon - 0.5^*\left\|M\vec{f} - \mathrm{b}\right\|_2^2\right)\right], \quad \mathrm{s.t.}\vec{f} \geq 0 \tag{9.13}$$

This equation is convex and can be solved using a gradient-based method. Equation 9.13 is actually equivalent to an optimization problem that contains data fidelity with an automatically tuned penalty weight and TV term. ABOCS needs a little longer time than the conventional least-square form method where the penalty is fixed. Despite this, it provides convenience to the practice to using consistent algorithm parameters.

A first-order Nesterov algorithm is used to solve this equation since it has monotonic convergence at the optimal rate among all the first-order approaches when the objective function has a Lipschitz continuous gradient (Nesterov, 1983). It is difficult to use the Nesterov algorithm directly since the Lipschitz constant (L) and the strong convex constants (σ) required by this method are difficult to compute. To solve this problem, the unknown parameter Nesterov (UPN) method is employed. The L value is estimated by applying a backtracking line search and σ is adjusted using a heuristic formula with each Nesterov iteration (Jensen et al., 2011). The backtracking strategy estimates the Lipschitz constant by iteratively increasing the estimate until the continuity of the objective function is satisfied. The decremental σ value is chosen when it satisfies the convexity condition of the objective function at each iteration.

Another difficulty to using the Nesterov algorithm directly is that the logarithmic term in the objective in Equation 9.13 is not differentiable at the location where $0.5^*M\vec{f} - b_2^2 = \varepsilon$. To solve this issue, the objective function is modified around and outside the singular point without affecting the solution of the optimization framework. As shown in Figure 9.11, a linear function (dashed line) with a large and finite positive slope is used to replace the segment of the logarithmic function (solid line) that has infinite values.

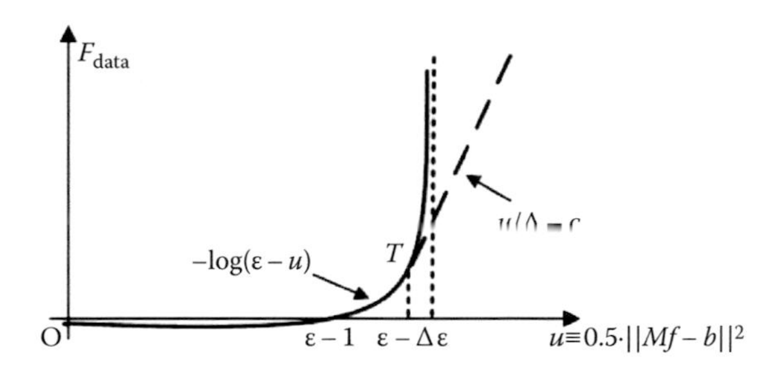

FIGURE 9.11
Illustration of the modified logarithmic term. (From T. Niu et al., *Phys. Med. Biol.* 59, 1801–1814, 2014.)

The modified logarithmic term is thus written as

$$F_{data}(u) = \begin{cases} -\log(\varepsilon - u) & \text{if } u \le \varepsilon - \Delta \\ \dfrac{u}{\Delta} - \log(\Delta) - \dfrac{\varepsilon - \Delta}{\Delta} & o.w. \end{cases} \tag{9.14}$$

where $u = \frac{1}{2} M\vec{f} - b_2^2$ and Δ is a user-defined parameter that controls the slope of the linear function.

The resultant objective function becomes

$$F(\vec{f}) = \left\|\vec{f}\right\|_{TV} + \begin{cases} -\log(\varepsilon - 0.5 * \left\|M\vec{f} - b\right\|_2^2), \text{ } if \text{ } 0.5 * \left\|M\vec{f} - b\right\|_2^2 \le \varepsilon - \Delta \\ \dfrac{0.5 * \left\|M\vec{f} - b\right\|_2^2}{\Delta} - \log(\Delta) - \dfrac{\varepsilon - \Delta}{\Delta} \quad o.w., \end{cases} \tag{9.15}$$

and the optimization problem (Equation 9.13) is rewritten as

$$\vec{f^*} = \arg\min F(\vec{f}) \text{ s.t., } \vec{f} \ge 0 \tag{9.16}$$

Equation 9.16 is then solved using the modified Nesterov algorithm.

Figure 9.12 shows the reconstruction results of the Catphan©600 phantom. With 60 projection views (about 17% of the total 362 projections from a short scan), the FBP reconstruction has severe view-aliasing artifacts, and the relative reconstruction error (RRE) compared with the full-view reconstruction is about 15% (Figure 9.12a and b). The ABOCS reconstruction using the modified Nesterov algorithm achieves a comparable full-view image quality (Figure 9.12c) as that of FBP, and effectively suppresses the view-aliasing artifacts when the projection views are reduced to 60 (Figure 9.12d).

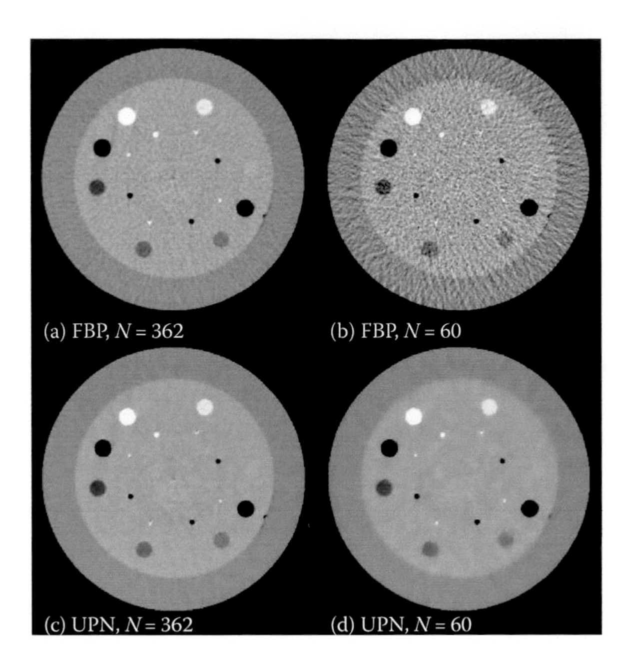

FIGURE 9.12
Reconstructed Catphan©600 images in a 200° short-scan mode: (a) FBP: using 362 views; (b) FBP: using 60 views; (c) ABOCS: using 362 views; and (d) ABOCS: using 60 views. Window level: [-600 400] HU. (From T. Niu et al., *Phys. Med. Biol.* 59, 1801–1814, 2014.)

9.4 Scatter/Shading Correction in CBCT

In CBCT, shading artifacts of dominantly low-frequency components are mainly caused by scatter contamination, uncorrected beam-hardening effects, and other nonlinearity conditions. These artifacts in CBCT will lead to inaccurate CT numbers, contrast loss, and spatial non-uniformity, which hamper the advanced clinical applications of CBCT. As the major factor of shading artifacts, scatter contamination has drawn a lot of attention. In this section, we briefly introduced several scatter correction algorithms.

The low-frequency errors are extracted using image processing techniques either in the image or projection domain. These methods are mainly based on prior knowledge or strong assumptions with generality. For instance, Brunner et al. (2011) proposed a prior constrained scatter correction method assuming a CT image consists of a series of weighted basis images. Li et al. (2011) proposed an image-domain correction algorithm based on the piecewise constant property of CT images. Here, the work of Niu et al. (2010) on scatter correction is taken as an instance for prior knowledge-based method.

Owing to its small inherent scatter signals and accurate detectors, diagnostic multidetector CT (MDCT) is widely applied in radiation treatment as planning CT (pCT). In Niu's work, the pCT images are used as "free" prior information to correct for the shading artifacts in CBCT (Niu et al., 2012). As shown in Figure 9.13, the method can be roughly divided into three steps. The first-pass CBCT images are reconstructed from the raw projections without correction. Then, the pCT images of the same patient are calibrated to translate the CT numbers to linear attenuation coefficients. In the third step, after the deformable registration between the calibrated pCT and the corrected CBCT, the deformed pCT can be filled with gas pockets optionally to match the geometry of gas pockets in CBCT images better. Then segmented couch from the corrected CBCT image is added into the pCT image, which is also optional. After the preparation, the correction for artifacts in CBCT can be implemented using the pCT-based correction algorithm. The iteration

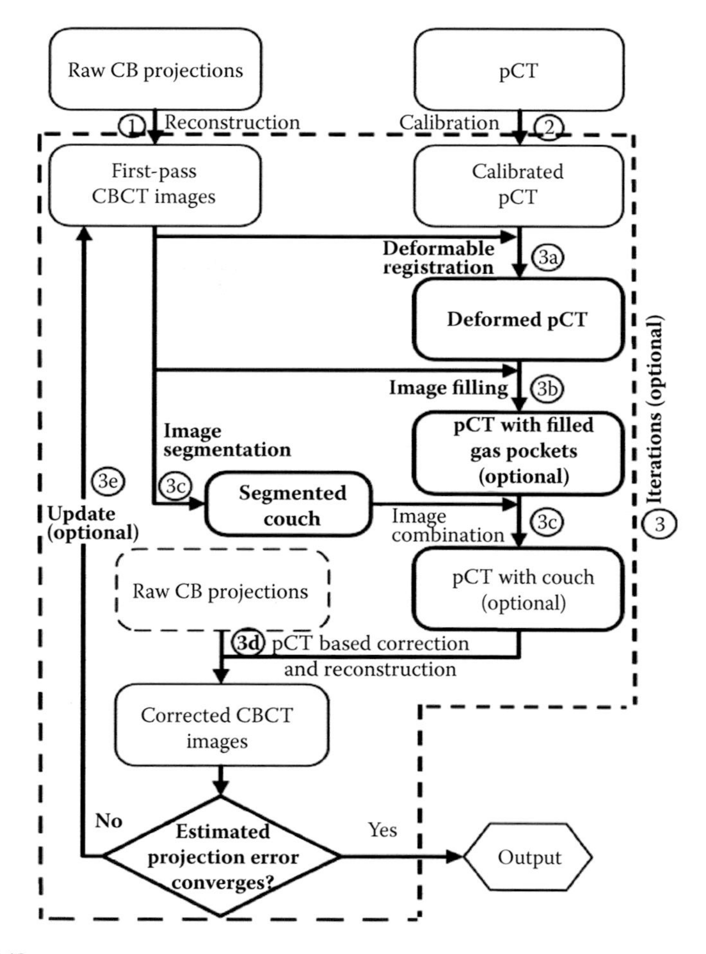

FIGURE 9.13
Improved workflow of the quantitative CBCT imaging scheme with the new components [bold]. (From T. Y. Niu et al., *Med. Phys.* 37, 5395–5406, 2010.)

of the pCT-based correction method is repeated till the estimated projection error converges to a stable value.

As shown in Figure 9.14, the raw projections are first reconstructed to CBCT images. Then the calibrated pCT images, where the affine function is used as calibration formula, are spatially registered to the CBCT images. The simulated projections are obtained via the forward projection of the registered pCT data. The subtraction of the simulated projection from the raw projection with the same projection angle contains the main low-frequency scatter signals and primary signal differences between the pCT and the CBCT scans.

Due to the registration errors from deformation, these primary signal differences are mostly high frequency. Thus, the scatter signals can be readily estimated from these blended components by low-pass filtering or smoothing. Finally, the estimated scatter is subtracted from the raw CBCT projections and scatter-corrected CBCT images are reconstructed. Note that, as a post-processing method, the proposed algorithm cannot estimate high-frequency statistical errors in the projections, and the scatter noise is, consequently, left in the corrected images.

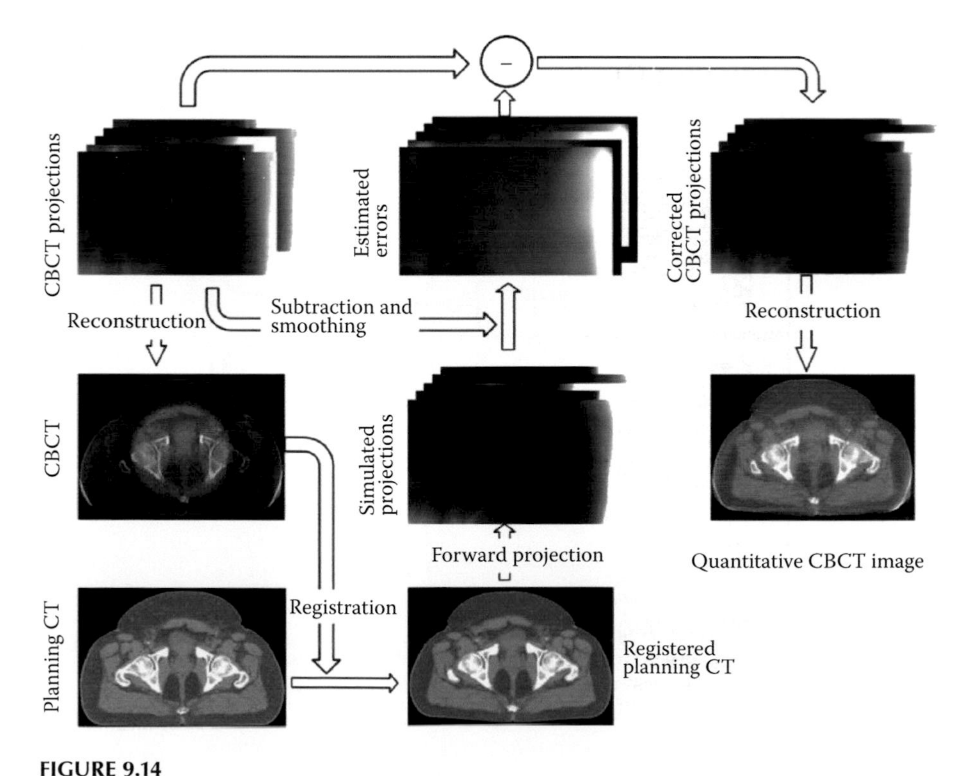

FIGURE 9.14
One iteration of scatter correction (mainly step 3d in Figure 9.13). (From T. Y. Niu et al., *Med. Phys.* 37, 5395–5406, 2010.)

The performance of the proposed pCT-based method is determined by the smoothing step of generating the estimated scatter. In principle, the small low-contrast discrepancy cannot be carried over at the process of registration and filtering. Thus, the implementation of the scatter correction can be effective as long as the high-contrast objects are well matched between the pCT and the CBCT images. The proposed correction method aims at not only scatter contamination but also the low-frequency components of other errors from beam-hardening effects, detector lag, and detector nonlinear gain.

Nevertheless, these methods are overly dependent on the prior knowledge, and the application is confined to certain clinical circumstances. For example, pCT of a patient is routinely scanned for treatment planning purpose. The patient position is not reproducible to a high degree of accuracy from pCT to CBCT scans and may cause a considerable amount of motion uncertainty in shading correction. More general shading correction schemes without prior knowledge are proposed. For instance, Wu et al. (2015) proposed an iterative CT shading correction method based on the general knowledge and obtained satisfactory results. Without depending on prior knowledge, the proposed method is only assisted by general anatomical information.

Figure 9.15 shows the workflow of the proposed iterative shading correction scheme (Wu et al., 2015). Reconstructed CT image or raw projections are chosen as the input data. After subtraction of the ideal template image from the reconstructed CT image, the residual image containing various

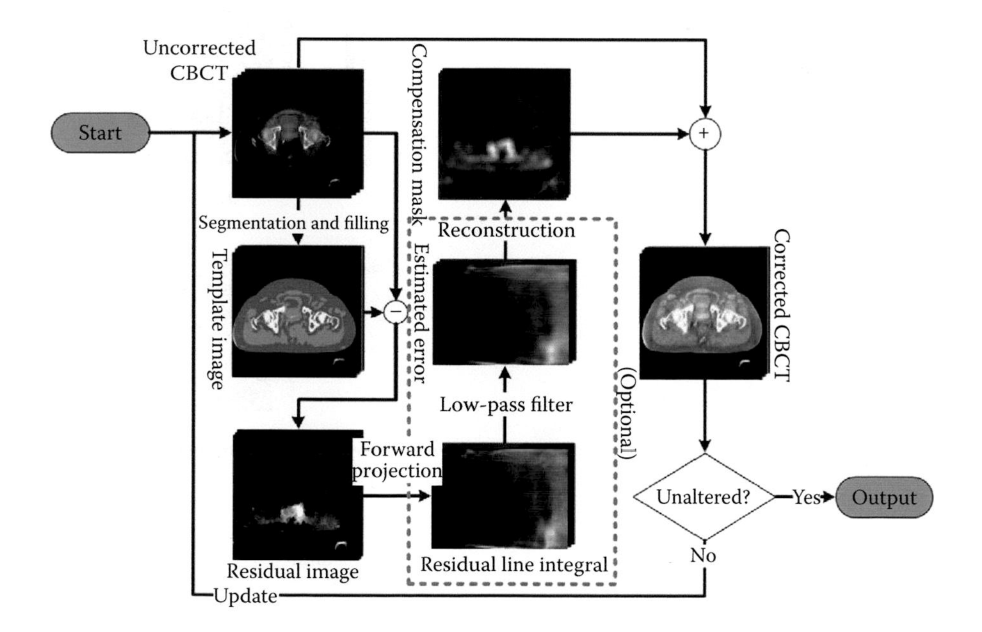

FIGURE 9.15
Workflow of the proposed iterative shading correcting scheme without relying on prior planning CT. (From P. Wu et al., *Phys. Med. Biol.* 60, 8437–8455, 2015.)

error sources is generated. Via the forward projection, the noncontinuous and low-frequency shading artifacts in the residual image turn into low-frequency signals with continuity in the line integral domain. Then, the shading errors are estimated by low-filtering the integral. By reconstruction, the compensation mask can be obtained from the estimated errors. The corrected image is obtained after adding the "demonstration" and the compensation mask. Until the variation of residual image is minimized, the iteration is processed.

Due to the severe shading artifacts in the CT image, tissue segmentation is inefficient or even impossible. Thus, an initial image with shading artifacts suppressing can be added into the iterative process as an option. Figure 9.16 shows the flow diagram of the initial image generation. In this study, pelvis patient data using a half-fan bowtie filter are selected as an example of severe shading artifacts around the periphery of the image. After the reconstruction from the modulated flat-field projections, we can get an initial bowtie mask. Due to the high performance on detection and elimination of the bright circle around the periphery of this mask, Hough transform is used to eliminate the artifacts outside the field of view. After the combination between the raw CBCT image reconstructed from the raw projections and the transformed bowtie mask, we can get the final corrected CBCT image.

The initial image without bowtie artifacts is obtained by

$$I_{init} = I_{uncorr} + \alpha M \tag{9.17}$$

where M is the bowtie mask, I_{uncorr} is the uncorrected CBCT image, and α is the scaling factor for each slice. Generally, the same tissue has comparable CT numbers, and a sharp peak usually exists in the histogram of a specific

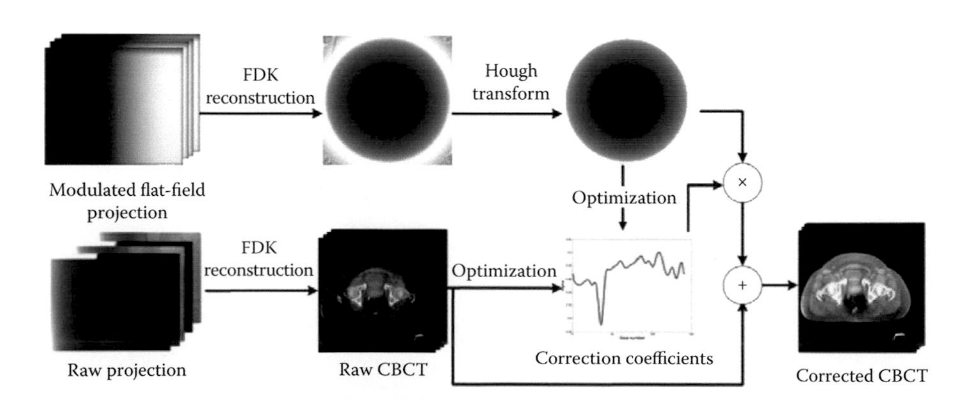

FIGURE 9.16
Workflow of generating the initial image. (From P. Wu et al., *Phys. Med. Biol.* 60, 8437–8455, 2015.)

tissue. Based on the above two points, the scaling factor is formulated as a mathematical optimization problem to minimize the peak value of the histogram, and is written as

$$\begin{aligned} &\alpha = \arg\min\left(-\max\left(\text{hist}(I_{uncorr} + \alpha M)\right)\right) \\ &\text{s.t.}\left|\alpha_{j+1} - \alpha_j\right| \le b(j = 1,2,\ldots,n) \end{aligned} \quad (9.18)$$

where $|\alpha_{j+1} - \alpha_j| \le b$ is the continuity constraint, which maintains a smooth transition of α_j from adjacent slice. b is the smoothness upper bound, which is empirically set as 0.05 in the study.

Without prior knowledge from high-quality MDCT images or strong assumptions on the CT image, the iterative correction method preserves the CT anatomical information well. Thus, the structure of the CT image is dependably kept up without bias toward the prior information. It is practical and attractive as a general solution to CT shading correction. Likewise, without prior knowledge of the imaged object, the scatter correction for full-fan volumetric CT using a stationary beam blocker achieves high scatter estimation accuracy (Niu et al., 2011).

In the conventional measurement-based method, the primary signal loss is inevitable. It cannot be applied into the clinical application well without the compensation for the primary data loss. In Niu's studies, similar to the conventional measurement-based method, the same imaging geometry is employed. And the sparse x-ray beam blockers are inserted between the x-ray source and the object (Niu et al., 2011). In a full-fan CT scan, one projection line through the object may be measured multiple times. Without considering the noise and measurement errors, we can block some of these redundant rays and maintain the reconstructed CT image quality.

If we can allocate the areas appropriately, the blocked data can only contain redundant rays. Figure 9.17 shows the geometry of the CBCT scanner and the blocker, which looks like a "crossing-finger" shape. Each half of the detector is respectively blocked in the vertical direction by horizontal strips. They are vertically shifted from the left to right sides circularly. Similarly, owing to the same half-blocking pattern, reconstruction from the blocked horizontal lines can also be precisely obtained. In the design of the blocker, for simplicity of description, two approximately redundant rays are referred as one conjugate ray pair, whether they are in the midplane or not.

As illustrated in Figure 9.18, the proposed reconstruction with the insertion of the "crossing finger" is coherently explained. After scatter correction, the line integral projection images p_m can be generated via interpolation on the blocked primary, p_0. And after being reconstructed using a conventional half-fan reconstruction algorithm, the left and the right half-fan CT images

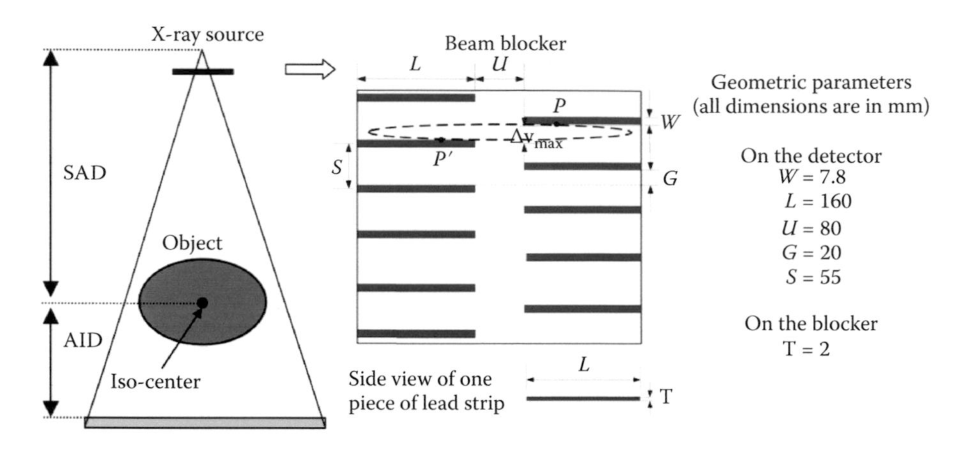

FIGURE 9.17
Geometry of the "crossing-finger" x-ray beam blocker used in the proposed method. (From T. Niu and L. Zhu, *Med. Phys.* 38, 6027–6038, 2011.)

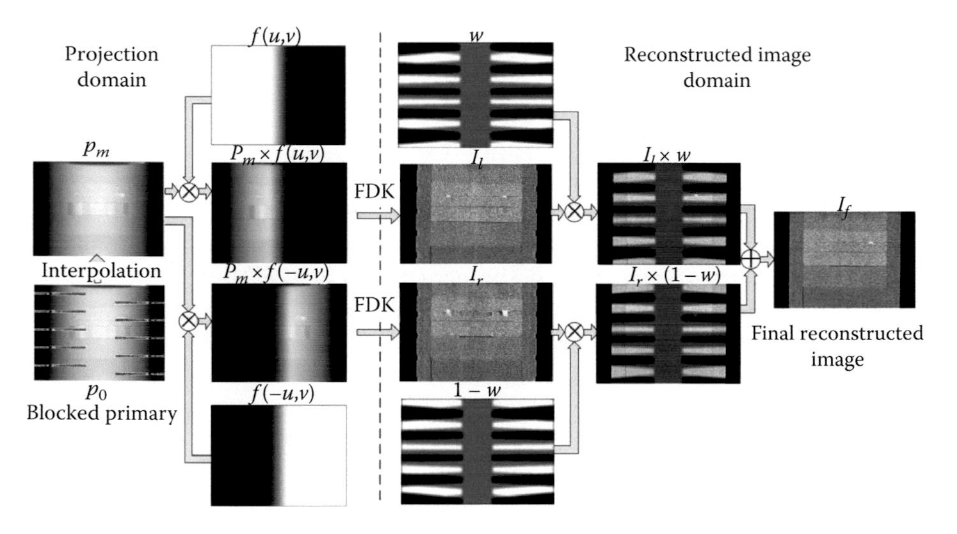

FIGURE 9.18
Workflow of the FDK-based reconstruction with the insertion of the "crossing-finger" blocker. (From T. Niu and L. Zhu, *Med. Phys.* 38, 6027–6038, 2011.)

are merged together finally. Note that the final image combination formula can be written as

$$I_f = I_l w + I_r(1 - w) \tag{9.19}$$

where I_f is the final merged image, I_l and I_r are the left and right half-fan CT images, respectively, and w is an intermediate function to quantify the image quality of two half-fan images.

Observably, the effective scatter correction and accurate reconstruction are achieved in this method. Compared to the existing methods, it enables simple and efficient scatter correction without increasing scan time or patient dose. Since the primary signals are considered to be approximately redundant in a full scan, the image quality is preserved even with primary signal loss.

9.5 Clinical Applications

9.5.1 Dental Practice

Dental CBCT generates the three-dimensional (3D) images as done by conventional CT. The 3D imaging capability will benefit clinicians in diagnosis, treatment planning, and evaluation. A literature review shows that CBCT has been widely utilized in oral and maxillofacial surgery, endodontics, implants, and orthodontics (Hadi et al., 2012).

9.5.1.1 Oral and Maxillofacial Surgery

Oral and maxillofacial surgery specializes in treating the diseases, injuries, and defects in the head, neck, face, jaws, and the hard and soft tissues of the oral and maxillofacial region.

CBCT imaging plays an important role in oral and maxillofacial surgery. In the past, Same-Lingual Opposite-Buccal Rule (SLOB Rule) was used to determine the location of a known or unknown object on intraoral radiographs by comparing two periapical radiographs taken from different directions. 3D images are more precise in the determination of not only the exact location and the extent of jaw pathologies but also the relationship of these teeth to vital structures. Additionally, the structural superimposition is avoided, which cannot be removed in panoramic images. The ability of accurate measurements of facial distance makes CBCT a prevalent tool in the planning of orthognathic and facial orthomorphic surgeries and the evaluation of the outline of the lip and bony regions of the palate.

9.5.1.2 Endodontics

Endodontics encompasses the study and practice of the basic and clinical sciences of the biology of the normal dental pulp. In addition, endodontics includes the etiology, diagnosis, prevention, and treatment of diseases and injuries of the dental pulp along with associated periradicular conditions.

CBCT imaging is a practical tool in detecting periapical lesions and viewing root fractures because of its high spatial resolution. The 3D scanning of a tooth allows for accurately measuring the depth of dentin fracture. Moreover, CBCT is also used presurgically to determine the proximity of a tooth to adjacent vital structures, execute precise measurement, and identify the anatomy of the area. When applied in diagnosing periapical lesions, CBCT imaging also has the ability of early detection of inflammatory root resorption that is almost impossible in conventional 2D radiographs.

9.5.1.3 Implant Dentistry

A dental implant is a surgical component that interfaces with the bone of the jaw or skull to support a dental prosthesis such as a crown, bridge, denture, or facial prosthesis, or to act as an orthodontic anchor.

CBCT images offer invaluable information when it comes to the assessment of implant dentistry. The 3D images are beneficial in developing treatment planning and preventing adjacent vital structures from superfluous damage. The proposed implant will also be the optimal selection in height and width due to the capacity of CBCT in quantifying the shape of the alveolus and measuring the accurate implant location. The fact that the dentists have a better 3D visualization than 2D radiography is very beneficial. As a result, the percentage of accidents in the implant surgery is significantly reduced. In addition, CBCT imaging also plays a role in postsurgical evaluations of bone grafts.

9.5.1.4 Orthodontics

Orthodontics deals primarily with the diagnosis, prevention, and correction of malpositioned teeth and the jaws.

CBCT imaging is the best instrument for evaluating airway function as a result of the causal relationship between airway disorders and malocclusion generating the adenoid facies. The CBCT images can also be applied to view the soft tissue and detect the vital structures adjacent to the impacted teeth, which may interfere with the orthodontic movement. Furthermore, 3D images help dentists to visualize from different angles or directions, some that cannot be available in panoramic or 2D radiography included. CBCT images are self-corrected for magnification, producing orthogonal images with a practical 1:1 measuring ratio (Hadi et al., 2012).

9.5.2 Radiation Therapy

Radiation therapy is a local treatment method that is designed to treat the defined tumor and spare the surrounding normal tissue from receiving the doses above specified dose tolerances. IGRT is the process of frequent 2D or 3D imaging to improve the radiation dose distribution to the tumor during the radiation treatment procedure.

It is a fact that structures may change in size, shape, and position, as well as the patient may react involuntarily to radiation therapy in the treatment procedure. Hence, obtaining the anatomical and positional information of patient online during the treatment is extremely vital.

In the past, patients were positioned by the skin marks determined at treatment planning. The accuracy of the position was guaranteed by comparing a secondary exposure projection image with the digital reconstructed radiography (DRR) generated from the planning CT images. Nowadays, patient positioning and error correction are more reliable in the IGRT session using CBCT. The CBCT scanner is integrated into the accelerator. A patient is initially positioned by skin marks with laser beams. Then CBCT images are acquired and compared with the planning CT. The images are imported into the planning system to update the treatment planning and to reposition the patient. Finally, the plan will be transferred to the treatment unit and translated to numerous different dimensions or locations in the treatment process. The difference between the portal images and the planning CT will be acquired and results in a correction using the 2D or 3D alignment. As a result, the 3D alignment is more accurate in patient positioning and provides the benefit of couch rotation, which cannot be implemented in the 2D alignment. In addition, CBCT imaging obtains the anatomical and positional information of the tumor and corrects for the errors online during the treatment procedure.

In the future, progress in computer technologies will enhance the efficiency of reconstruction algorithms and image processing. The advances in CBCT image quality and reduction in imaging dose will benefit the clinicians and patients respectively, allowing more frequent imaging.

Digital subtraction angiography (DSA) is a fluoroscopy technique to provide excellent blood vessel visualization in a dense soft tissue environment. DSA images are obtained using contrast medium images subtracting the "pre-contrast image" or the mask image after the contrast medium has been injected into the blood vessels.

DSA is recognized as the "gold standard" in imaging neck and head blood vessels, owing to its high specificity, accuracy, and sensibility in cerebrovascular disease detection. Nevertheless, conventional DSA images may attain unsatisfactory visualization in the cases of multiple vessels overlapping and improper photography angle. Furthermore, slimy vessels may be obscured, especially when they are adjacent to large chaotic areas. CBCT technology, with its 3D visualization nature and advanced postprocessing algorithm, is superior to 2D DSA in slimy blood vessels and complex vascular structures. The 3D DSA images are reconstructed from two data sets of rotational 2D DSA and a 3D model. The first rotation provides the subtraction mask, and the second is performed during the administration of contrast medium.

9.5.3 Head and Neck

CBCT imaging technology has become a preferred option in the intraoperative and diagnostic applications in head and neck regions. CBCT imaging obtains high spatial resolution coupled with a low dose under the background of the high sensitivity to radiation of the head structures. Applications in paranasal sinus, temporal bone, and skull base imaging have been explored.

9.5.3.1 Paranasal Sinus

Relatively low-dose features and high-quality bony structure visualization makes CBCT the best option for the paranasal sinuses detection. Up to now, very few studies were performed on the comparison of the image quality in paranasal sinus using CBCT and conventional multidetector CT (MDCT). Alspaugh et al. (2007) directly compared the spatial resolution on the paranasal sinuses obtained by CBCT scans and the 64-slice MDCT scans. The effective dose is as low as 0.17 mSv using the CBCT scanner and as high as 0.87 mSv using the 64-slice conventional CT scanners. The 64-slice conventional CT scanners had a transaxially spatial resolution of 7 lp/mm (standard filter) and 11 lp/cm (bone filter). The CBCT scanner had isotropic spatial resolution of 12 lp/cm.

9.5.3.2 Temporal Bone

The temporal bone scan is one of the earliest targets for head and neck CBCT imaging. The images show a significantly higher structural visualization compared with MDCT, especially for fine and tiny structures, such as the ossicular chain, bony labyrinth of the inner ear, internal cochlear anatomy, and the facial nerve.

In addition to the high resolution, CBCT images present superiority in the evaluations of middle and inner ear implants. A reduction in metal artifacts is practical compared with conventional CT images. As a result, the combination of high spatial resolution and reduced metal artifacts in CBCT imaging facilitate the postsurgical evaluation of reconstructed middle and inner ears. However, with the poor soft-tissue contrast resolution, CBCT imaging in general diagnostic imaging of the temporal bone is still hampered.

9.5.3.3 Skull Base

Bony structures and neurovascular anatomy structures of the skull base are extremely subtle and complex, leading to unsatisfactory visualization of conventional CT images. The CBCT imaging becomes an attractive option in skull base detection due to the characteristic high spatial resolution of CBCT images.

Due to the inferior soft-tissue contrast resolution in CBCT imaging, the existing practice in oncologic imaging of the skull base relies on MDCT and MR imaging for their osseous and soft-tissue definitions. There are still several clinical groups working to explore the potential use of CBCT during surgeries in the skull base.

9.5.4 C-arm CBCT Imaging to Intracranial Hemorrhage

Intracranial hemorrhage (ICH) is a serious medical emergency, since the buildup of blood in the skull may bring a sharp increase in intracranial pressure. The increased pressure will crush delicate brain tissue or limit its blood supply. As a result, ICH is closely related to a number of severe neurological diseases and injuries.

Conventional CT imaging is an effective tool in detection, diagnosis, and monitoring of ICH, coupled with advantages of high speed and sensitivity to intracranial bleed. Considering the serious emergency of ICH, patient transportation between a sickroom to a scanner suite is associated with terribly high mortality. Mobile C-arm cone-beam CT offers an ideal solution to this problem for its mobility and convenience. It has the capability of 2D radiographic as well as 3D CBCT imaging and compatibility with the environment. Nevertheless, imaging performance of the current status cannot suit for diagnostic request satisfactorily. When used in IGRT, CBCT image quality is credited for positioning and correcting error tasks. As for skull and soft tissue imaging, the contrast and the spatial resolution are not sufficient for diagnosis. In recent years, numerous works have developed in promoting the CBCT imaging quality, including the scanner and detector designs, the reconstruction and artifact correction algorithms, etc. Jennifer et al. (2016) used high-resolution FBP reconstruction and artifact correction for scatter, beam hardening, and image lag. In reality, the quality of their images is improved to a large extent, including an excellent delineation of bone feature and a high contrast-to-noise ratio.

Acknowledgments

The work was supported by the Zhejiang Provincial Natural Science Foundation of China (Grant No. LR16F010001), National High-tech R&D Program for Young Scientists by the Ministry of Science and Technology of China (863 Program, 2015AA020917), Natural Science Foundation of China (NSFC Grant No. 81201091, 51305257), National Key Research Plan by the Ministry of Science and Technology of China (Grant No. 2016YFC0104507), Key Laboratory of Diagnosis and Treatment of Neonatal Diseases of Zhejiang Province (2016-ZJKD-ND-004), and Zhejiang Province "151 Talent" Program. Yu Kuang was supported in part by an NIH/NIGMS grant (P20GM103440).

References

Alspaugh J, Christodoulou E, Goodsitt M et al., Dose and image quality of flat-panel detector volume computed tomography for sinus imaging, *Med. Phys.* **43**, (2007).

Brunner S, Nett BE, Tolakanahalli R, and Chen G-H, Prior image constrained scatter correction in cone-beam computed tomography image-guided radiation therapy, *Phys. Med. Biol.* **56**, 1015 (2011).

Chen L, Research on several weight factor models of ART, *Optic Instrum.* **36**, 142–146 (2014).

Grimmer R and Kachelriess M, Empirical binary tomography calibration (EBTC) for the precorrection of beam hardening and scatter for flat panel CT, *Med. Phys.* **38**, 2233–2240 (2011).

Hadi Mohammed Alamri, BDS Mitra Sadrameli, DMD Mazen Abdullah Alshalhoob, et al., Applications of CBCT in dental practice: A review of the literature, *Gen Dentistry*, (2012).

Hsieh J, Molthen RC, Dawson CA, and Johnson RH, An iterative approach to the beam hardening correction in cone beam CT, *Med. Phys.* **27**, 23–29 (2000).

Jensen TL, Jorgensen JH, Hansen PC, and Jensen SH, Implementation of an optimal first-order method forstrongly convex total variation regularization, *BIT Numer. Math.* **51**, 1–28 (2011).

Kak AC and Slaney M, *Principles of Computerized Tomographic Imaging*, PA: SIAM, (2001).

Li X, Li T, Yang Y, Heron DE, and Huq MS, A novel image-domain-based cone-beam computed tomography enhancement algorithm, *Phys. Med. Biol.* **56**, 2755 (2011).

Mori I, Machida Y, Osanai M, and Iinuma K, Photon starvation artif acts of x-ray CT: Their true cause and a solution, *Radiol. Phys. Technol.* **6**, 130–141 (2013).

Nesterov Y, A method for unconstrained convex minimization problem with the rate of convergence O(1/k2), *Doklady AN USSR* (translated as Soviet Math Docl), (1983).

Niu T, Al-Basheer A, and Zhu L, Quantitative cone-beam CT imaging in radiation therapy using planning CT as a prior: First patient studies, *Med. Phys.* **39**, 1991–2000 (2012).

Niu TY, Sun MS, Star-Lack J, Gao HW, Fan QY, and Zhu L, Shading correction for on-board cone-beam CT in radiation therapy using planning MDCT images, *Med. Phys.* **37**, 5395–5406 (2010).

Niu T, Ye X, Fruhauf Q, Petrongolo M, and Zhu L, Accelerated barrier optimization compressed sensing (ABOCS) for CT reconstruction with improved convergence, *Phys. Med. Biol.* **59**, 1801–1814 (2014).

Niu T and Zhu L, Scatter correction for full-fan volumetric CT using a stationary beam blocker in a single full scan, *Med. Phys.* **38**, 6027–6038 (2011).

Niu T and Zhu L, Accelerated barrier optimization compressed sensing (ABOCS) reconstruction for cone-beam CT: Phantom studies, *Med. Phys.* **39**, 4588–4598 (2012).

Perkin Elmer Inc, XRD 1611 Digital X-ray Detector Reference Manual, **Rev.02**, (2014).

Siewerdsen JH, Daly MJ, Bakhtiar B, Moseley DJ, Richard S, Keller H, and Jaffray DA, A simple, direct method for x-ray scatter estimation and correction in digital radiography and cone-beam CT, *Med. Phys.* **33**, 187–197 (2006).

Tanabe K, Projection method for solving a singular system, *Numer. Math.* **17**, 203–214 (1971).

Tanaka R, Ichikawa K, Mori S, Dobashi S, Kumagai M, Kawashima H, Minohara S, and Sanada S, Investigation on effect of image lag in fluoroscopic images obtained with a dynamic flat-panel detector (FPD) on accuracy of target tracking in radiotherapy, *J. Radiat. Res.* **51**, 723–731 (2010).

Varian Medical System Inc, RAD-94 Rotating Anode X-ray Tube Datasheet, **Rev I**, (2007).

Wu P, Sun X, Hu H, Mao T, Zhao W, Sheng K, Cheung, AA, and Niu T, Iterative CT shading correction with no prior information, *Phys. Med. Biol.* **60**, 8437–8455 (2015).

Xu J, Sisniega A et al., Technical assessment of a prototype cone-beam CT system for imaging of acute intracranial hemorrhage. *Med. Phys.* **43**, (2016).

10

Cadmium (Zinc) Telluride 2D/3D Spectrometers for Scattering Polarimetry

Rui Miguel Curado da Silva, Ezio Caroli, Stefano del Sordo, and Jorge M. Maia

CONTENTS

10.1 Introduction

The semiconductor detectors technology has dramatically changed the broad field of x- and γ-rays spectroscopy and imaging. Semiconductor detectors, originally developed for particle physics applications, are now widely used for x/γ-rays spectroscopy and imaging in a large variety of fields, among which, for example, x-ray fluorescence, γ-ray monitoring and localization, noninvasive inspection and analysis, astronomy, and diagnostic medicine. The success of semiconductor detectors is due to several unique characteristics as the excellent energy resolution, the high detection efficiency, and the possibility of development of compact and highly segmented detection systems (i.e., spectroscopic imager). Among the semiconductor devices, silicon (Si) detectors are the key detectors in the soft x-ray band (<15 keV). Si-PIN diode detectors (Pantazis et al. 2010) and silicon drift detectors (SDDs; Lechner et al. 2004), operated with moderate cooling using small Peltier cells, show excellent spectroscopic performance and good detection efficiency below 15 keV. On the other side, germanium (Ge) detectors are unsurpassed for high-resolution spectroscopy in the hard x-ray energy band (>15 keV) and will continue to be the first choice for laboratory-based high-performance spectrometers system (Eberth and Simpson 2006).

However, in the last decades, there has been an increasing demand for the development of room-temperature detectors with compact structure having the portability and convenience of a scintillator, but with a significant improvement in energy resolution and/or spatial resolution. To fulfill these requirements, numerous high-Z and wide bandgap compound semiconductors have been exploited (Owens and Peacock 2004; Sellin 2003). As demonstrated by the impressive increase in the scientific literature and technological development, cadmium telluride (CdTe) and cadmium zinc telluride (CZT) based devices are today dominating the room-temperature semiconductor applications scenario, being widely used for the development of x/γ-ray instrumentation (Lebrun et al. 2003; Lee et al. 2010; Ogawa and Muraishi 2010) in different application fields.

In particular, applications that require imaging capabilities with high spatial resolution possibly coupled with good spectroscopic performance (at room temperature) are certainly the field in which CdTe/CZT sensors technology can exploit all its potential and advantages. In fact, the possibility to easily segment the charge collecting electrodes into strips and/or arrays, as well as to assemble mosaics of even small sensitive units (i.e., crystals), allow one to obtain devices with excellent bi-dimensional spatial resolution (down to tens of microns). According to the type of readout electronics, these devices allow the accurate measure of the energy released by the interaction of photons within the material (Limousin et al. 2011; Watanabe et al. 2009; see also other chapters in this book).

One quite new and challenging application field for CdTe/CZT spectro-imagers is x- and γ-rays polarimetry. This type of measurement is becoming increasingly important in high-energy astrophysics. Until now, polarimetry in high-energy astrophysics has been an almost unexplored field due to the inherent difficulty of the measurement and also to the complexity of the required detection, electronic, and signal processing systems, since celestial x/γ-ray sources are only observable from space. To date, x- and γ-ray cosmic source emissions have been studied exclusively through traditional spectral and timing analysis, and imaging of the measured fluxes.

Polarization measurements will increase the number of observational parameters of the same x/γ-ray source by two: the polarization angle and the level of linear polarization. These additional parameters should allow a better discrimination between different emission models characterizing the same object. Polarimetric observations can provide fundamental information about the geometry, the magnetic field, and the active emission mechanisms of cosmic-ray sources, helping to solve several hot scientific issues. For these reasons, the high-energy polarimetric capability is currently recognized as an essential requirement for the next generation of space telescopes.

In the range 10–1000 keV, effective polarization measurements can be performed by using the properties of Compton scattering for polarized photons. A spectroscopic imager made of CZT/CdTe offers a suitable and high-performance solution to build a scattering polarimeter (Caroli et al. 2000). Furthermore, this solution offers the capability to perform polarization measurements simultaneously with those of spectroscopy, imaging, and timing. This represents a major advantage for new space instruments, both for the optimization of payload and inflight resources utilization and for the scientific return, because the various observational parameters on the same source can be correlated without problems due to the time variability of the sources itself and/or background.

The chapter is divided into two main parts. The first part, comprising Sections 10.2 and 10.3, gives a summary of room-temperature semiconductor principle and CZT/CdTe development for the realization of detectors for x- and γ-rays suitable for building two- and three-dimensional spectroscopic imagers (Caroli and Del Sordo 2015). The second part is dedicated to addressing a very hot and challenging topic: the use of CdTe/CZT spectroscopic imagers as scattering polarimeters for high-energy astrophysics applications and is mainly based on the results obtained by the authors and colleagues both by experiments and by Monte Carlo simulations.

10.2 X- and γ-Rays Spectroscopy with CZT/CdTe Sensors

The typical operation of semiconductor detectors is based on collection of the charges, created by photon interactions, through the application of an external electric field. The energy range of interest mainly influences the choice of the best semiconductor material for a radiation detector. Among the various interaction mechanisms of x- and γ-rays with matter, three play an important role in radiation measurements: photoelectric absorption, Compton scattering, and pair production (Leo 1994). In photoelectric absorption, the photon transfers all its energy to an atomic electron, while a photon interacting through Compton process transfers only a fraction of its energy to an outer electron, producing a hot electron and a degraded photon. In pair production, a photon with energy above a threshold energy of 1.02 MeV interacts within the Coulomb field of the nucleus, producing an electron–positron pair. Neglecting the escape of characteristic x-rays from the detector volume (the so-called escape fluorescent lines), only the photoelectric effect results in the total absorption of the incident energy and thus gives the correct information on the impinging photon energy. The interaction cross sections are highly dependent on the atomic number. In photoelectric absorption, it varies as Z^{4-5}, Z for Compton scattering and Z^2 for pair production.

10.2.1 CdTe/CZT as X- and γ-Rays Spectrometer

An optimum spectroscopic detector must favor photoelectric interactions, and so semiconductor materials with a high atomic number are preferred. Figure 10.1a shows the linear attenuation coefficients, calculated by using tabulated interaction cross-section values (Berger et al. 2010), for photoelectric absorption and Compton scattering of Si, CdTe, HgI_2, NaI, and BGO; NaI and BGO are solid scintillator materials typically used in radiation measurements. As shown in Figure 10.1a, photoelectric absorption is the main process up to about 200 keV for CdTe. The efficiency for CdTe detectors versus detector thickness for various typical photon energies is reported in Figure 10.1b. A 10 mm thick CdTe detector ensures good photoelectric efficiency at 140 keV (>95%), while a 1 mm thick CdTe detector is characterized by a photoelectric efficiency of 100% at 40 keV. It is important to note for the scope of this chapter that for all high-Z semiconductors, the Compton cross section becomes comparable with the photoelectric one over 200 keV.

Semiconductor detectors for x- and γ-rays spectroscopy behave as solid-state ionization chambers operated in pulse mode. The simplest configuration is a planar detector, i.e., a slab of a semiconductor material with metal electrodes on the opposite faces of the semiconductor (Figure 10.2a). Photon interactions produce electron–hole pairs in the semiconductor volume through the above-discussed interactions. The interaction is a two-step process where the electrons created in the photoelectric or Compton process lose their energy through electron–hole pair production. The number of electron–hole pairs is proportional to the released photon energy. If E is the released photon energy, the number of electron–hole pairs N is equal to E/w, where w is the average energy for pair creation. The generated charge cloud is $Q_0 = eE/w$. The electrons and holes move toward the opposite electrodes, anode, and cathode for electrons and holes, respectively (Figure 10.2a).

The movement of the electrons and holes causes variation ΔQ of induced charge on the electrodes. It is possible to calculate the induced charge ΔQ by the Shockley–Ramo theorem (Cavalleri et al. 1997; He 2001), which makes use of the concept of a weighting potential defined as the potential that would exist in the detector with the collecting electrode held at unit potential, while holding all other electrodes at zero potential. According to the

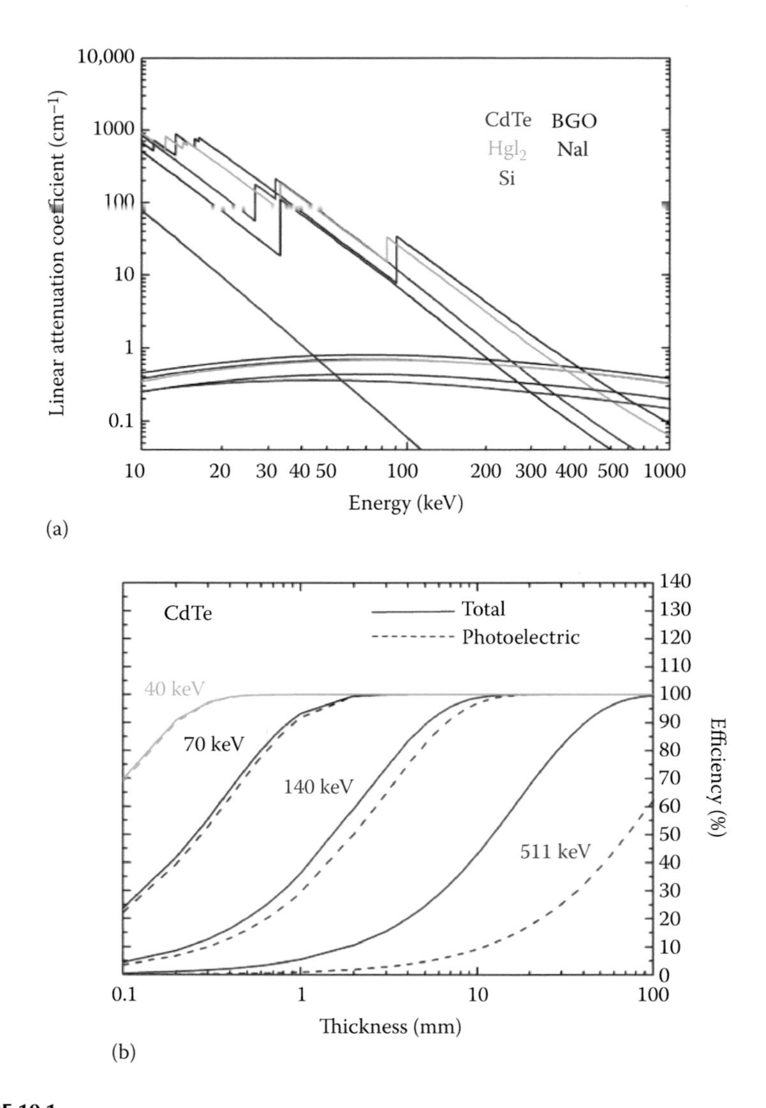

FIGURE 10.1
(a) Linear attenuation coefficients for photoelectric absorption and Compton scattering of CdTe, Si, HgI_2, NaI, and BGO. (b) Intrinsic efficiency of CdTe detectors as function of detector thickness at various photon energies.

Shockley–Ramo theorem, the induced charge by a carrier q, moving from x_i to x_f, is

$$\Delta Q = -q[\varphi(x_f) - \varphi(x_i)] \tag{10.1}$$

where $\varphi(x)$ is the weighting potential at position x. The analytical solution of the Laplace equation inside the detector enables calculating the weighting

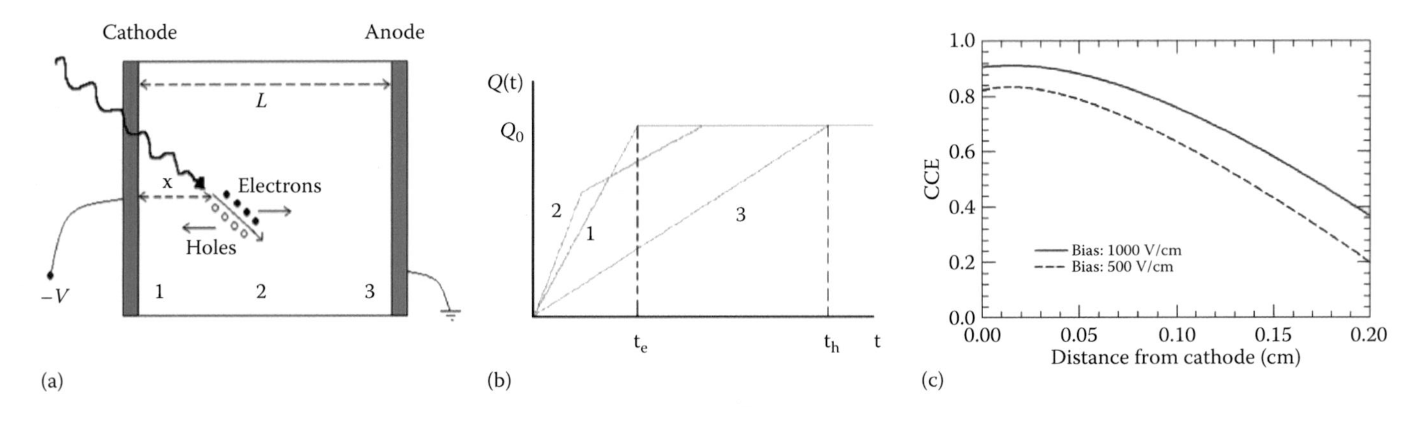

FIGURE 10.2
Planar configuration of a semiconductor detector. (a) Electron–hole pairs, generated by radiation, are swept toward the appropriate electrode by the electric field. (b) Time dependence of the induced charge for three different interaction sites in the detector (positions 1, 2, and 3). The fast rising part is due to the electron component, while the slower component is due to the holes. (c) CCE of a CZT planar detector (2 mm thick) evaluated by the Hecht equation for two different bias (i.e., electric field values) and for $\mu_e\tau_e$ and $\mu_h\tau_h$ equal to 1×10^{-3} cm^2/V and 8×10^{-5} cm^2/V, respectively.

potential (Eskin et al. 1999). In a semiconductor, the total induced charge is given by the sum of the induced charges due both to the electrons and holes.

Charge trapping and recombination are typical effects in compound semiconductors and may prevent full charge collection. For a planar detector, having a uniform electric field, neglecting charge detrapping, the charge collection efficiency (CCE), i.e., the induced charge normalized to the generated charge (Figure 10.2c), can be evaluated by the Hecht equation (Hecht 1932) and derived models (Zanichelli et al. 2013) and is strongly dependent on the photon interaction position. This dependence coupled with the random distribution of photon interaction points inside the sensitive volume increases the fluctuations on the induced charge and produces peak broadening in the energy spectrum as well as the characteristic low tail asymmetry in the full energy peak shape observed in planar CdTe/CZT sensors.

The charge transport properties of a semiconductor, expressed by mobility-lifetime products for holes and electrons ($\mu_h\tau_h$ and $\mu_e\tau_e$), are key parameters in the development of radiation detectors. Poor mobility-lifetime products result in short mean drift length λ, and therefore small λ/L ratios, which limit the maximum thickness and energy range of the detectors. Compound semiconductors, generally, are characterized by poor charge transport properties due to charge trapping. Trapping centers are mainly caused by structural defects (e.g., vacancies), impurities, and irregularities (e.g., dislocations, inclusions). In compound (CdTe and CZT) semiconductors, $\mu_e\tau_e$ is typically of the order of 10^{-5}–10^{-3} cm²/V, while $\mu_h\tau_h$ is usually much worse with values around 10^{-6}–10^{-4} cm²/V. Therefore, the corresponding mean drift lengths of electrons and holes are 0.2–20 mm and 0.02–2 mm, respectively, for typical applied electric fields of 2000 V/cm (Sato et al. 2002).

The charge collection efficiency is a crucial property of a radiation detector and affects the spectroscopic performances and in particular the energy resolution. High charge collection efficiency ensures good energy resolution, which also depends on the statistics of the charge generation and on the noise of the readout electronics. Three contributions mainly affect the energy resolution (FWHM) of a radiation detector:

$$\Delta E = \sqrt{(2.355)^2(F \cdot E \cdot w) + \Delta E_{el}^2 + \Delta E_{coll}^2} \tag{10.2}$$

The first contribution is the noise due to the statistics of the charge carrier generation, where F represents the Fano factor. In semiconductors, F is much smaller than unity (0.06–0.14) (Devanathan et al. 2006). The second contribution is the electronic noise, which is generally measured directly using a precision pulser, while the third term takes into account the contribution of the charge collection process. Different semi-empirical relations have been proposed for the charge collection contribution evaluation of different detectors (Kozorezov et al. 2005).

Figure 10.3 shows the typical spectroscopic system based on a semiconductor detector. The detector signals are read out by a charge sensitive preamplifier (CSP) and then shaped by a linear amplifier. A multichannel analyzer (MCA), which samples and records the shaped signals, finally acquires and records the deposited energy spectrum.

As will be pointed out later, poor holes transport properties of CdTe and CdZnTe materials are a critical issue in the development of x- and γ-rays detectors. Hole trapping reduces the charge collection efficiency of the detectors and produces asymmetry and a long tail in the photopeaks in the measured spectra (holes tailing). In order to minimize this effect, several methods have been used. Some techniques concern the particular irradiation configuration of the detectors (Figure 10.4a). *Planar parallel field* (PPF) is the classical configuration used in overall planar device. In this

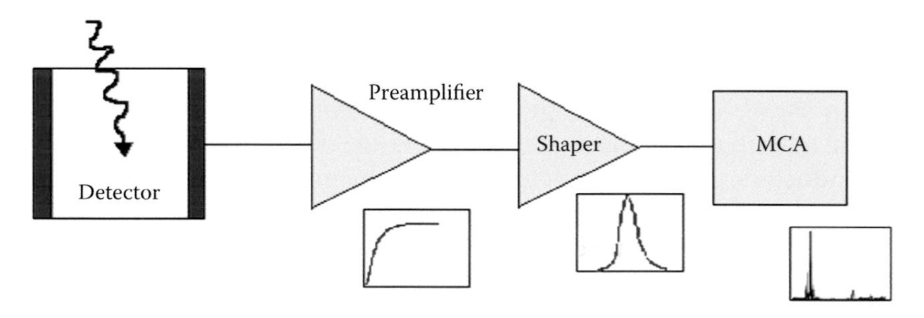

FIGURE 10.3
Block diagram of a standard spectroscopic detection system for x- and γ-rays.

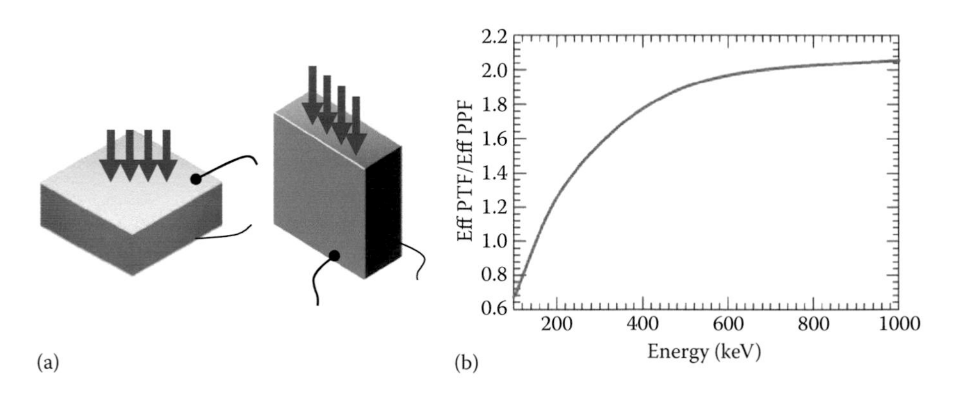

FIGURE 10.4
(a) Usual irradiation configuration in which photons impinge (arrows) the detector through the cathode surface (PPF: planar parallel field) and the PTF (planar transverse field), one in which the photons impinge on the sensor orthogonally with respect to the charge collecting field. (b) Ratio between PTF and PPF efficiency for impinging photon energies from 50 to 1000 keV assuming the PTF thickness equal to 10 mm and the distance between electrodes (i.e., the PPF absorption thickness) 2.5 mm.

configuration, the irradiation of the detector is through the cathode electrode, thus minimizing the hole trapping probability. In an alternative configuration, denoted as *planar transverse field* (PTF) (Casali et al. 1992), the irradiation direction is orthogonal (transverse) to the electric field. In this irradiation condition, different detector thicknesses can be chosen to fit the detection efficiency required, without modifying the interelectrodes distance and then the charge collection properties of the detectors. This technique is particularly useful in developing detectors with high detection efficiency in the γ-ray energy range. In Figure 10.4b, the ratio is plotted between the efficiency achievable by a CdTe spectrometers with lateral sides of 10 mm and a distance between electrodes of 2.5 mm (Caroli et al. 2008). This plot shows that the PTF irradiation configuration starts to be convenient in terms of detection efficiency above 200 keV.

10.2.2 Spectroscopic Performance Improvement Techniques

To compensate for the trapping effects in CdTe/CZT semiconductor detectors, and therefore to improve their spectroscopic performance and increase their full energy efficiency, different methods have been proposed. The most used methods rely on the possibility of avoiding the contribution of holes on the formation of the charge signal and therefore using the CZT/CdTe detector as single charge sensing devices. In this configuration, only electrons are collected and, because their mobility-lifetime product, the effect of trapping is limited and can be even more efficiently compensated by using simple signal manipulation. There are several techniques to realize single charge carrier (namely electrons) sensing CdTe and CdZnTe detectors (unipolar detectors). Some of these techniques are based on electronic methods (e.g., pulse rise time discrimination; Jordanov et al. 1996) and bi-parametric readout (Richter and Siffert 1992; Auricchio et al. 2005). While others rely on particular electrode design (e.g., Frisch-grid—McGregor et al. 1998; Bolotnikov et al. 2006; pixels—Barrett et al. 1995; Kuvvetli and Budtz-Jørgensens 2005; coplanar grids—Luke 1995; strips—Shor et al. 1999; Perillo et al. 2004; multiple electrodes—Lingren et al. 1998; Kim et al. 2004; Abbene et al. 2007). Figure 10.5 shows some electrode designs used for CdTe and CdZnTe unipolar detectors. Within the proposed unipolar electrode configuration, particularly interesting for their intrinsic imaging properties, pixels and microstrips sensors (Figure 10.5b, c) are also characterized by unipolar properties, when the ratio between charge collection distance and the pixel/strip pitch is large ($\gg$1), i.e., the so-called small pixel effect.

In general, the almost unipolar characteristics of these detector configurations are due to the particular shape of the weighting potential: it is low near the cathode and rises rapidly close to the anode. According to this characteristic, the charge induced on the collecting electrode is proportional to the weighting potential, as stated by the Shockley–Ramo theorem, and its major contribution comes from the drift of charge carriers close to the anode,

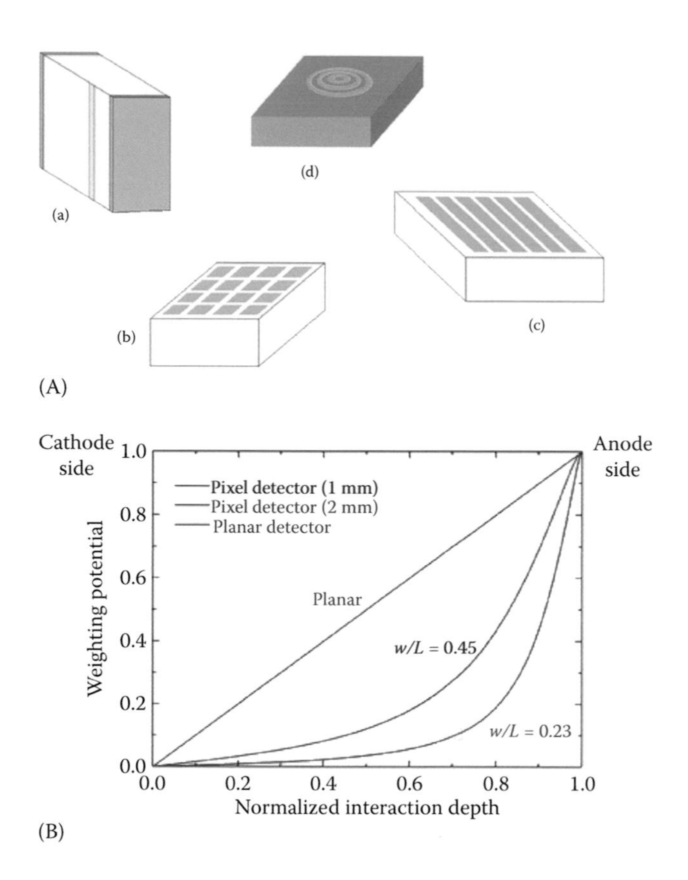

FIGURE 10.5
(A) Single charge sensing electrode configurations widely used in CdTe and CdZnTe detectors: (a) parallel strip Frisch grid, (b) pixel, (c) strip, and (d) multiple electrodes. (B) Weighting potential of a pixel detector, compared to a planar detector. It is possible to improve the unipolar properties of pixel detectors by reducing the w/L ratio (i.e., pixel size to detector thickness), according to the theory of small pixel effect.

i.e., the electrons. On the contrary, the linear shape of the weighting potential of a planar detector makes the induced charge sensitive to both electrons and holes, as discussed above.

In particular, the introduction of coplanar-grid noncollecting electrodes in the anode side design of sensors provides an important additional feature that is fundamental to realize 3D sensing spectrometers, that is, the position information of the radiation interactions point inside the sensitive volume (Luke 1995). In fact, for these electrode configurations, the induced charge on the planar cathode Q_p increases roughly as a linear function of the distance D of γ-ray interaction location from the coplanar anodes ($Q_p \propto D \cdot E$) because it is proportional to the drift time of electrons. On the other hand, the coplanar anode signal Q_s is only approximately proportional to the γ-ray deposited energy ($Q_s \propto E$). Therefore, the interaction depth can be estimated by reading both Q_p and Q_s signal amplitude for each interaction through their ratio (also called depth parameter): $d = Q_p/Q_s, \propto D$ (He et al. 1997).

10.3 CZT/CdTe Spectrometers with 3D Spatial Resolution

In this section, we focus on a particular type of detector based on sensitive elements of CZT/CdTe, namely spectrometers with spatial resolution in three dimensions. These devices represent the new frontiers for applications in different fields that require increasing performance such as high-energy astrophysics, environmental radiation monitoring, medical diagnostics with PET, and inspections for homeland security (Vetter et al. 2007; Gu et al. 2011; Whal et al. 2015).

A 3D spectrometer is a detector divided into volume elements (voxels), each operating as an independent spectroscopic detector. The charge produced in each voxel by the interaction of an incoming x/γ photon is converted into a voltage signal proportional to the released energy. If the readout electronics of the detection system implements a coincidence logic, it will be possible to determine to some extent (depending on the voxel dimensions and the time coincidence window) the history of the incident photon inside the sensitive volume by associating the energy deposits in more voxels to the same incident photon. These capabilities are of fundamental importance for applications requiring high-detection efficiency even at high energies (>200 keV), i.e., in the Compton scattering regime, as well as a wide-field localization of the direction of incidence and a uniform spectroscopic response throughout the sensitive volume. In fact, the possibility to reconstruct the photon interaction position in 3D will allow correcting from signal variations due to charge trapping and material non-uniformity and therefore will increase the sensitive volume of each detector unit without degrading the spectroscopic performance. A straightforward application of 3D spectro-imagers in hard x- and γ-rays is the realization of advanced Compton detectors that use the interaction position reconstruction with energy determination of each hit to evaluate the incoming photon direction through the Compton kinematics (Du et al. 2001; Mihailescu et al. 2007).

In the field of hard x-ray and soft γ-ray astrophysics (10–1000 keV), there are promising developments of new focusing optics operating for up to several hundreds of kiloelectron volts, through the use of broadband Laue lenses (Frontera et al. 2013; Virgilli et al. 2015) and new generation of multilayer mirrors (Della Monica Ferreira et al. 2013; Blozer et al. 2016). These systems make it possible to drastically improve the sensitivity of a new generation of high-energy space telescopes at levels far higher (i.e., 100 times) than current instrumentation. To obtain the maximum return from this type of optics up to megaelectron volts, focal plane detectors with high performance are required. These detectors should guarantee at the same time high efficiency (>80%, at least) even at higher energies, fine spectroscopic resolution (1% at 500 keV), and also accurate localization (0.1–1 mm) of the interaction point of the photons used for the correct attribution of their direction of origin in the sky.

In fact, we should point out that the 3D spectro-imager represents a promising way to realize highly efficient scattering polarimeters. This capability can finally open the hard x- and γ-rays polarimetry windows to space astronomy, making the measurement of the polarization of cosmic sources a standard observational mode, as it is now for imaging, spectroscopy, and timing, in the next generation of high-energy space telescopes.

The realization of 3D spectrometers by a mosaic of single CdTe/CZT crystals is not as easy as the case of bi-dimensional (2D) imagers. These difficulties are mainly due both to the small dimension of each sensitive unit necessary to guarantee the required spatial resolution and to the packaging of such 3D sensor units, requiring an independent spectroscopic readout electronics chain. A solution is the realization of a stack of 2D spectroscopic imagers (Watanabe et al. 2002; Judson et al. 2011). This configuration, while very appealing for large area detectors, has several drawbacks for applications requiring fine (<0.5 mm) spatial resolution in 3D and compactness (as focal plane detectors). Indeed, the distance between each 2D layer of the stack limits the accuracy and the sampling of the third spatial coordinate. Furthermore, passive materials normally required for mechanical support of each detection layer could introduce large amount of unwanted scatter.

To solve this kind of problem, in the last 10–15 years, several groups have focused their activity on the development of sensor units based on high-volume (1–10 cm^3) single crystals of CZT/CdTe capable of intrinsically operating as 3D spectrometers. The main target of these developments is to fulfill the requirements for a given application with only one high-performance sensor, and/or to make more efficient and easy the realization of 3D detectors based on matrices of these basic units. The main benefits of such approaches run from limitation of readout channel numbers to achieve the required spatial resolution to packing optimization with reduction of passive material between sensitive volumes. The adopted electrode configurations play a key role in these developments. As already seen in the previous section, various electrode configurations have been proposed and studied to improve both the spectroscopic performance and the uniformity of response of CZT/CdTe detectors. In fact, these electrode configurations, with the implementation of appropriate logical reading of the signals, make the sensors intrinsically able to determine the position of interaction of the photon in the direction of the collected charge (depth sensing) and therefore are particularly suited to the realization of 3D monolithic spectrometers without requiring a drastic increase in the electronics readout chains.

In the following sections, we describe, as examples, a couple of configurations currently proposed and under development for the realization of 3D spectrometers based on single large volume crystals of CdTe/CZT. Within other undergoing developments (Cui et al. 2008; Bale and Szeles 2006; Owens et al. 2006; Dish et al. 2010; Macri et al. 2002, 2003; Luke 2000; Matteson et al. 2003), we report only on these two configurations which are intrinsically capable to fulfill requirements for fine spatial resolution in all three dimensions coupled with high and uniform spectroscopic response.

10.3.1 Pixel Spectrometers with Coplanar Guard Grid

By combining a pixelated anode array, already providing good energy resolution because of the small pixel effect introduced in Section 10.2.2, and an interaction depth sensing technique for electron trapping corrections, it is possible to build CdZnTe γ-ray spectrometers with intrinsic 3D position sensing capability over a quite large volume (1–3 cm^3) of bulk crystals (He et al. 1999).

The first prototype was based on a 10 × 10 × 10 cm^3 CZT crystal with an 11 × 11 pixel anode array and a single cathode electrode on the opposite surface (Stahle et al. 1997). The 2D sensing of γ-ray interactions is provided by the pixel (x, y) anode where electrons are collected. Instead of using an array of simple square pixel anodes, each collecting anode is surrounded by a common noncollecting grid (Figure 10.6a, b). The pixel pitch had a dimension of 0.7 × 0.7 mm^2, with a collecting anode of 0.2 × 0.2 mm^2 at the center surrounded by a common noncollecting grid with a width of 0.1 mm. Since the noncollecting grid is biased at lower potential relative to that of the collecting anodes, electrons are forced toward the collecting pixel anodes. Even more important, the dimension of the pixel anode is small with respect to the anode–cathode distance and smaller than the geometrical pixel dimension enhancing the small pixel effect and minimizing any induced signal from the holes movement. To guarantee a good electron collection, the bias voltage between anodes and the planar cathode is in the 1.5–2 kV range, while the voltage difference between anodes and the noncollecting common grid is typically of few tens of volt (30–50 V).

The ratio between the cathode and the anode signals allow determining the γ-ray interaction depth between the two electrodes planes. With a simple coincidence logic between cathode and anode signals, this technique can provide the depth (z) of the photon interaction for single-site events, and only the centroid depth for multiple-site interactions (e.g., Compton scattered events). The identification of individual hit depths for multiple-site events requires the readout, through a charge sensitive preamplifier, of the signals from the noncollecting grid. When electrons generated by an energy deposit are detected toward the collecting pixel anode near the anode surface, a negative pulse is induced on the noncollecting grid as shown in Figure 10.6c. This signal is differentiated, generating positive pulses corresponding to the slope inversion points of the noncollecting grid signal. Finally, a threshold circuit uses the differential output to provide a logic pulse when it is

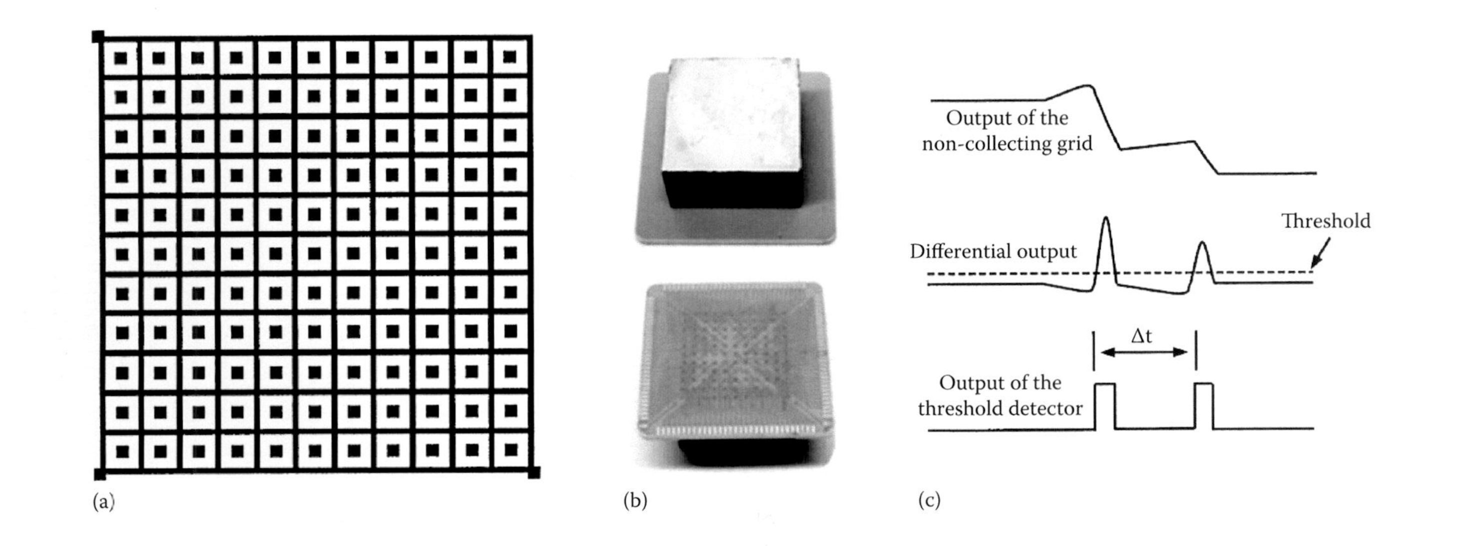

FIGURE 10.6
(a) Scheme of the anode side of the 10 × 10 × 10 mm^3 CZT prototype. (Reprinted from *Nuclear Instruments and Methods in Physics Research Section A: Accelerators, Spectrometers, Detectors and Associated Equipment*, 422, 1–3, Z. He, W. Li, G.F. Knoll, D.K. Wehe, J. Berry, C.M. Stahle, 3-D position sensitive CdZnTe gamma-ray spectrometers, 173–178, Copyright 1999, with permission from Elsevier.) (b) Photos of the detector (15 × 15 × 10 mm^3) with the ceramic substrate facing up (top) and with the cathode facing down (bottom), where, through the thin ceramic substrate, the anode bonding pads array is visible. (F. Zhang, Improved resolution for 3-D position sensitive CdZnTe spectrometers, *IEEE Transactions on Nuclear Science*, 51(5), © 2004 IEEE.) (c) Scheme of the depth sensing logic used for multiple-site events handling. (Reprinted from *Nuclear Instruments and Methods in Physics Research Section A: Accelerators, Spectrometers, Detectors and Associated Equipment*, 422, 1–3, Z. He, W. Li, G.F. Knoll, D.K. Wehe, J. Berry, C.M. Stahle, 3-D position sensitive CdZnTe gamma-ray spectrometers, 173–178, Copyright 1999, with permission from Elsevier.)

above a defined threshold (Li et al. 1999). These logic pulses provide start and stop signals to a time-to-amplitude converter (TAC) that measure the electrons drifting time intervals.

By combining the centroid depth, pulse amplitudes from each pixel anode, and the depth interval between energy deposits derived from the measure of electrons drifting time, it is possible to obtain the depth of each hit (Figure 10.7a). Although the differential circuit could identify multiple hits of the same incoming photon, the TAC limits the number of interactions to two. Therefore, the original system was able to provide interaction depths for only single- and two-site (double) events. Events having more than two interactions can be identified using the number of triggered anode pixels, but only the centroid interaction depth can be obtained. While the single event low-energy threshold was small (~10 keV), the threshold for double events results is relatively high, because their detection depends on the non-collecting grid signal threshold being in the first measurements ~100 keV. The reconstructed interaction depth accuracy using this technique becomes worse with decreasing energy (Li et al. 2000) and is ~0.25 mm for single events and ~0.4 mm for double ones at 662 keV.

Since the first realization, the same groups have made several improvements on both the CZT sensor configuration and the readout and processing electronics allowing to increase, in particular, the sensitive volume of each CZT device up to 6 cm^3 (i.e., $2 \times 2 \times 1.5$ cm^3) (Zhang et al. 2004, 2012). This sensor can achieve very impressive spectroscopic performance (Figure 10.7b) for all the event types for energy up to several megaelectron volts. One of the main problems operating in this energy regime (>500 keV) is represented by the dimension of the electron cloud, generated at each photon interaction point, that becomes larger than the pixel lateral size (>1 mm) as the energy deposit increases (Figure 10.7c). This effect tends to degrade the spatial resolution because transient signals are collected by several anode pixels around the central one (charge sharing) and, in the direction of charge collection, increase the depth reconstruction accuracy. The geometrical spatial resolution in the anode plane of the 6 cm^3 sensor was only 1.8 mm. However, with a custom-designed digital readout scheme, handling the charge shared signals out from the eight neighboring pixels of the triggered one, it has been demonstrated that a Δx of 0.23–0.33 (FWHM) mm can be achieved for 662 keV single interaction (Zhu et al. 2011).

10.3.2 PTF Microstrip with Drift Configuration

Another way to build 3D spectroscopic sensors relies on the use of CZT crystals in the PTF configuration. The drawback of the PTF irradiating geometry is that all the positions between the collecting electrodes are uniformly hit by impinging photons leading to a stronger effect of the difference in charge collection efficiency and then in the spectroscopic performance with respect to the standard irradiation configuration through the

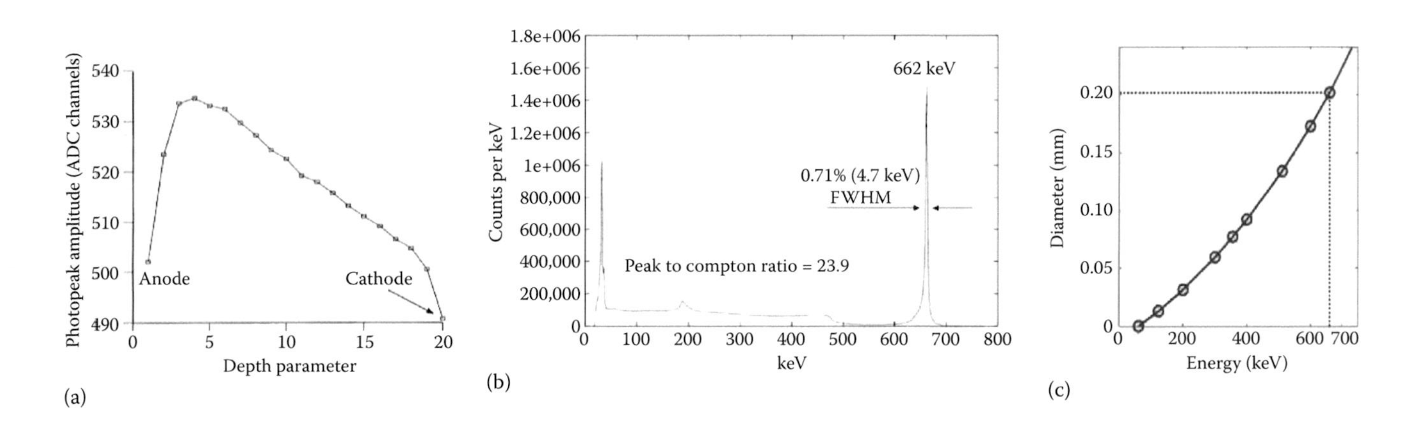

FIGURE 10.7
(a) Typical dependence of the centroids of ^{137}Cs photopeak from the interaction depth parameter (1 depth step = 0.5 mm) for one pixel. (Reprinted from *Nuclear Instruments and Methods in Physics Research Section A: Accelerators, Spectrometers, Detectors and Associated Equipment*, 422, 1–3, Z. He, W. Li, G.F. Knoll, D.K. Wehe, J. Berry, C.M. Stahle, 3-D position sensitive CdZnTe gamma-ray spectrometers, 173–178, Copyright 1999, with permission from Elsevier.) (b) ^{137}Cs Spectrum measured by one pixel after compensation for interaction depth for all event multiplicity. The measured resolution is 0.71% (FWHM). (F. Zhang, Characterization of the H3D ASIC readout system and 6.0 cm^3 3-D Postion Sensitive CdZnTe detectors, *IEEE Transactions on Nuclear Science*, 59(1) © 2012 IEEE.) (c) Diameter of the electron cloud generated by the photon interaction vs energy. (Reprinted from Nuclear Instruments and Methods in Physics Research Section A: Accelerators, Spectrometers, Detectors and Associated Equipment, 654, 1, Kim, J.C. ,Anderson, S.E., Kaye, W., Zhang, F., Zhu, Y., Kaye, S.J. He, Z., Charge sharing in common-grid pixelated CdZnTe detectors, 233–243, Copyright 2011, with permission from Elsevier.)

cathode (PPF). Therefore, worst spectroscopic performances are expected in PTF with respect to standard PPF irradiation configuration (Auricchio et al. 1999). To recover this spectroscopic degradation and to improve the CZT sensitive unit performance, an array of microstrips in a drift configuration can be used instead of a simple planar anode (Figure 10.8): the anode surface is made of a thin collecting anode strip surrounded by guard strips that are biased at decreasing voltages. This anode configuration (Gostilo et al. 2002) allows the detector to become almost a single charge carrier device. This avoids the degradation of the spectroscopic response by the charge loss due to the holes trapping and provides a more uniform spectroscopic response (i.e., independent from the distance of the interaction from the collecting electrodes; Caroli et al. 2010). The spectroscopic resolution of this type of sensor ranges from 6% at 60 keV down to 1.2% at 662 keV, without any correction for the interaction depth. In fact, similarly to the previous configuration presented above, it will be possible to perform a compensation of the collected charge signals using the photon interaction position in between the metalized surfaces that can be inferred by the ratio between the cathode and the anode strip signals (Kuvvetli et al. 2010a).

The achievable spatial resolution in this direction is a function of energy, depending on the dimension of the charge-generated cloud. The measurements have given (Kuvvetli et al. 2010b) a value around 0.2 mm (FWHM) up to 500 keV. Further segmentation of a cathode into an array of metallic strips, in the direction orthogonal to the anode ones, can provide the third hits coordinate, i.e., the 3D sensitivity for the photon interaction position (Figure 10.8c). Of course, with the described configuration, the spatial resolution along the anode surface is defined geometrically by the collecting anode and cathode strip pitch.

Both anode and cathode strips are read out by standard spectroscopic electronics chains, and therefore, the segmentation of cathode and anode surfaces will set the number of readout channels. In fact, ongoing developments on this sensor configuration are demonstrating that with a readout logic able to weight the signal between strips, the achievable spatial resolution along the anode and the cathode strip sets can result finer than the geometrical one. For a CZT sensor, in which the cathode is segmented in 2 mm pitch strips, the final spatial resolution can be as low as 0.6 mm (FWHM, up to 600 keV) weighing the cathode strips signals. In fact, along the anodic strips set, the effective resolution can be further improved to a small fraction (1/5–1/10) of the geometrical pitch between collecting strips by implementing an appropriate readout of the drift strips signal similar to the one suggested by Luke et al. (2000) for 3D coplanar grid detectors (Kuvvetly et al. 2014). This expectation has been confirmed by recent tests on a sensor implementing the PTF drift strip configuration on a 20 × 20 × 5 mm^3 single CZT spectroscopic crystal made at the ESRF (Grenoble) with a fine (50 μm) high-flux collimated monochromatic beam (Figure 10.9). The CZT sensor is characterized by an anode and a cathode pitch of 1.6 and 2 mm, respectively.

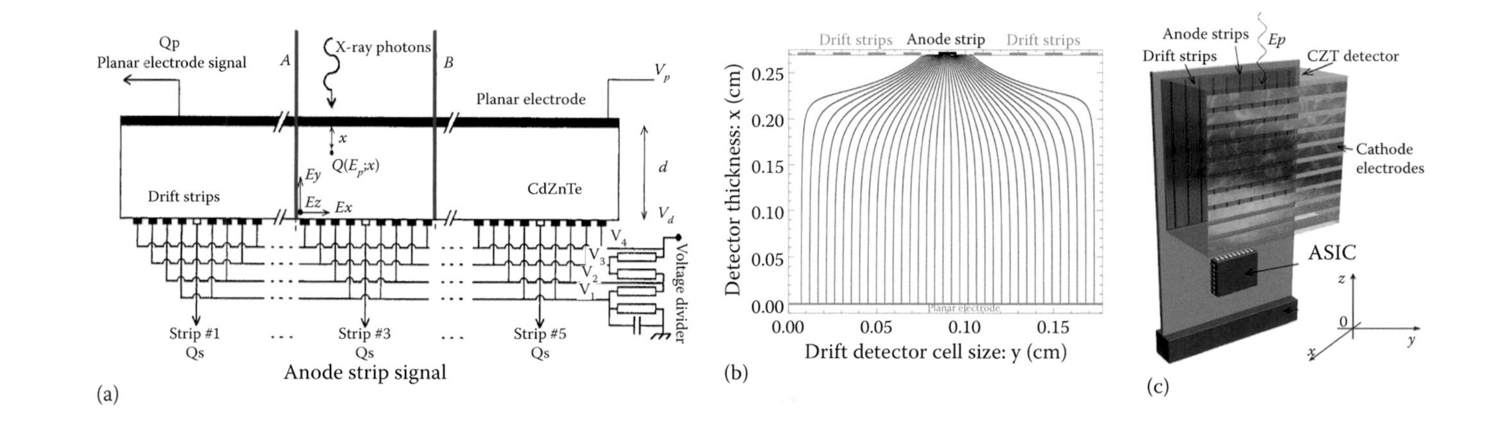

FIGURE 10.8
(a) Schematic configuration of a CZT drift strip sensor. The two thick vertical lines A and B define the volume of the drift strip cell. The drift strips and the planar electrodes are biased in such a way that the electrons move to the anode collecting strips (central white strips). (b) Shape of the charge collecting electric field calculated for a drift strip cell on a CZT sensor: the anode–cathode bias is set at 150 V, and the difference between each adjacent strip pairs is $\Delta V = -25$ V. (c) Schematic view of a 3D PTF drift strip CZT sensor with segmented cathode.

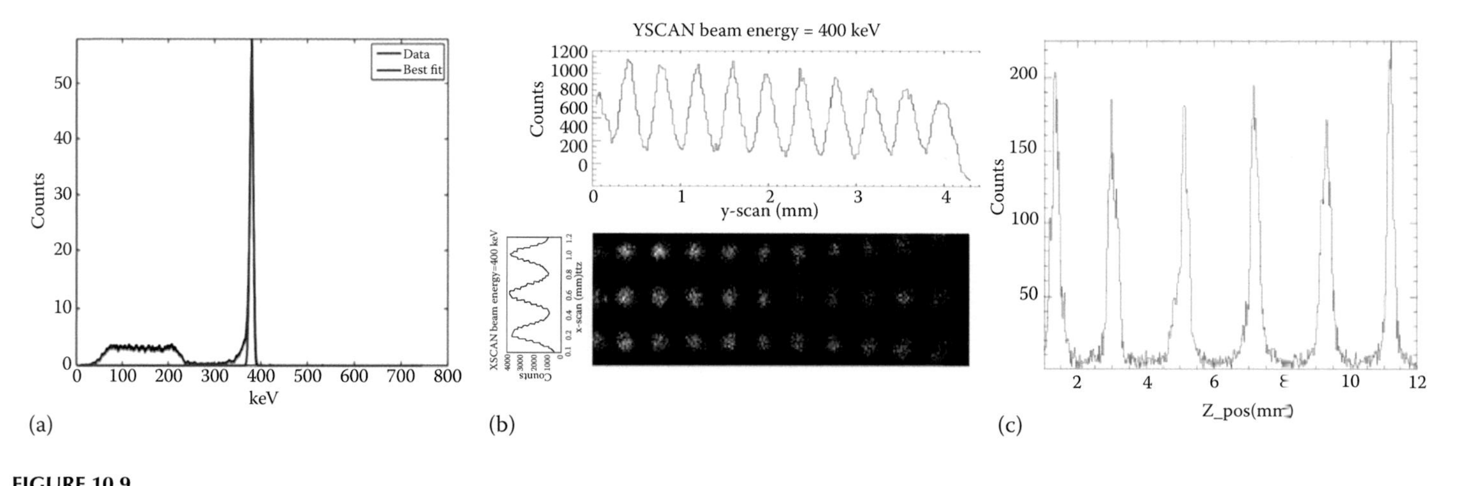

FIGURE 10.9
Performance of a 3D PTF drift strip CZT sensor of 20 × 20 × 5 mm^3 using a fine collimated (0.05 × 0.05 mm^2 spot) x-ray monochromatic beam (ID15A) at ESRF (Grenoble, France). (a) Single events spectrum for 400 keV beam exhibiting an energy resolution of 1.4% (FWHM). (b) Reconstruction of photon (400 keV) interaction position across the impinging surface. The vertical axis is across the anode strips for a total of 1 drift cell (i.e., 1.[illegible] mm), while the horizontal one is parallel to the interelectrode distance (5 mm length). (c) Reconstruction of 400 keV photon interaction position in z direction (across the cathode strips), showing a resolution (FWHM) of 0.65 mm.

Using the mentioned technique of drift strips readout and signal weighing, the beam tests have demonstrated very good performance, both in spectroscopy (e.g., 1.4% FWHM @ 400 keV) and in 3D position reconstruction, achieving in all the three directions spatial resolution (FWHM) at a submillimeter level ($\Delta x = 0.15$ mm, $\Delta y = 0.26$ mm, $\Delta z = 0.65$ mm).

Because of the use of the PTF configuration, the dimensions of the 3D sensor unit can reach up to 20–30 mm in the lateral sizes and up to 5 mm as charge collecting distance, allowing one to limit the high-bias voltage required to have a high charge collection efficiency to values below 500 V.

While the electrical field intensity between the cathode and the anode is typically 100 V/mm, the drift strips, to be effective in shaping the charge collection electric field and to minimize dead volume, is biased at decreasing relative voltage with respect to the cathode strips of $\Delta V = 20–30$ V. These values depend, in particular, on the thickness (distance between cathode and anode surfaces) of the sensor tile and the best bias voltage scheme needs to be optimized. Using such PTF CZT drift strip sensor units (Auricchio et al. 2012), large-volume 3D spectrometers can be built by packaging several units as shown in Figure 10.10, in which CZT 3D sensors are bonded on thin high resistivity support layers (e.g., Al_2O_3) forming linear modules that provide the electrical interface both for readout electronics and bias circuits.

A PTF drift strip sensor unit, like the one discussed above, has the great advantage, with respect to pixel spectrometers with coplanar guard grid implementation, represented by the few readout channels (~30) required to obtain a sensor segmentation equivalent to ~8×10^4 "virtual" voxels in a sensitive volume of 2 cm^3. This characteristic is quite important, in particular, for applications with limited power resources, like space astronomy, and opens the possibility to implement efficiently new readout systems based on the use of fast digitizers to record the original charge sensitive preamplifier (Abbene et al. 2015).

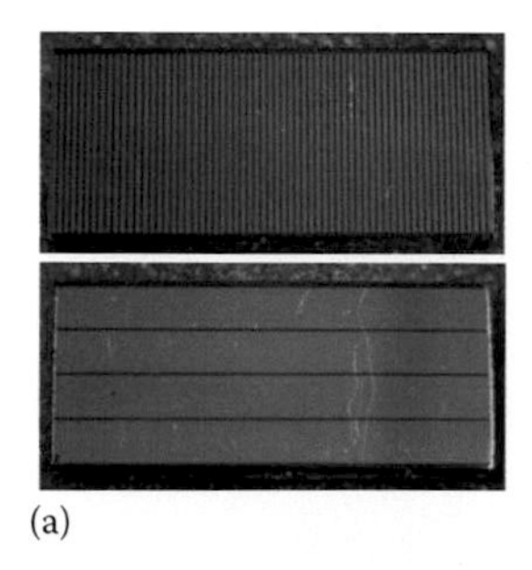
(a)

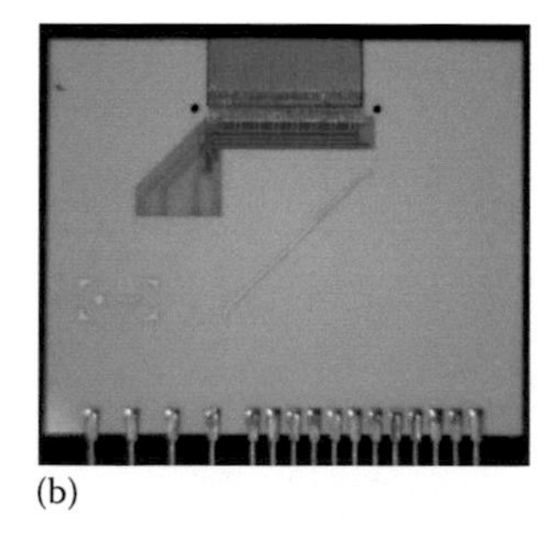
(b)

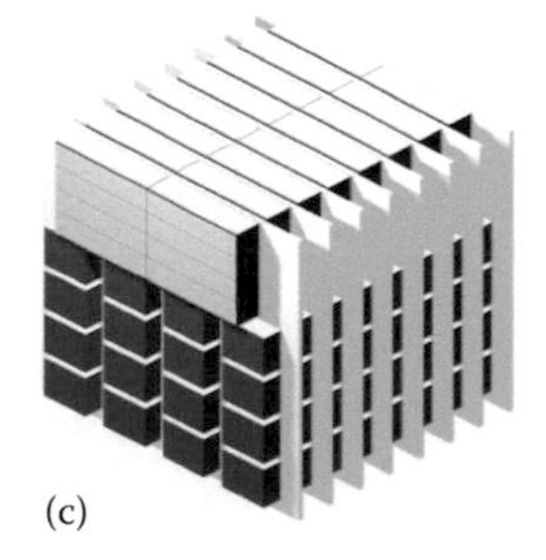
(c)

FIGURE 10.10
(a) Drift strip CZT sensor ($18 \times 8 \times 2.5$ mm^3): (top) anode side with 64 (0.15 mm wide) strips set; (bottom) cathode side with 4 (2 mm wide) strips set. (b) Linear module prototype seen from anode side: this constitutes the basic element for building a large-volume 3D sensor. (c) Suitable packaging scheme of eight linear modules, each supporting two CZT drift 3D sensors of $20 \times 20 \times 5$ mm^3 to obtain a spectroscopic imager of 32 cm^3 sensitive volume.

10.4 CZT/CdTe Spectro-Imagers for Compton Polarimetry in Astrophysics

High-energy polarized emissions are expected in a wide variety of gamma-ray sources such as pulsars, solar flares, active galactic nuclei, galactic black holes, and gamma-ray bursts (Lei et al. 1997; Bellazzini et al. 2010; McConnell et al. 2009), but polarimetry in this energy regime is still a completely unexplored field mainly due to two facts. In the first place, the expected polarized hard x/γ-rays flux from cosmic sources is, in general, only a small percentage of the already low incoming flux (a few to 10–20%), and only in a few cases can represent a large fraction of it (>40%), requiring very high sensitivity instruments to be detected. Second, x/γ-ray polarimetric measurements require the implementation of the complex of detection, electronic, and signal processing systems, onboard to high-altitude balloon or satellite missions in space. Therefore, until a few years ago, no dedicated hard x/γ-ray polarimetric missions have been launched into space, and x- and γ-ray source emissions have been studied almost exclusively through spectral and timing analysis of the measured fluxes and by using imaging techniques. On the other hand, polarization measurements will increase the number of observational parameters of a γ-ray source by two: the polarization angle and the level of linear polarization. These additional parameters should allow a better discrimination between different emission models characterizing the same object. Polarimetric observations can provide important information about the geometry, the magnetic field, the composition, and the emission mechanisms. In the soft γ-ray domain (0.1–1 MeV), only a few polarimetric measurements were performed by the SPI and IBIS instruments onboard the INTEGRAL (INTErnational Gamma-Ray Astrophysics Laboratory) mission (Winkler et al. 2003; Ubertini et al. 2003), on the Crab Pulsar, on the galactic black-hole Cygnus X-1, and on some high flux gamma-ray bursts (Dean et al. 2008; Forot et al. 2008; Laurent et al. 2011; Götz et al. 2009).

Today, the importance of high-energy polarimetry is largely recognized, and several research groups are involved in the development of dedicated instruments (Kole et al. 2016; Chauvin et al. 2016; Kislat et al. 2017). In any case, the next generation of space telescopes should certainly provide polarimetric observations, contemporaneously with spectroscopy, timing, and imaging. These multipurpose instrument types were proposed in recent high-energy (100 keV–1 GeV) space mission concepts submitted to ESA Cosmic Vision calls where our groups were proposal partners, such as the Gamma-Ray Imager (GRI), DUAL, and e-ASTROGAM (Knödlseder et al. 2007; von Ballmoos et al. 2010; Tatischeff et al. 2016). In the framework of these space mission proposals, different configuration detection planes suitable to high-energy polarimetry are under study and development.

A pixel/voxel detector inherently offers the possibility to operate as a scattering (Compton) polarimeter if equipped with a readout logic that allows to manage events with two (double events) or more (multiple events) interactions in coincidence (Figure 10.11a). Furthermore, a polarimeter based on a pixel/voxel detector permits an optimal use of the entire sensitive volume, since each element operates in the same time as a scatterer and as an absorber one. Another important advantage for the use of segmented detectors, such as 2D/3D spectro-imager as a Compton polarimeter, is to allow the use of the same detector to make contemporary spectroscopy, timing, and imaging measurements. This capability allows overcoming problems linked to the inherent time variability of both cosmic sources flux and instrumental background, making it possible to directly correlate the various types of measurement for the same observation.

The choice of CZT/CdTe spectroscopic imager as a scattering polarimeter, mainly to optimize the detection efficiency, for the high Z of the material, and simultaneously ensure good spectroscopic performance and high spatial resolution (2D or 3D), obviously implies a limitation on the low-energy threshold useful for polarimetric measurement. As in these materials, the Compton cross section becomes significant only above 100 keV; by equating the photoelectric one approximately at 200 keV, CZT/CdTe spectro-imagers can work efficiently as scattering polarimeters above 100 keV and depending on the thickness up to energies of a few megaelectron volts.

10.4.1 Compton (or Scattering) Polarimetry Principle

The polarimetric performance of a high-energy detection plane is determined by the fundamental concepts associated with polarized Compton interactions and by its design. The Compton scattering of a polarized photon beam generates non-uniformity in the azimuthal angular distribution of the scattered photons. The scattered photon's angular direction depends on its initial polarization angle. If the scattered photon goes through a new interaction inside the detector, the statistical distribution of the photon's angular directions defined by the two interactions (double event) provides a modulation curve from which the degree and polarization direction of the incident beam can be derived. The angular distribution of the scattered photons is given by the Klein–Nishina differential cross section for linearly polarized photons:

$$\frac{d\sigma}{d\Omega} = r_0^2 \left(\frac{E'}{E}\right)^2 \left[\frac{E'}{E} + \frac{E}{E'} - 2\sin^2\theta\cos^2\phi\right] \tag{10.3}$$

where r_0 is the classical electron radius, E and E' are, respectively, the energies of the incoming and outgoing photons, θ is the angle of the scattered photons, and ϕ is the angle between the scattering plane (defined by the incoming and outgoing photon directions) and the incident polarization plane (defined by the polarization direction and the direction of the incoming photon). As can

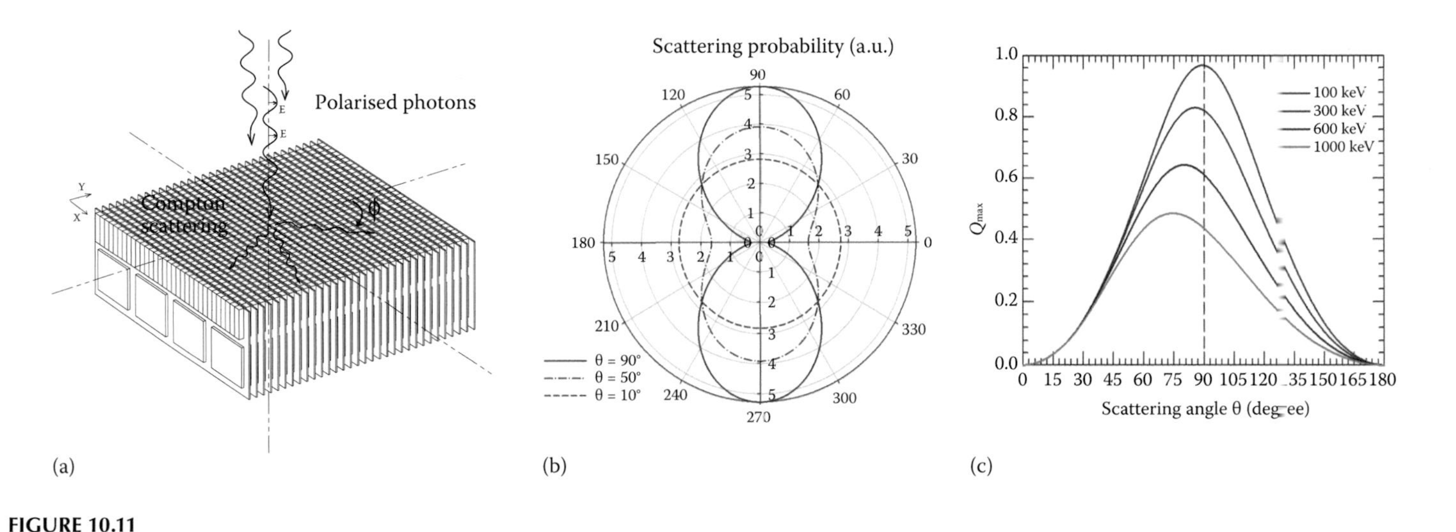

FIGURE 10.11
(a) Scheme of a CZT pixel detector used as a scattering polarimeter. The polarization vector direction is along the y-axis. (b) Azimuthal angle (ϕ) probability distribution for a given Compton scattering angle (θ) of linearly polarized photons at E = 200 keV derived from Equation 10.3. The direction of the polarization is parallel to the horizontal axis of the polar plot. The probability distribution of the scattering angle will present a maximum along the orthogonal direction to the polarization plane and a minimum in the parallel direction. (c) Maximum modulation expected (Q_{max}) as a function of the scattering angle for different impinging photon energy.

be seen from Equation 10.3, after fixing all the other parameters, the scattering probability varies with the azimuthal angle ϕ and its maximum and minimum arises for orthogonal directions (Figure 10.11b). For $\phi = 0°$, the cross section reaches a minimum and for $\phi = 90°$, the cross section reaches a maximum. However, this relative difference reaches a maximum for a scattering angle θ_M, dependent on the incident photon energy (Lei et al. 1997). For hard x/γ-rays (0.1–1 MeV), the θ_M value is 90° at 100 keV slowly decreasing down to ~75° at 1 MeV (Figure 10.11c). Note that E and E' are related through

$$\frac{E'}{E} = \frac{1}{1 + \frac{E}{m_0 c^2}(1 - \cos\theta)} \tag{10.4}$$

where c is the speed of light in free space, and m_o is the electron rest mass.

The modulation factor, Q_{100}, of double-event distribution generated by a 100% polarized beam provides the evaluation of the polarimetric performance of an instrument. For the case of a planar pixelated detector, Q_{100} can be calculated from the modulation curve resulting from a double-event angular distribution around a central irradiated pixel:

$$Q_{100} = \frac{N_{\parallel} - N_{\perp}}{N_{\parallel} + N_{\perp}} \tag{10.5}$$

where $N_{\parallel}$ and $N_{\perp}$ are the double events integrated over two orthogonal directions defined on the detector plane along the maxima and minima of the modulation curve (Suffert et al. 1959).

For a given polarimeter, another parameter is of fundamental importance to quantify its final performance, once implemented in a particular instrument: the minimum detectable polarization (i.e., MDP). MDP indicates when one may be confident that polarization is detected, i.e., that the source is not unpolarized. The expected MDP should be significantly smaller than the degree of polarization to be measured. For a space polarimeter in a background noise environment, the following relation estimates the MDP at 99% (3σ significance) confidence level (Weisskopf et al. 2009):

$$MDP_{99\%} = \frac{4.29}{A \cdot \varepsilon \cdot S_F \cdot Q_{100}} \sqrt{\frac{A \cdot \varepsilon \cdot S_F + B}{\Delta T}} \tag{10.6}$$

where Q_{100} is the modulation factor for a 100% polarized source, ε is the double-event detection efficiency, A is the polarimeter detection area in cm^2, S_F is the source flux (photons s^{-1} cm^{-2}), B is the background count rate (counts/s), and ΔT is the observation time in seconds.

10.4.2 Polarimetry Modulation in CdTe/CZT Pixel Spectrometers

To optimize the polarimetric performances of future high-energy space proposals, a series of experiments based on CZT/CdTe pixel detector prototypes were carried out at the European Synchrotron Radiation Facility (ESRF), where a ~99% polarized gamma-ray beam is available (Curado da Silva et al. 2004, 2008, 2011, 2012; Caroli et al. 2009; Antier et al. 2015). The main purpose of these experiments, denominated as POLCA (POlarimtery with Cadmium Telluride Arrays) series, was to assess the performance of a CZT/CdTe focal plane as a polarimeter up to 750 keV. Monte Carlo simulations were also performed, implementing in the code the same CZT/CdTe detector prototype design irradiated under analogous conditions. The Monte Carlo simulation code was based on the GEANT4 (GEometry ANd Tracking, Allison et al. 2016) a very suitable and efficient tool. The simulation code implemented two main functions: (a) the modules implementing the physics of the electromagnetic interactions of polarized photons, in particular, for the Compton scattering; and (b) the detection system with the definition of the beam characteristics, the detection plane design (geometry and material), and the readout logic.

The POLCA experimental system was composed of four functional subsystems: the synchrotron beamline optical system, the CdZnTe detection system (Figure 10.12), the shaping and coincidence electronic system, and the control and data acquisition workstation.

10.4.2.1 Synchrotron Beamline Optical System

The ID 15A beamline optical system allows tuning the energy of the monochromatic photon beam between 100 keV and up to 1 MeV, with a beam spot

(a)

(b)

FIGURE 10.12
(a) Setup inside the experimental hutch of the ID15A beamline at the ESRF. The large ring provides the rotation around directions parallel to the beam axis. (b) In its center, the CZT pixelized prototype detector system is visible with its readout cables.

of about 500 μm in diameter and a linearly polarized component at the beam center higher than 99% (ESRF 2017).

10.4.2.2 CdZnTe Detection System

Several types of CZT/CdTe detectors (Eurorad, IMARAD, and ACRORAD) were tested under POLCA experiments. Herein, we concentrate on the results obtained with the most tested model during these experiments: the IMARAD detector. This polarimeter prototype was based on an IMARAD 5 mm thick CZT mosaic of four units with anodes segmented to obtain a total of 16 × 16 pixels, each with 2.5 × 2.5 mm^2 area. Due to limitations in our back-end electronics (only 128 channels available), only 11 × 11 pixels have been connected for a total sensitive area of ~8 cm^2. The CZT unit was installed on a supporting layer that contains the readout application-specific integrated circuit (ASIC) supplied by eV Products, Pennsylvania, USA (De Geronimo et al. 2003), the bias circuit, and the connectors for the back-end electronics (Figure 10.12b). The device sensitivity is determined from the energy selectable from 1.2 to 7.2 mV/keV and a peaking time variable between 0.6 and 4 μs.

10.4.2.3 Shaping and Coincidence Electronic Subsystems

The signals were processed by a custom multi-parametric system consisting of 128 independent channels with filters, coincidence logic, and analog-to-digital converter (ADC) units (Guazzoni et al. 1991). When operating in coincidence mode, all signals exceeding the lower energy threshold occurring in the same coincidence time window (2 μs) are analyzed as generated by the same event. The typical irradiated pixel count rate was about 10^4 counts/s.

10.4.2.4 Data Acquisition Unit

This unit was based on a commercial data acquisition card PXI DAQ-6533 provided by National Instruments connected to a personal computer and controlled by a LabView application. For each event, we obtained information about the number of hits, the triggered pixels, and the energy deposited in each hit (Caroli et al. 2002). The recorded data are analyzed offline by an *interactive data language* (IDL 2017) s/w custom tool, which allows the selection of single, double, and multiple events (photons undergoing at least three interactions in the detection plane).

The CZT prototype was tested under a 500 μm diameter monochromatic linearly polarized beam from 150 up to 750 keV in steps of 100 keV. The experimental procedure adopted in order to minimize several factors that might introduce errors in the calculation of Q, such as the non-uniformity of the detection efficiency of the pixels that compose the 11 × 11 CZT matrix and the misalignment of the beam with respect to the irradiated pixel center, consisted of four steps for each energy:

1. The photon beam was aligned with respect to four pixels (2 × 2) by displacing the mechanical system until the number of events in the four pixels became almost uniform. This identified the centroid of the 2 × 2 pixels.
2. The beam was aligned with the center of each pixel in turn, because our beam had a maximum spot diameter of 500 μm, and a slight deviation from this position could be responsible for an undesirable artificial asymmetric distribution due to a different mean free path for scattered photon in the neighboring pixels.
3. Each of the 11 × 11 CdZnTe pixels was irradiated by the polarized beam by moving the detector in the *x* and *y* directions with 2.5 mm steps.
4. The detector was rotated by 90° with respect to its initial position and the steps from 1 to 3 were repeated in order to confirm the 90° double-event scattering distribution symmetry.

The single events obtained in each of the directly irradiated pixels allowed us to determine the relative detection efficiency map of the 11 × 11 pixels. The data were used to perform the correction of the non-uniformities in the response of the CZT detector pixels. The true double-event counts N_{true} for each pixel becomes

$$N_{true} = \frac{N_{pol}}{N_{non}} N_{max} \tag{10.7}$$

where N_{pol} is the number of double events detected (that depend on the beam polarization), N_{non} is the number of single events of the response map obtained when the pixel is directly irradiated, and N_{max} is the maximum value among all the matrix pixels N_{non} (Lei et al. 1997). By applying this method to the pixels around the irradiated pixel, the error introduced by the non-uniformity of the detector matrix response is minimized, and the double-event distributions obtained allow improving the precision of the modulation *Q* factor of the CZT prototype, which is given by Equation 10.5.

Figure 10.13 shows false color maps resulting from double-event distributions generated by a 511 keV monochromatic beam with polarization angles of 0° and 45°. As can be seen, double events are not uniformly distributed around the irradiated pixel for a polarization angle of 0°. As expected from theory, a maximum number of Compton photons were detected in the pixels along the direction defined by the top–center–bottom of the matrix. Polarization direction is perpendicular to the maximum intensity direction,

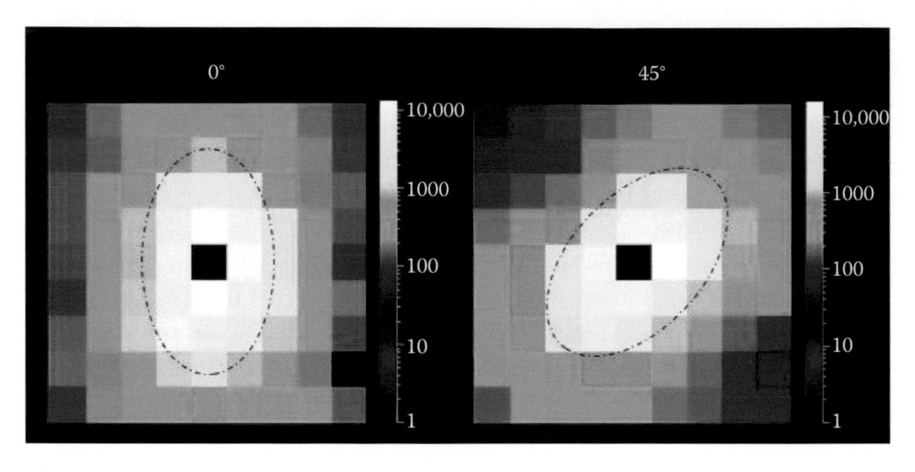

FIGURE 10.13
Double-event maps obtained for a 100% polarized beam at 511 keV by rotating the polarization by an angle φ of 0° and 45°. The 511 keV beam was directed to a CZT matrix central pixel, in black at the center. The dashed ellipse superposed at the center of each double-event distribution is represented to guide the visualization of polarization angle rotation over the matrix. Note that the major ellipse axis is oriented with the pixels along the direction where higher numbers of Compton photons were recorded. The minor axis is perpendicular and is aligned with the incoming beam polarization direction.

represented by the major axis of the ellipse (represented only for guideline purposes) superposed to the double-event distribution. Auxiliary ellipse minor axis takes the incoming polarization direction. When the CZT matrix is rotated by 45°, the projection of the polarization in the detector plane is also rotated by the same amount. It is noticeable in Figure 10.13 that the direction traced by the ellipse's major and minor axis rotates according to the polarization angle apparent rotation.

Figure 10.14a shows the modulation factor Q calculated for the CdZnTe prototype as a function of the polarized photon beam energy. These values were obtained after the correction for the non-uniformity in the response of the detector throughout its pixelated volume by using the method explained above where *true* double events are given by Equation 10.7. For comparison, Monte Carlo simulations were performed with a 5 mm thick CdZnTe matrix similar to the POLCA prototype under analogous conditions. The modulation factor Q values obtained from these simulations are shown in Figure 10.14.

The experimental modulation factor Q obtained is about 0.35 or higher up to 350 keV. It decreases to about 0.15 for 650 keV, since for higher energies the probability of Compton interactions occurring with a scattering angle θ lower than 90° is higher than in the 150–350 keV band. Lower scattering angles provide poorer polarization information; the optimum scattering angle θ_M is about 90° for soft γ-rays and hard x-rays. Furthermore, a lower θ also means that a higher fraction of Compton scattered photons escape the CdZnTe without interacting a second time in the crystal. The fraction of photons that cross the CdZnTe matrix without interaction also increases with the beam energy. As can be seen in Figure 10.14, the CdZnTe prototype performances obtained up to 450 keV are in good agreement with the Monte Carlo simulation results performed with a GEANT4-based code. From 550 keV to higher energies, a secondary synchrotron beam (due to a gap in the beam collimator

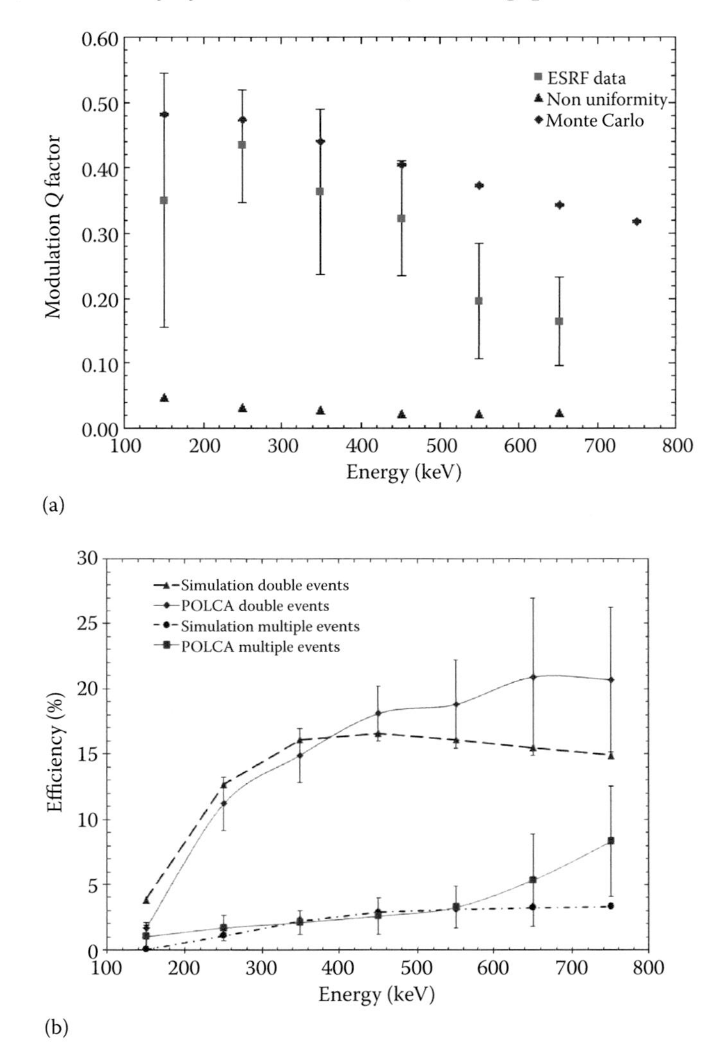

FIGURE 10.14
(a) Q factor as a function of the energy for a 5 mm CdZnTe prototype when irradiated by a monochromatic ~99% polarized photon beam. Monte Carlo simulation results obtained in similar conditions are shown for comparison. The modulation generated by the non-uniformity of matrix pixels response is also represented (triangle). The simulated residual modulation obtained for an unpolarized beam in the same energy range was lower than 0.01 [8]. (b) Experimental and Monte Carlo double and multiple events' relative efficiencies (double events/total detected photons and multiple events/total detected photons) as a function of the γ-ray beam energy.

shield) was projected onto the CdZnTe active surface area, which introduced a substantial error component in the Q factor calculation. For 750 keV, the secondary was so dramatically close to the main beam (a few pixels) that the double-event distributions of the two beams overlapped, and it was not possible to perform the polarimetric analysis of our prototype.

The double and multiple events' relative efficiencies (double events/detected photons and multiple events/detected photons) obtained over the experiment energy range are shown in Figure 10.14b together with efficiencies obtained from Monte Carlo simulations. The absolute efficiency (events/incident photons) was not determined since the auxiliary instruments of the ID 15 beamline were not stable and did not accurately measure the count rate of the photon beam. As can be seen in Figure 10.14b, up to 550 keV, the double-event relative efficiency increases with the energy in agreement with the Monte Carlo data, up to about 18%. However, from 550 keV up, experimental efficiency values diverge from the Monte Carlo relative efficiencies, attaining about 20% for 750 keV, while Monte Carlo simulations show a slight diminution of the efficiency for higher energies. Since Compton scattering probability increases with energy, the double events detected increase up to 550 keV, and then lower Compton scattering angles favor escape Compton photons that leave the CdZnTe block without undergoing a second interaction, which explains why the efficiency decreases slightly as the beam energy is increased. The experimental divergence for higher energies is explained by the difficulty to exclude coincidence events generated by the simultaneous projection of the main and secondary beams in the detection plane that occurred from 550 keV up. The multiple-event efficiency increases with energy, since the original photon energy becomes sufficiently high in order to increase noticeably the probability to generate two successive Compton scatterings. However, comparison between Monte Carlo generated and experimental multiple-event relative efficiencies shows similar divergence to the double-event curve, confirming that simultaneous beam detection in the CdZnTe plane artificially increases the efficiency of events measured in coincidence. This problem could be solved if the distance between the detector and the beam output window is increased. Unfortunately, the rack where the mechanical system was mounted was already at its maximum distance from the beam window. Excluding this anomaly for higher energies due to multiple beam detection, both experimental relative efficiency results are in good agreement with the results obtained by the Monte Carlo simulation code.

10.4.3 Polarimetry Optimization of CdTe/CZT Pixel Detector

In order to optimize a CdZnTe focal plane for γ-ray polarimetry in astrophysics, we tested several CZT/CdTe pixel prototypes in a series of experiments covering various aspects, from polarimetric performance to possible sources of systematic error. Several factors limit the performance of a polarimeter when measurements take place under conditions that are not ideal. One of the most important is the angle between the polarimeter detection plane and the direction of the incoming polarized photons. If the direction of the incoming photons is not orthogonal with respect to the detector plane, the observed modulation of the Compton events distribution is distorted. The degradation of polarimetric measurements will be more important as the angle of the off-axis source increases. The optical system employed to collect photons will influence the direction of the incoming photons. In the case of coded mask telescope, the tilt angle is the same for all photons from one source, but when Laue lenses are used, photons are diffracted at different angles, but typically at less than 1° tilt angle with respect to the optical axis. The effect of impinging photon beam inclination on the measured polarization is also dependent on the pixel size, because this characteristic influences the separation of the hits in a scattered event. Therefore, it is important to determine the maximum tilt angle for which the real polarization modulation is only faintly affected. Another important factor is how the polarimetric sensitivity of the detector depends on the polarization level of beam polarization (i.e., the minimum percentage of polarized photons that the detector is able to detect), since its configuration (mainly spatial resolution and geometry) might limit the capacity to recognize a weakly polarized source. Because of the "square" geometry of scattering elements in a typical pixel detector, systematic effects are introduced in the polarimetric modulation when the incident polarization plane angle is not parallel to one of the detector pixel sides. In fact, square pixels introduce a quantization effect in the distribution of the polarized scattered photons that limits the angular resolution of the polarimeter when considering pixels at different distances from that which scattered the incoming photon. Herein we extend this investigation, obtaining a finer response to the polarization angle direction, not only testing the detector response to a wider set of angles, but also by carefully choosing angles that are not redundant when considering the matrix double-event distribution. The double-event spread pattern in a square pixel matrix repeats itself every 45° (10° is equivalent to 80°, 20° to 70°, 30° to 60°, etc.); therefore, we tested our polarimeter in a 0° to 45° angle range at 5°, 10°, 20°, 30°, 40°, and 45°.

Firstly, the CZT detector central pixel was irradiated by a polarized beam forming different inclination angles with the optical detector axis: 0°, 0.5°, 1°, 1.5°, 2°, 3°, 4°, 5°, and 10°. These measurements at different tilt angles were repeated for different energies (200, 300, 400, and 511 keV) and for polarization vector directions parallel to both the detector plane axis: x and y. Then the modulation factor Q as a function of the inclination angle Θ was calculated from the double-event distributions obtained from each measurement.

Figure 10.15a shows the Q factors as a function of the tilt angle. Up to 2° tilt angle, the Q factor is not significantly affected by the beam inclination. However, from 3° up to 10° tilt angles, the Q factor dramatically increases when polarization and inclination add their effects and decreases when these effects partially cancel each other. These results confirm previous simulation

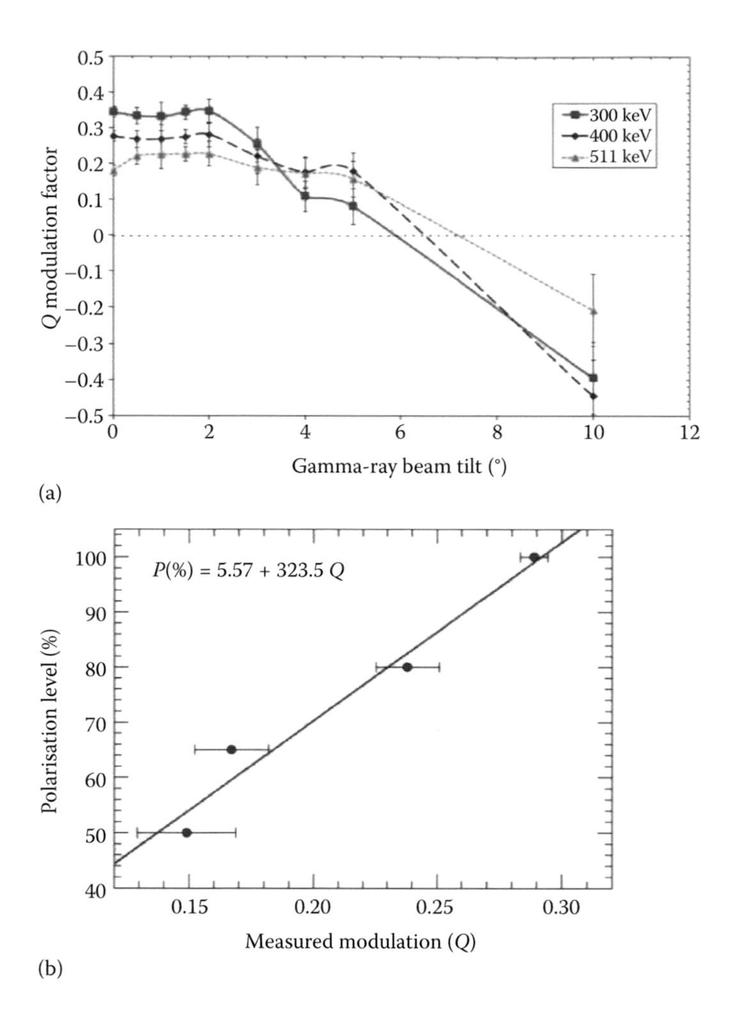

FIGURE 10.15
(a) The modulation factor Q as a function of the tilt angle of the ESRF 100% polarized beam for different energies. In these measurements, the polarization vector was parallel to the x-axis. (b) Measured factor Q as a function of the polarization level from 100% to 50% of a 400 keV photon beam. The error associated to each point was obtained by averaging the measured Q for a set of measurements at the same beam polarization level. These results show a good linearity of the polarimetric response of the used CZT pixel detector to the levels of the γ-ray beam polarization.

studies performed by a Monte Carlo simulation program based on GEANT4 (Curado da Silva et al. 2003). Both experimental and simulation results show that during an observation period onboard a γ-ray satellite, it is essential that polarized sources are no more than 2° off-axis in order that polarization measurements are not affected. This study shows the importance of a pointing system with accuracy better than 1° for an instrument designed for polarimetry. This accuracy should be sufficient so that double-event distributions can be read directly without further data correction methods.

Afterward, the polarimetric sensitivity of a CZT prototype, as a function of the polarization fraction of the incoming γ-ray beam, was tested. Tests were limited to one energy because of time limitations resulting from the very low flux at lower polarization degrees. During each measurement, a polarized 400 keV monochromatic beam irradiated one of the four central pixels of the CZT matrix. The polarization angle was fixed in a parallel direction to the detector *x*-axis. Measurements were performed for different beam polarization degrees: 100%, 80%, 65%, and 50%. The measurement live time was tuned to acquire, in the pixels situated near the irradiated one, a number of double events of the order of 10^4, so that a good polarimetric sensitivity could still be achieved.

The modulation factor *Q* was calculated from the double-event distributions generated by each measurement performed at different beam polarization degrees. The fraction of double and multiple events recorded by the detector was ~20% and ~3%, respectively (Curado da Silva et al. 2004, 2008). Multiple events do not enter into our calculations since the data handling system cannot determine the order of each hit. For double events, we know which is the first interaction because this is coincident with the position of the pixel irradiated by the collimated beam. Therefore, during the analysis, we exclude double events that do not have at least one interaction in the target pixel, e.g., chance coincidence events due to noise and flaring pixels and/or triple events in which the first interaction in the target pixel was under the low-energy threshold (~30 keV). Furthermore, because the impinging beam was monochromatic, we applied a further simple selection of double events using the energy deposited in each hit. Knowing the beam energy, we have selected as good double events only those in which the sum of the two interactions is within a window centered at the selected beam energy within 3σ derived from the expected energy resolution at that energy—evaluated by a simple square root relation derived from calibration data with radioactive sources. Energy resolution (FWHM) ranged between ~8% and 7% for single events and between 16% and 15% for double events in the 200–400 keV band. Although energy resolution was relatively high, this was not a critical factor for polarization performance analysis in the adopted experimental setup.

Figure 10.15b shows a good linear relation between the polarization level of the beam and the measured factor *Q*. At least, down to 50%, this CZT prototype exhibits a good sensitivity to the beam polarization degree. The error associated to each point of Figure 10.15b was obtained by averaging the *Q* values for a set of measurements at the same beam polarization degree. Previous measurements revealed a background noise residual factor *Q* of about 0.02 (Curado da Silva et al. 2011) that would limit our polarimetric sensitivity in the described experimental conditions (setup and acquisition time and CZT pixel detector configuration) to about 12% when extrapolating the linear fitting.

In order to evaluate the accuracy of the polarization angle measurement of a planar CZT matrix prototype, additional tests were performed using the ESRF ID15 beamline. A central pixel was irradiated by a 100% polarized monochromatic beam at different energies (200, 300, 400, and 511 keV), and the support ring of the detector prototype was rotated by an azimuthal angle φ of 5°, 10°, 20°, 30°, 40°, and 45° (Figure 10.16a). It is relevant to choose a set of nonsymmetrical angles relative to 45° when testing a square pixel matrix; otherwise, double events generate similar distributions. As mentioned before, in previous experiments, other authors tested square matrices rotating the polarization angle only by 30° and 60° (Xu et al. 2005)—angles whose double-event distribution is going to spread in a symmetrical pattern throughout the whole array—and by 45° (Kroeger et al. 1996).

Figure 10.16b shows the measured polarization angle (φ_{obs}) as a function of the effective ESRF beam polarization angle (φ_{beam}) at 200, 300, 400, and 511 keV. The linear fits calculated for each energy are also represented. Overall analysis of these results shows a good agreement between measured polarization angle and the effective beam polarization angle. The error bars of most of the measured polarization angles lie within a few degrees.

The best agreement between φ_{obs} and φ_{beam} was found for the set of measurements performed at 300 keV. Actually, it is at approximately 300 keV that better polarization sensitivity is obtained ($Q \sim 0.4$), as shown in a previous study performed with the same CZT prototype and at the same beamline (Curado da Silva 2008). Furthermore, errors associated with polarization angle observations show that systematic effects due to the square pixels generate higher uncertainties for angles near 45°. This is an expected result, since for these angles, the pixels that correspond to the maximum and minimum directions of the double-event distribution lie close to the diagonal of the detector plane axis, where the effect due to square pixels is more pronounced. When square pixels or parallelepiped voxels are the only technical solutions available, another way to minimize this problem consists of employing pixels of shorter lateral size, improving the spatial (and angular) resolution of the double-event distribution, and therefore reducing the systematic effects.

In order to study the optimal pixel size of the Laue lens telescope focal plane, the MDP was calculated as a function of pixel lateral size dimensions under the same irradiation conditions as explained before. Since the expected point spread function is of about 30 mm, a pixel scale of a few millimeters (1–3 mm) would be enough to have a good sampling from the imaging and source detection point of view. A smaller pixel scale would allow a better sensitivity to the polarized emission, but it means an increase in the focal plane complexity (a large number of channels require more complex electronics and more resources). Therefore, we limited our study to pixel lateral dimensions between 0.5 and 2 mm. Figure 10.16c shows the factor Q and the MDP (for 10^6 s observation time) obtained for a broad band Laue lens in the 120 to 200 keV energy band-pass combined with a 10 mm thickness CdTe focal plane. Since a 32×32 CdTe matrix is always simulated, for lateral pixel sizes smaller than 1.0 mm, its volume is smaller and therefore a fraction

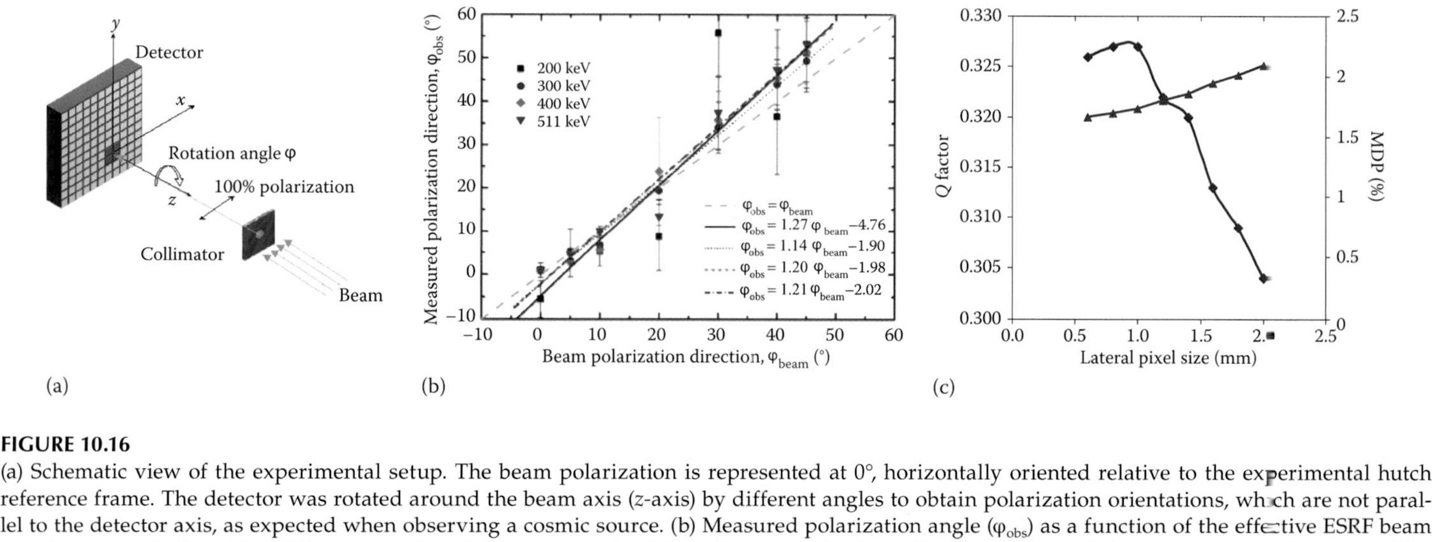

FIGURE 10.16
(a) Schematic view of the experimental setup. The beam polarization is represented at 0°, horizontally oriented relative to the experimental hutch reference frame. The detector was rotated around the beam axis (z-axis) by different angles to obtain polarization orientations, which are not parallel to the detector axis, as expected when observing a cosmic source. (b) Measured polarization angle (φ_{obs}) as a function of the effective ESRF beam polarization angle (φ_{beam}) for different energies. The errors associated to each measured polarization angle were obtained by averaging the angle of the two maxima and the two minima of each double-event modulation curve applying a 90° independent partial fitting, since the polarization is 90° symmetric. A linear fitting was applied for each energy. (c) Factor Q (diamonds) and the MDP (triangles) for a 10^6 s observation time as a function of pixel lateral size obtained for a broad band Laue lens with a 10 mm thick CdTe focal plane in the 120–200 keV energy band.

of Compton photons escape from the detection plane before having a second interaction, which explains why Q decreases for lateral sizes smaller than 1.0 mm. From 1.0 mm up, this effect becomes residual and smaller lateral dimensions result in higher factor Q values due to a higher rate of second interactions occurring inside pixels further from the central pixel, which contributes to an improvement in the double-event distribution angular resolution. However, the net gain observed in the MDP for pixel dimensions lower than 1 mm does not compensate for the technical difficulties associated with its production. Therefore, focal plane pixel lateral dimensions of about 1 up to 2 mm provide a good trade-off between focal plane complexity and polarimetric performance. Furthermore, we point out that the improvement in the MDP achievable with the smaller pixel scale might be obtained with less expensive background noise reduction techniques such as optimizing the shielding and/or applying event selection procedures based on Compton kinematics.

10.5 Consideration on CZT/CdTe Spectroscopic-Imager Applications and Perspective for Scattering Polarimetry

The development of CZT/CdTe spectrometers with high 2D/3D spatial resolution and fine spectroscopy represents a challenge to the realization of a new class of high-performance instruments, for hard x/γ-rays, able to fulfill the current and future requirements in several applications fields.

Such detectors can achieve very good detection efficiency at high energy (up to a few megaelectron volts; Boucher et al. 2011), without significant loss of spectroscopic performance and response uniformity. These characteristics, together with room-temperature operability, are appealing for application in radiation monitoring and identification (Wahl and He 2011), in industrial noninvasive controls, in nuclear medicine, and in hard x/γ-ray astronomy instrumentation.

Furthermore, 3D CZT/CdTe spectro-imagers, because of the fine spectroscopy (few % at 60 keV and <1% above 600 keV) and the high 3D spatial resolution (0.2–0.5 mm FWHM) achievable, allow operation not only in full energy mode but also as Compton scattering detectors if equipped with appropriate electronics providing a suitable coincidence logic to handle multihit events. These possibilities imply that these sensors are suitable to realize wide field detector for γ-ray sources (>100 keV) localization and detection both in ground and space applications (Xu et al. 2004). Evaluation done using a single thick 3D CZT sensor (Section 10.3.1) as a 4π Compton imager has demonstrated the possibility to obtain an angular resolution

~15° (FWHM) at 662 keV. This is really an excellent result in the small distance scale used to reconstruct the Compton events kinematics and can be achieved only because the good 3D and spectroscopic performance of the CZT proposed sensor units.

As seen in Section 10.4, the possibility operating 2D/3D CZT/CdTe spectrometers as Compton scattering detectors relies on the appealing opportunity to use these devices for hard x/γ-rays polarimetry. Today, this type of measurement is recognized for its fundamental importance in high-energy astrophysics and is one of the most demanding requirements for next space mission instrumentation in this energy band (10–1000 keV). This capability is well described in Section 10.4 by using both the experimental results and Monte Carlo evaluations obtained by authors for 2D CZT/CdTe pixel detector (2D spectro-imager).

In fact, a 3D spectrometer able to handle properly scattered events in three dimensions over the entire sensitive volume can offer even better performance as a scattering polarimeter. In the case of 3D spectrometer devices, such as described in Section 10.3, each single sensor unit could be operated as a Compton polarimeter (Xu et al. 2005). The quality (modulation factor) of a scattering polarimeter strictly depends on both spatial and spectroscopic resolution, because these characteristics affect the capability of Compton kinematics reconstruction and good event selection (Curado da Silva et al. 2011; Antier et al. 2015).

The development of 3D CZT/CdTe spectroscopic imagers in the coming years represents a great opportunity for the implementation of high-performance detectors operated as high-efficiency scattering polarimeters. This development can definitely open the polarimetric dimension in hard x/γ-rays astronomy, making polarimetry the new standard observation mode in the next space instrumentation. Compared to the pixel detectors, the determination of the 3D position of each hit in scattering events represents a great advantage in the measurement of polarization as it allows a more accurate reconstruction of the Compton kinematics and therefore a more efficient selection of the events to optimize the response to the polarization modulation. For example, a better Compton kinematics reconstruction allows implementing reliable methods to recognize good events (i.e., events form the source) with respect to chance coincidence ones and background events, improving the signal-to-noise ratio of the detection. The 3D spatial resolution capability can help also to handle some typical systematics that can negatively affect polarization measurements, like the one introduced by incoming flux direction angle (Section 10.4.3). Furthermore, the possibility to select events within thin layers of the sensitive volume, thanks to the 3D segmentation of the detector (i.e. close to 90° scattering direction) improves the modulation factor and therefore the reliability of the polarimetric measurements (Caroli et al. 2015).

References

Abbene, L. et al. (2007), "Spectroscopic response of a CdZnTe multiple electrode detector," *Nucl. Instr. Methods Phys. Res. A*, Vol. 583, p. 324.

Abbene, L. et al. (2015), "Digital performance improvements of a CdTe pixel detector for high flux energy-resolved X-ray imaging," *Nucl. Instr. Methods Phys. Res. A*, Vol. 777, p. 54.

Allison, J. et al. (2016), "Recent developments in GEANT4," *Nucl. Instr. Methods Phys. Res. A*, Vol. 835, p. 186. GEANT 4 home page.

Antier, S. et al. (2015), "Hard X-ray polarimetry with Caliste, a high performance CdTe based imaging spectrometer," *Exp. Astron.*, Vol. 39, p. 233.

Auricchio N. et al. (1999), "Investigation of response behaviour in CdTe detectors versus inter-electrode charge formation position," *IEEE Trans. Nucl. Sci.*, Vol. 46, p. 853.

Auricchio, N. et al. (2005), "Twin shaping filter techniques to compensate the signals from CZT/CdTe detectors," *IEEE Trans. Nucl. Sci.*, Vol. 52, p. 1982.

Auricchio, N. et al. (2012), "Development status of a CZT spectrometer prototype with 3D spatial resolution for hard X ray astronomy," *Proc. SPIE*, Vol. 8453, p. 84530S.

Bale, D.S., and Szeles, C. (2006), "Design of high performance CdZnTe quasi-hemispherical gamma-ray CAPture™ plus detectors," *Proc. SPIE*, Vol. 6319, p. 63190B.

Barrett, H.H. et al. (1995), "Charge transport in arrays of semiconductor gamma-rays detectors," *Phys. Rev. Lett.*, Vol. 75, p. 156.

Bellazzini, R. et al. (2010), *X-ray Polarimetry: A New Window in Astrophysics*, Cambridge University Press.

Berger, M.J. et al. (2010), "XCOM: Photon Cross Sections Database," http://www.nist.gov/pml/data/xcom/index.cfm.

Blozer, P.F. et al. (2016), "A concept for a soft gamma-ray concentrator using thin-film multilayer structures," *Proc. SPIE*, Vol. 9905, p. 99056L.

Bolotnikov, A.E. et al. (2006), "Performance characteristics of Frisch-ring CdZnTe detectors," *IEEE Trans. Nucl. Sci.*, Vol. 53, p. 607.

Boucher, Y.A. et al. (2011), "Measurements of gamma rays above 3 MeV using 3D position-sensitive 20 × 20 × 15 mm^3 CdZnTe detectors," 2011 *IEEE Nucl. Sci. Symp. Conference Rec.*, p. 4540.

Caroli, E. et al. (2000), "CIPHER: coded imager and polarimeter for high-energy radiation," *Nucl. Instr. Methods Phys. Res. A*, Vol. 448, p. 525.

Caroli, E. et al. (2002), "PolCA (Polarimetria con CdTe Array)," IASF/CNR Internal Report n. 345.

Caroli, E. et al. (2008), "A three-dimensional CZT detector as a focal plane prototype for a Laue Lens telescope," *Proc. SPIE*, Vol. 7011, p. 70113G.

Caroli, E. et al. (2009), "A polarimetric experiment with a Laue Lens and CZT pixel detector," *IEEE Trans. Nucl. Sci.*, Vol. 56, p. 1848.

Caroli, E. et al. (2010), "Development of a 3D CZT detector prototype for Laue Lens telescope," *Proc. SPIE*, Vol. 7742, p. 77420V.

Caroli, E., and Del Sordo, S., (2015), "CdTe/CZT Spectrometers with 3-D Imaging Capabilities," in *Solid-State Radiation Detectors—Technology and Applications*, Awadalla, S. and Iniewski, K. (eds.), CRC Press (Boca Baton, London, New York), ISBN: 978-1-4822-6221-6.

Caroli, E. et al. (2015), "Monte Carlo evaluation of a CZT 3D spectrometer suitable for a hard X- and soft-γ rays polarimetry balloon borne experiment," 2015 *IEEE Conf Records.*

Casali, F. et al. (1992), "Characterization of small CdTe detectors to be used for linear and matrix arrays," *Nucl. IEEE Trans. Nucl. Sci.,* Vol. 39, p. 598.

Cavalleri, G. et al. (1971), "Extension of Ramo theorem as applied to induced charge in semiconductor detectors," *Nucl. Instr. Methods Phys. Res.,* Vol. 92, p. 137.

Chauvin, M. et al. (2016), "Optimising a balloon-borne polarimeter in the hard X-ray domain: From the PoGOLite Pathfinder to PoGO+," *Astroparticle Phys.,* Vol. 82, p. 99.

Cui, Y. et al. (2008), "Hand-held gamma-ray spectrometer based on high-efficiency Frisch-ring CdZnTe detectors," *IEEE Trans. Nucl. Sci.,* Vol. 55, p. 2765.

Curado da Silva, R.M. et al. (2003), "CIPHER, a polarimeter telescope concept for Hard X-ray Astronomy," *Exp. Astron.,* Vol. 15, p. 45.

Curado da Silva, R.M. et al. (2004), "Hard-X and soft gamma-ray polarimetry with CdTe array prototypes," *IEEE Trans. Nucl. Sci.,* Vol. 51, p. 2478.

Curado da Silva, R.M. et al. (2008), "Polarimetric performance of a Laue lens gamma-ray CdZnTe focal plane prototype," *J. Appl. Phys.,* Vol. 104, p. 084903.

Curado da Silva, R.M. et al. (2011), "Polarimetry study with a CdZnTe focal plane detector," *IEEE Trans. Nucl. Sci.,* Vol. 58, p. 2118.

Curado da Silva, R.M. (2012), "Polarization degree and direction angle effects on a CdZnTe focal plane performance," *IEEE Trans. Nucl. Sci.,* Vol. 59, p. 1628.

Dean, A.J. et al. (2008), "Polarized gamma ray emission from the CRAB," *Science,* Vol. 321, no 5893, p. 1183.

De Geronimo, G. et al. (2003), "Advanced-readout ASICs for multielement CdZnTe detectors," *Proc. SPIE,* Vol. 4784, p. 105, 2003.

Della Monica Ferreira, D. et al. (2013), "Hard X-ray/soft gamma-ray telescope designs for future astrophysics missions," *Proc. SPIE,* Vol. 886, p. 886116-1

Devanathan, R. et al. (2006), "Signal variance in gamma-ray detectors—A review," *Nucl. Instr. Methods Phys. Res. A,* Vol. 565, p. 637.

Dish, C. et al. (2010), "Coincidence measurements with stacked (Cd,Zn)Te coplanar grid detectors," *IEEE Nucl. Sci. Symp. Conference Record,* p. 3698.

Du, Y.F. et al. (2001), "Evaluation of a Compton scattering camera using 3-D position sensitive CdZnTe detectors," *Nucl. Instr. Methods Phys. Res. A,* Vol. 457, p. 203.

Eberth, J. and Simpson, J. (2006), "From Ge(Li) detectors to gamma-ray tracking arrays—50 years of gamma spectroscopy with germanium detectors," *Prog. Particle Nucl. Phys.,* Vol. 60, p. 283.

Eskin, J.D. et al. (1999), "Signals induced in semiconductor gamma-ray imaging detectors," *J. Appl. Phys.,* Vol. 85, p. 647.

ESRF (2017), European Synchrotron Research Facility (Grenoble, France): http://www.esrf.eu/UsersAndScience.

Forot, M. et al. (2008), "Polarization of the crab pulsar and nebula as observed by the INTEGRAL/IBIS telescope," *Ap. J.,* Vol. 688, p. L29.

Frisch, O. (1944), British Atomic Energy Report, BR-49.

Frontera, F. et al. (2013), "Scientific prospects in soft gamma-ray astronomy enabled by the LAUE project," *Proc. SPIE,* Vol. 8861, p. 886106-1.

Gostilo, V. et al. (2002), "The development of drift-strip detectors based on CdZnTe," *IEEE Trans. Nucl. Sci.,* Vol. 49, p. 2530.

Götz, D. et al. (2009), "Variable polarization measured in the prompt emission of GRB 041219A using IBIS on board INTEGRAL," *Astrophys. J. Lett.*, Vol. 695, p. 2.
Gu, Y. et al. (2011), "Study of a high-resolution, 3D positioning cadmium zinc telluride detector for PET," *Phys. Med. Biol.*, Vol. 56, p. 1563.
Guazzoni, G. et al. (1991), "A mixer unit for data acquisition," *Nucl. Instr. Methods Phys. Res. A*, Vol. 305, p. 442
He, Z. (2001), "Review of the Shockley–Ramo theorem and its application in semiconductor gamma ray detectors," *Nucl. Instr. Methods Phys. Res. A*, Vol. 463, p. 250.
He, Z. et al. (1997), "Position-sensitive single carrier CdZnTe detectors," *Nucl. Instr. Methods Phys. Res. A*, Vol. 388, p. 180.
He, Z. et al. (1999), "3-D position sensitive CdZnTe gamma-ray spectrometers," *Nucl. Instr. Methods Phys. Res. A*, Vol. 422, p. 173.
Hecht, K. (1932), "Zum Mechanismus des lichtelektrischen Primärstromes in isolierenden Kristallen," *Z. Phys.*, Vol. 77, p. 235.
IDL 2017, Interactive Data Language from Harris Geospatial Solution: http://www.exelisvis.it/ProdottieServizi/IDL.aspx.
Jordanov, V.T. et al. (1996), "Compact circuit for pulse rise-time discrimination," *Nucl. Instr. Methods Phys. Res. A*, Vol. 380, p. 353.
Judson, D.S. et al. (2011), "Compton imaging with the PorGamRays spectrometer," *Nucl. Instr. Methods Phys. Res. A*, Vol. 652, p. 587.
Kim, H. et al. (2004), "Investigation of the energy resolution and charge collection efficiency of Cd(Zn)Te detectors with three electrodes," *IEEE Trans. Nucl. Sci.*, Vol. 51, p. 1229.
Kislat, F. et al. (2017), "Design of the telescope truss and gondola for the balloon-borne X-ray polarimeter X-Calibur," *J. Astronom. Instrument.*, Vol. 6, p. 1740003.
Knödlseder, J. et al. (2007), "GRI: Focusing on the evolving violent universe," *Proc. SPIE*, Vol. 6688, p. 668806.
Kole, M. et al. (2016), "POLAR: Final calibration and in-fligth performance of a dedicated GRB polarimeter," *IEEE NSS/MC Conf. in Strasbourg (F)*, 29 Oct.–5 Nov. 2016, arXiv:1612.04098v1.
Kozorezov, A.G. et al. (2005), "Resolution degradation of semiconductor detectors due to carrier trapping," *Nucl. Instr. Methods Phys. Res. A*, Vol. 546, p. 207.
Kroeger, R.A. et al. (1996), "Gamma-ray instrument for polarimetry, spectroscopy, and imaging (GIPSI)," *Proc. SPIE*, Vol. 2806, p. 52.
Kuvvetli, I., and Budtz-Jørgensen, C. (2005), "Pixelated CdZnTe drift detectors," *IEEE Trans. Nucl. Sci.*, Vol. 52, p. 1975.
Kuvvetli, I. et al. (2010a), "CZT drift strip detectors for high energy astrophysics," *Nucl. Instr. Methods Phys. Res. A*, Vol. 624, p. 486.
Kuvvetli, I. et al. (2010b), "Charge collection and depth sensing investigation on CZT drift strip detectors," *IEEE Nucl. Sci. Symp. Conference Record*, p. 3880.
Kuvvetly, I. et al. (2014), "A 3D CZT high resolution detector for x-and gamma-ray astronomy," *Proc. SPIE*, Vol. 9154, p. 91540X-1.
Laurent, P. et al. (2011), "Polarized gamma-ray emission from the galactic black hole Cygnus X-1," *Science*, Vol. 332, p. 438.
Lebrun, F. et al. (2003), "ISGRI: The INTEGRAL soft gamma-ray imager," *Astron. Astrophys.*, Vol. 411, p. L141.
Lechner, P. et al. (2004), "Novel high-resolution silicon drift detectors," *X Ray Spectrometr.*, Vol. 33, p. 256.

Lee, K. et al. (2010), "Development of X-ray and gamma-ray CZT detectors for Homeland Security Applications," *Proc. SPIE*, Vol. 7664, p. 766423-1.

Lei, F. et al. (1997), "Compton polarimetry in gamma-ray astronomy," *Space Sci. Rev.*, Vol. 82, p. 309.

Leo, W.R., *Techniques for Nuclear and Particle Physics Experiments*, Springer-Verlag (Berlin, Heidelberg), ISBN: 978-3-642-57920-2 (1994).

Li, W. et al. (1999), "A data acquisition and processing system for 3-D position sensitive CZT gamma-ray spectrometers," *IEEE Trans. Nucl. Sci.*, Vol. 46, p. 1989.

Li, W. et al. (2000), "A modeling method to calibrate the interaction depth in 3-D position sensitive CdZnTe gamma-ray spectrometers," *IEEE Trans. Nucl. Sci.*, Vol. 47, p. 890.

Limousin, O. et al. (2011), "Caliste-256: A CdTe imaging spectrometer for space science with a 580 μm pixel pitch," *Nucl. Instr. Methods Phys. Res. A*, Vol. 647, p. 46.

Lingren, C.L. et al. (1998), "Cadmium-zinc telluride, multiple-electrode detectors achieve good energy resolution with high sensitivity at room-temperature," *IEEE Trans. Nucl. Sci.*, Vol. 45, p. 433.

Luke, P.N. (1995), "Unipolar charge sensing with coplanar electrodes—Application to semiconductor detectors," *IEEE Trans. Nucl. Sci.*, Vol. 42, p. 207.

Luke, P.N. (2000), "Coplanar-grid CdZnTe detector with three-dimensional position sensitivity," *Nucl. Instr. Methods Phys. Res. A*, Vol. 439, p. 611.

Macri, J.R. et al. (2002), "Study of 5 and 10 mm thick CZT strip detectors," 2002 *IEEE Nucl. Sci. Symp. Conference Record*, p. 2316.

Macri, J.R. et al. (2003), "Readout and performance of thick CZT strip detectors with orthogonal coplanar anodes," 2003 *IEEE Nucl. Sci. Symp. Conference Record*, p. 468.

Matteson, J.L. et al. (2003), "CZT detectors with 3-D readout for gamma-ray spectroscopy and imaging," *Proc. SPIE*, Vol. 4784.

McConnell, M. et al. (2009), "X-ray and gamma-ray polarimetry," Astro2010: The Astronomy and Astrophysics Decadal Survey, Science White Papers, no. 198.

McGregor, D.S. et al. (1998), "Single charge carrier type sensing with a parallel strip pseudo-Frisch-Grid CdZnTe semiconductor radiation detector," *Appl. Phys. Lett.*, Vol. 12, p. 192.

Mihailescu, L. et al. (2007), "SPEIR: A Ge Compton camera," *Nucl. Instr. Methods Phys. Res. A*, Vol. 570, p. 89.

Ogawa, K., and Muraishi, M. (2010), "Feasibility study on an ultra-high-resolution SPECT with CdTe detectors," *IEEE Trans. Nucl. Sci.*, Vol. 57, p. 17.

Owens, A., and Peacock, A. (2004), "Compound semiconductor radiation detectors," *Nucl. Instr. Methods Phys. Res. A*, Vol. 531, p. 18.

Owens, A. et al. (2006), "Hard X- and g-ray measurements with a large volume coplanar grid CdZnTe detector," *Nucl. Instr. Methods Phys. Res. A*, Vol. 563, p. 242.

Pantazis, T. et al. (2010), "The historical development of the thermoelectrically cooled X-ray detector and its impact on the portable and hand-held XRF industries," *X-Ray Spectrosc.*, Vol. 39, p. 90.

Perillo, E. et al. (2004), "Spectroscopic response of a CdTe microstrip detector when irradiated at various impinging angles," *Nucl. Instr. Methods Phys. Res. A*, Vol. 531, p. 125.

Richter, M., and Siffert P. (1992), "High resolution gamma ray spectroscopy with CdTe detector systems," *Nucl. Instr. Methods Phys. Res. A*, Vol. 322, p. 529.

Sato, G. et al. (2002), "Characterization of CdTe/CdZnTe detectors," *IEEE Trans. Nucl. Sci.*, Vol. 49, p. 1258.
Sellin, J. P. (2003), "Recent advances in compound semiconductor radiation detectors," *Nucl. Instr. Methods Phys. Res. A*, Vol. 513, p. 332.
Shor, A. et al. (1999), "Optimum spectroscopic performance from CZT γ- and X-ray detectors with pad and strip segmentation," *Nucl. Instr. Methods Phys. Res. A*, Vol. 428, p. 182.
Stahle, C.M. et al. (1997), "Fabrication of CdZnTe strip detectors for large area arrays," *Proc. SPIE*, Vol. 315, p. 90,
Suffert, P.M. et al. (1959), "Polarization measurements of proton capture gamma rays," *Physica*, Vol. 25, p. 659.
Tatischeff, V. et al. (2016), "The e-ASTROGAM gamma-ray space mission," *Proc. SPIE*, Vol. 9905, p. 99052N.
Ubertini, P. et al. (2003), "IBIS: The Imager on-board INTEGRAL," *Astron. Astr.*, Vol. 411, p. L131.
Vetter, K. et al. (2007), "High-sensitivity Compton imaging with position-sensitive Si and Ge detectors," *Nucl. Instr. Methods Phys. Res., A*, Vol. 579, p. 363.
Virgilli, E. et al. (2015), "Hard x-ray broad band Laue lenses (80–600 keV): Building methods and performances," *Proc. SPIE*, Vol. 9603, p. 960308-1.
Von Ballmoos, P. et al. (2010), "A DUAL mission for nuclear astrophysics," *Nucl. Instr. Methods*, Vol. A623, p. 431.
Wahl, C.G., and He, Z. (2011), "Gamma-ray point-source detection in unknown background using 3D-position-sensitive semiconductor detectors," *IEEE Trans. Nucl. Sci.*, Vol. 58, p. 605.
Watanabe, S. et al. (2002), "CdTe stacked detectors for gamma-ray detection," *IEEE Trans. Nucl. Sci.*, Vol. 49, p. 1292.
Watanabe, S. et al. (2009), "High energy resolution hard X-ray and gamma-ray imagers using CdTe diode devices," *IEEE Trans. Nucl. Sci.*, Vol. 56, p. 777.
Weisskopf, M.C. et al. (2009), "The prospects for X-ray polarimetry and its potential use for understanding neutron stars," *Neutron Stars Pulsars Astrophys. Space Sci. Lib.*, Vol. 357, p. 589.
Whal, C.G. et al. (2015), "The Polaris-H imaging spectrometer," *Nucl. Instr. Methods Phys. Res. A*, Vol. 784, p. 377.
Winkler, C. et al. (2003), "The integral mission," *Astron. Astr.*, Vol. 411, p. L1.
Xu, D. et al. (2004), "4π Compton imaging with single 3D position sensitive CdZnTe detector," *Proc. SPIE*, Vol. 5540, p. 144.
Xu, D. et al. (2005), "Detection of gamma ray polarization using a 3-D position-sensitive CdZnTe detector," *IEEE Trans. Nucl. Sci.*, Vol. 52, p. 1160.
Zanichelli, M. et al. (2013), "Charge collection in semi-insulator radiation detectors in the presence of a linear decreasing electric field," *J. Phys. D: Appl. Phys.*, Vol. 46, p. 365103.
Zhang, F. et al. (2004), "Improved resolution for 3-D position sensitive CdZnTe Spectrometers," *IEEE Trans. Nucl. Sci.*, Vol. 51, p. 2427.
Zhang, F. et al. (2012), "Characterization of the H3D ASIC readout system and 6.0 cm 3-D position sensitive CdZnTe detectors," *IEEE Trans. Nucl. Sci.*, Vol. 59, p. 236.
Zhu, Y. et al. (2011), "Sub-pixel position sensing for pixelated, 3-D position sensitive, wide band-gap, semiconductor, gamma-ray detectors," *IEEE Trans. Nucl. Sci.*, Vol. 58, p. 1400.

Index

D

E

F

G

H

I

J

K

L

M